Melanges:
Their Nature, Origin, and Significance

Edited by

Loren A. Raymond
Department of Geology
Appalachian State University
Boone, North Carolina 28608

THE GEOLOGICAL SOCIETY OF AMERICA · 1888 ·

SPECIAL PAPER

198

Published by The Geological Society of America, Inc.
3300 Penrose Place, P.O. Box 9140, Boulder, Colorado 80301

Printed in U.S.A.

Library of Congress Cataloging in Publication Data
Main entry under title:

Melanges : their nature, origin, and significance,

Special paper / Geological Society of America ; 198)
Papers presented at a symposium organized for the
1981 annual meeting of the Geological Society of America.
Includes bibliographies.
1. Melanges (Petrology)—Congresses. I. Raymond,
Loren A., 1943- . II. Series: Special paper
(Geological Society of America) ; 198.
QE471.15.M44M45 1984 552 85-761
ISBN 0-8137-2198-9

Contents

Preface

Melanges were the focus of a one-half day, 14-paper symposium I organized for the 1981 Annual Meeting of the Geological Society of America. The theme for that symposium was the same as the title of this volume. Although the number of papers presented was greater than the number included herein, this volume presents the wide range of views expressed by speakers during the symposium. Eight of the 13 authors or coauthors of papers in this volume were a part of the 1981 symposium, and four of the remaining five presented papers dealing with melanges during the 1981 Annual Meeting. With this volume, we hope to focus attention on the variability in nature, origin, and significance of melanges and to encourage future research on these subjects.

Numerous individuals, in addition to the Geological Society of America and agencies that provided support for individual studies, have aided in the preparation of this volume. Book editors Campbell Craddock and W.G.E. Caldwell of the Geological Society of America provided advice and encouragement during volume preparation. Reviewers included Ken Aalto, Richard Abbott, James O. Berkland, Dave Bieler, William Bosworth, William Brown, Tim Byrne, Mark Cloos, Robert Jacobi, Brenna Lorenz, Frank K. McKinney, Robert McLaughlin, Gregory F. Moore, Lucian Platt, Nicholas Rast, Loren A. Raymond, Ghulam Sarwar, M. B. Underwood, Fred Webb, Jr., and Pinar Yilmaz. To them the authors and editor are indebted.

The papers in this volume were written and submitted in late 1981 and early 1982. Final revisions were received between late 1982 and early 1984. References cited and the texts of the papers reflect those dates.

Geological Society of America
Special Paper 198
1984

Prologue
The melange problem—a review

Loren A. Raymond
Tom Terranova
Department of Geology
Appalachian State University
Boone, North Carolina 28608

Masses of mixed rock called melanges (Fr. for mixture) have been recognized widely in the geologic record during the past 15 years. Greenly, in 1919, first used the term as a geological name. Yet in spite of many recent studies on rock bodies similar to those to which Greenly applied the term, the only agreement about melanges that exists is that melanges are bodies of mixed rocks.

Melange-like rock bodies, which are distributed widely, have been described from all continents except Antarctica (Fig. 1). Nearly all of the melanges reported in the literature occur in Phanerozoic orogenic belts. Many melanges are found in zones paralleling convergent plate margins, a circumstance that led many workers to the assumption that melanges only form along such margins. Along the west coast of North American from Alaska to California and in southern Europe from Italy to the Caucasus Mountains, melanges, in fact, do occur in zones interpreted to be post-Paleozoic subduction or overthrust zones. Paleozoic melanges in northeastern North America and northwestern Europe are related similarly to former zones of plate convergence. Nevertheless, melanges form in other environments, as emphasized in this volume (Saleeby, 1984; Jacobi, 1984; Phipps, 1984; Yilmaz and Maxwell, 1984; Sarwar and DeJong, 1984; and Vollmer and Bosworth, 1984). Whatever their petrogenesis, it is clear that melanges are common features in orogenic belts and additional melanges will probably be discovered in all of the Phanerozoic orogenic belts. How extensive melanges may be in Precambrian terranes is an open question.

Melanges form through processes of fragmentation and mixing; this is a widely accepted view (e.g., Hsu, 1968; Byrne, 1984; Cloos, 1984; Phipps, 1984; Vollmer and Bosworth, 1984). Yet several fundamental problems remain to be solved. These are: (1) what is a melange and how do we define the term; (2) how do melanges form; i.e., what processes are responsible for fragmentation and mixing in bodies or rock; and (3) what is the significance of melanges in the geologic record?

The first question is important because it relates to problems of communication between geologists. These communication problems have hampered progress on melange studies. For example, at the Penrose Conference on Melanges (see Silver and Beutner, 1980), unanimous agreement could not be reached even on the statement that "melanges consist of blocks in a matrix." More important, however, the responses to questions (2) and (3) depend on the response to question (1).

The definition of the term *melange* is important in considering melange origins; that is clear from various examples. As an illustration, those who consider all melanges to be tectonic in origin eliminate from consideration, *a priori,* other possible origins. They also may be eliminating the type melanges, the Gwna melanges of Anglesey, from the category. Wood (1974, personal communication, 1978) has argued that the Gwna melanges owe their fragmentation and mixing to sedimentary processes. Because various workers have assigned tectonic, sedimentary, diapiric, and polygenetic origins to melange-like rock bodies, it seems prudent at this point to use the term in a descriptive way. Some workers do not agree. Melange terminology and its various applications are reviewed in this volume by Raymond (1984).

Inasmuch as a variety of origins has been proposed, definitive criteria are needed to distinguish between possible origins. General criteria considered to date include:

(1) the nature of contacts (depositional or structural);
(2) the relationships to surrounding units, including paleogeographic considerations;
(3) the nature of the matrix;
(4) the internal structure of the rock body;
(5) the presence or absence of features indicative of soft sediment deformation; and
(6) the presence or absence of deformation and metamorphism within inclusions.

However, these criteria have not yet resolved the questions of origin, as all but depositional contacts (1) have been ascribed to both tectonic and sedimentary processes of formation. Some progress has been made through detailed analysis of internal structures and sequences (e.g., Cowan, 1982; Naylor, 1982; Byrne, 1984-this volume).

Recognition of melanges of various types (sedimentary, diapiric, tectonic, and polygenetic) in the geologic record also will depend on development of that set of definitive criteria. Subse-

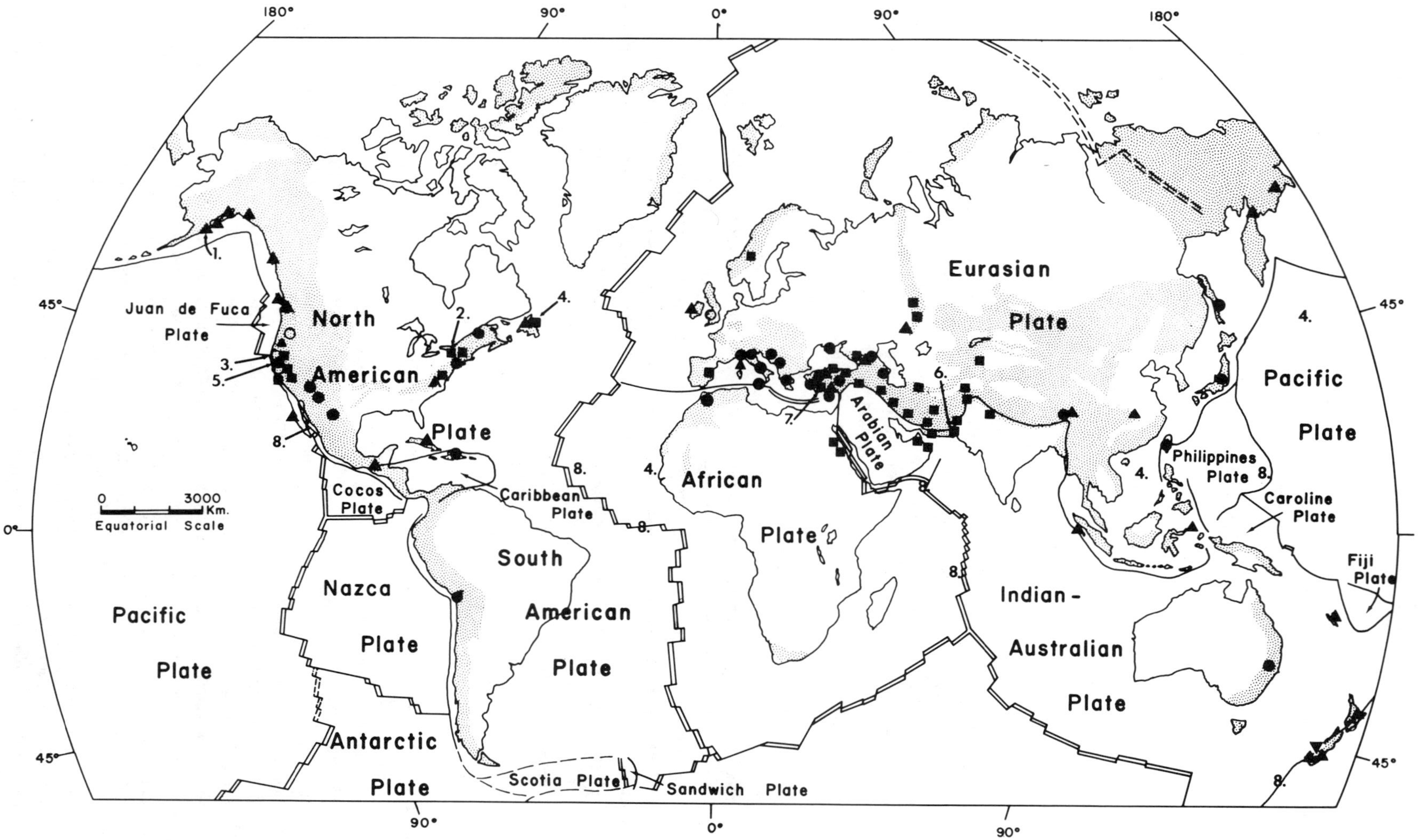

Figure 1. Global tectonic map (after Lowman, 1980) showing locations of selected melanges of various types reported in the literature. Triangles with point up represent tectonic melanges, triangles with point down are diapiric melanges, filled circles are olistostromes, open circles are melanges of controversial origin, and solid squares are polygenetic melanges. Stippled areas are Phanerozoic orogenic belts (after Condie, 1982), which are likely to contain additional melanges. Numbers refer to selected sites discussed in papers in this volume: 1-Ghost Rocks Formation, Alaska—Byrne; 2-Taconic Foreland, New York—Vollmer and Bosworth; 3-Franciscan Complex, California—Cloos; 4-Dunnage Formation, Newfoundland, Atlantic margin of Africa, NW Pacific basin, and South China Sea—Jacobi; 5-Great Valley Group, California—Phipps; 6-Kanar Melange, Pakistan—Sarwar and DeJong; 7-Alakir Cay Unit, Turkey—Yilmaz and Maxwell; and 8-various transform fault zones, the Macquarie Ridge, and the Mariana arc-trench system—Saleeby. See reference list for papers on which map data are based.

quent subdivision of melanges into the above categories will allow a more thorough understanding of the significance of melanges in geologic history. Toward that end, the papers in this volume provide images of the nature of melanges, discussion of possible origins, and comments on the perceived significance of melanges. Specifically, Raymond addresses various uses of the term *melange* and the history of classification. Byrne describes the internal fabric of the Ghost Rocks Formation, a subduction zone unit in Alaska, and discusses its deformation history. Vollmer and Bosworth describe the flysch-melange transition and structural history of melange rocks in a foreland basin in New York. Subduction-related flow melanges, tectonic mixing, and diapirism are the subject of Cloos' report. In contrast, Jacobi describes geologically young submarine sediment slides that may be analogous to olistostromes in the ancient geologic record, then notes the significance of such slides to melange studies, and applies these and other observations to a study of the Dunnage melange of Newfoundland. Submarine slides also provide the mechanism of melange petrogenesis for California Coast Range units described by Phipps. These units are considered to have formed in a fore-arc basin. Sarwar and DeJong describe a little-known melange, the Kanar Melange in Pakistan, and suggest an origin involving oblique convergence and obduction along a transform fault zone. A polygenetic origin involving ophiolite obduction is also suggested for the Alakir Cay unit of the Antalya Complex of Turkey, described by Yilmaz and Maxwell. Finally, Saleeby focuses on the role of serpentinite in melange genesis and emphasizes polygenetic histories that may involve transform faulting, accretion in subduction zones, diapirism, and olistostromal recycling.

The significance of individual melanges in the geologic record will be best realized when processes and products of melange formation in all possible tectonic sites are understood. Thus, it is important to continue the study of melanges in subduction zones (e.g., Franciscan Complex-Cowan, 1978, 1982; Aalto, 1982), fore-arc and back-arc basins (e.g., Lichi melange - Page and Suppe, 1981), transform and overthrust fault zones (e.g., Taconic allochthon - Vollmer and Bosworth, 1984; Pulaski Thrust - Schultz, 1981; transform faults - Saleeby, 1984), continental margin sedimentary wedges (e.g., northwest African continental margin - Jacobi, 1984), continental interior sedimentary wedges (e.g., Kearny, Arizona - Krieger, 1977), and elsewhere. The fact that evidences of melange-forming processes and their products can be, and often are, obscured by later deformation and metamorphism compounds the problem.

ACKNOWLEDGMENTS

We wish to thank F. Webb, R. N. Abbott and an anonymous reviewer for reviews of an earlier version of this paper and Susan Wilson and Elizabeth Stevens for technical assistance in its preparation. Appalachian State University provided support.

REFERENCES CITED

Aalto, K. R., 1981, Multistage melange formation in the Franciscan Complex, northernmost California: Geology, v. 9, p. 602–607.

——1982, The Franciscan Complex of northernmost California: sedimentation and tectonics, *in* Leggett, J. K., ed., Trench-forearc geology: Sedimentation and tectonics on modern and ancient active plate margins: Oxford, Blackwell Scientific Publications, p. 410–432.

Abbate, E., Bortolotti, V., and Passerini, P., 1970, Olistostromes and olistoliths: Sedimentary Geology, v. 4, p. 521–557.

Alekseyev, E. S., 1981, The Kuyul' serpentinite melanges and Talovsk-Mayna Zone (Koryak Highlands): Geotectonics, v. 26, no. 1, p. 68–78.

Bachman, S. B., 1978, A Cretaceous and early Tertiary subduction complex, Mendocino Coast, northern California, *in* Howell, D. G., and McDougall, K. A., eds., Mesozoic paleogeography of the western United States: Los Angeles, Pacific Section Society of Economic Paleontologists and Mineralogists, p. 419–430.

Bailey, E., and McCallien, W. J., 1950, The Ankara Melange and the Anatolian thrust: Nature, v. 166, p. 938–940.

Beneo, E., 1956a, Accumuli terziari da risedimentazione (olisthostroma) nell'Appennino centrale e frane sottomarine: Bollettino Servizio Geologia Italia, v. 78, p. 291–319.

——1956b, The results of the studies on petroleum exploration in Sicily: Bollettino Servizio Geologia Italia, v. 78, p. 27–50.

Berkland, J. O., Raymond, L. A., Kramer, J. C., Moores, E. M., and O'Day, M., 1972, What is Franciscan?: American Association of Petroleum Geologists Bulletin, v. 56, p. 2295–2302.

Bishop, D. G., Bradshaw, J. D., Landis, C. A., and Turnbull, I. M., 1976, Lithostratigraphy and structure of the Caples Terrane of the Humboldt Mountains, New Zealand: New Zealand Journal of Geology and Geophysics, v. 19, p. 827–848.

Black, P. M., and Brothers, R. N., 1977, Blueschist ophiolites in the melange zone in northern New Caledonia: Contributions to Mineralogy and Petrology, v. 65, p. 69–78.

Blake, M. C., Jr., and Jones, D. L., 1974, Origin of Franciscan melanges in northern California: Society of Economic Paleontologists and Mineralogists Special Publication No. 19, p. 345–357.

Boccaletti, M., Bortolotti, V., and Sagri, M., 1966, Ricerche sulle ofioliti delle Catene Alpine. I-Osservazioni sull'Ankara Melange nella zona di Ankara: Bollettino Societa Geologia Italia, v. 85, p. 485–508.

——and Sagri, M., 1967, Olistostromi e olistoliti di Maiolica negli Scisti Policromi della Val di Lima (Provincia di Lucca): Bollettino Societa Geologia Italia, v. 86, no. 3, p. 525–536.

Borukayev, B., Rastsvetayev, L. M., and Shcherba, I. G., 1982, Mesozoic and Cenozoic olistostromes on the southern slope of the Western Caucasus: International Geology Revue, v. 9, p. 1009–1018.

Bosworth, W., and Vollmer, F. W., 1981, Structures of the medial Ordovician flysch of eastern New York: deformation of synorogenic deposits in an overthrust environment: Journal of Geology, v. 89, p. 551–568.

Byrne, T., 1982, Structural evolution of coherent terranes in the Ghost Rocks Formation, Kodiak Island, Alaska, *in* Leggett, J. K., ed., Trench-forearc geology: Sedimentation and tectonics on modern and ancient active plate margins: Oxford, Blackwell Scientific Publications, p. 229–242.

——1984, Early deformation in melange terranes of the Ghost Rocks Formation, Kodiak Islands, Alaska, *in* Raymond, L. A., ed., Melanges: Their nature, origin, and significance: Geological Society of America Special Paper 198 (this volume).

Cerda, M. E., and Prat, P.F.S., 1974, Deslizamientos gravitacionales y olistostromas miocenos en el bajo Gaia y Campo de Tarragona: Acta Geologica Hispanica: v. IX, p. 117–126.

Cloos, M., 1982, Flow melanges: Numerical modeling and geologic constraints on their origin in the Franciscan subduction complex, California: Geological

Society of America Bulletin, v. 93, p 330–345.
—— 1984, Flow melanges and the structural evolution of accretionary wedges, *in* Raymond, L. A., ed., Melanges: Their nature, origin, and significance: Geological Society of America Special Paper 198 (this volume).
Cobiella, J. L., 1978, Una melange en Cuba oriental: La Mineria en Cuba, v. 4, no. 4, p. 46–51.
Cohen, C. R., Schweickert, R. A., and Odom A. L., 1981, Age of emplacement of the Schistes Lustres Nappe, Alpine Corsica: Tectonophysics, v. 73, p. 267–283.
Condie, K. C., 1982, Plate tectonics and crustal evolution: 2nd ed.: New York, Pergamon Press, 310 p.
Connelly, W., 1978, Uyak Complex, Kodiak Islands, Alaska: A Cretaceous subduction complex: Geological Society of America Bulletin, v. 89, p. 755–769.
Cowan, D. S., 1974, Deformation and metamorphism of the Franciscan subduction zone complex northwest of Pacheco Pass, California: Geological Society of America Bulletin, v. 85, p. 1623–1634.
—— 1978, Origin of blueschist-bearing chaotic rocks in the Franciscan Complex, San Simeon, California: Geological Society of America Bulletin, v. 89, p. 1415–1423.
—— 1982, Deformation of partly dewatered and consolidated Franciscan sediments near Piedras Blancas Point, California, *in* Leggett, J. K., ed., Trench-Forearc geology: Sedimentation and tectonics on modern and ancient active plate margins: Oxford, Blackwell Scientific Publications, p. 439–457.
—— and Page, B. M., 1975, Recycled Franciscan material in Franciscan melange west of Paso Robles, California: Geological Society of America Bulletin, v. 86, p. 1089–1095.
Craw, D., 1979, Melanges and associated rocks, Livingstone Mountains, Southland, New Zealand: New Zealand Journal of Geology and Geophysics, v. 22, no. 4, p. 443–454.
Crawford, K. E., 1975, The Geology of the Franciscan Tectonic Assemblage near Mount Hamilton, California [Ph.D. thesis]: Los Angeles, University of California, 137 p.
DeJong, K., 1974, Melange (olistostrome) near Lago Titicaca, Peru: The American Association of Petroleum Geologists Bulletin, v. 58, p. 729–741.
Dimitrijevic, M. D., and Dimitrijevic, M. N., 1973, Olistostrome melange in the Yugoslavian Dinarides and Late Mesozoic plate tectonics: Journal of Geology, v. 81, p. 328–340.
Drake, A. A., and Morgan, B. A., 1981, The Piney Branch Complex—A metamorphosed fragment of the central Appalachian ophiolite in northern Virginia: American Journal of Science, v. 281, p. 484–508.
Duffield, W. A., and Sharp, R. V., 1975, Geology of the Sierra foothill melanges and adjacent areas, Amador County, California: United States Geological Survey Professional Paper 827, 30 p.
Elter, P., and Trevisan, L., 1973, Olistostromes in the tectonic evolution of the northern Apennines: *in* DeJong, K. A., and Scholton, R., eds., Gravity and Tectonics: John Wiley and Sons, New York, p. 175–188.
Feary, D. A., and Pessagno, E. A., Jr., 1980, An Early Jurassic age for chert within the Early Cretaceous Oponae Melange (Torlesse Supergroup), Raukumara Peninsula, New Zealand: New Zealand Journal of Geology and Geophysics, v. 23, p. 623–628.
Gansser, A., 1974, The ophiolite melange, a world-wide problem on Tethyan examples: Ecologae Geologicae Helvetiae, v. 67, p. 469–507.
Gasanov, T. A., Gadzhiyev, T. G., and Ali-zade, S. A., 1980, Structures of the serpentinite melange in Tutkhun River basin (Lesser Caucasus): International Geology Revue, v. 22, p. 797–804.
Greenly, E., 1919, The geology of Anglesey: Great Britain Geological Survey Memoir, v. 1, 980 p.
Hall, R., 1980, Unmixing a melange: the petrology and history of a disrupted and metamorphosed ophiolite: Journal of the Geological Society, v. 137, p. 195–206.
Horne, G. S., 1969, Early Ordovician chaotic deposits in the Central Volcanic Belt of northeastern Newfoundland: Geological Society of America Bulletin, v. 80, p. 2451–2464.
—— 1979, Melange in the Trondheim Nappe suggests a new tectonic model for central Norwegian Caledonides: Nature, v. 281, p. 267–270.
Hsu, K. J., 1968, Principles of melanges and their bearing on the Franciscan-Knoxville paradox: Geological Society of America Bulletin, v. 79, p. 1063–1074.
Isayev, B. M., Gadzhiyev, T. D., Ali-zade, S. A., and Kengerli, T. N., 1981, Tectonic sheets and olistostrome complexes in the southeastern Caucasus: Geotectonics, v. 15, p. 45–53.
Jacobi, R. D., 1984, Modern submarine sediment slides and their implications for melange and the Dunnage Formation in north-central Newfoundland, *in* Raymond, L. A., ed., Melanges: Their nature, origin, and significance: Geological Society of America Special Paper 198 (this volume).
Johns, D. R., 1978, Mesozoic carbonate rudites, megabreccias and associated deposits from central Greece: Sedimentology, v. 25, p. 561–573.
Jones, D. L., Blake, M. C., Jr., Bailey, E. H., and McLaughlin, R. J., 1977, Distribution and character of Upper Mesozoic subduction complexes along the west coast of North America: Tectonophysics, v. 47, p. 207–222.
Kano, K., 1982, Gravity sliding of the Mesozoic strata and its related structures in the Mino-Ryoke Belt, central Japan: Geoscience Reports Shizuoka University, v. 7, p. 9–33.
Karig, D. E., 1980, Material transport within accretionary prisms and the "knocker" problem: Journal of Geology, v. 88, p. 27–39.
Kay, M., 1976, Dunnage melange and subduction of the Protacadic Ocean, northeast Newfoundland: The Geological Society of America Special Paper 175, 49 p.
Khvorova, I. V., and Il'inskaya, M. N., 1981, Devonian olistostromes in the southern Ural Mountains: Academy of Sciences of the USSR, Institute of Geology, Lithology and Mineral Resources (USSR), v. 16, no. 6, p. 558–568.
Kienast, J. R., and Rangin, C., Mesozoic blueschists and melanges of Cedros Island (Baja California, Mexico): A consequence of nappe emplacement or subduction?: Earth and Planetary Science Letters, v. 59, p. 119–138.
Kilmetz, M. P., 1983, Mesozoic accretionary events in south-eastern Asia [abs.]: EOS (American Geophysical Union Transactions), v. 64, p. 320.
Knipper, A. L., 1971, Development of serpentinite melange in the Lesser Caucasus: Geotectonics, v. 6, p. 384–390.
Kopteva, V. V., 1981, Sedimentary-tectonic breccia of "Gadilevo Bed," South Urals: Academy of Science of the USSR, Institute of Geology, Lithology and Mineral Resources (USSR), v. 16, no. 1, p. 39–49.
Krieger, M. H., 1977, Large landslides, composed of megabreccia, interbedded in Miocene basin deposits, southeastern Arizona: United States Geological Survey Professional Paper 1008, 25 p.
Kurenkov, S. A., 1978, The serpentinite melange and olistostrome complexes of the Alai Range (Southern Tien Shan): Geotectonics, v. 12, p. 376–382.
Landis, C. A., 1974, Stratigraphy, lithology, structure and metamorphism of Permian, Triassic, and Tertiary rocks between the Mararoa River and Mount Snowdon, western Southland, New Zealand: Journal of the Royal Society of New Zealand, v. 4, p. 229–251.
Leitch, E. C., and Cawood, P. A., 1980, Olistoliths and debris flow deposits at ancient consuming plate margins: an eastern Australian example: Sedimentary Geology, v. 23, p. 5–22.
Lindsley-Griffin, N., 1977, Paleogeographic implications of ophiolites: The Ordovician Trinity Complex, Klamath Mountains, California: *in* Stewart, J. H., Stevens, C. H., and Fritsche, A. E., eds., Paleozoic Paleogeography of the Western United States: Pacific Section, Society of Economic Paleontologists and Mineralogists, p. 409–420.
Lowman, P. D., 1980, Global tectonic and volcanic activity of the last one million years [map]: Greenbelt, Maryland, Goddard Space Flight Center.
McBride, E. F., 1978, Olistostrome in the Tesnus Formation (Mississippian-Pennsylvanian), Payne Hills, Marathon region, Texas: Geological Society of America Bulletin, v. 89, p. 1550–1558.
Moore, G. F., Kadarisman, D., Evans, C. A., and Hawkins, J. W., 1981, Geology of the Talaud Islands, Molucca Sea collision zone, northeast Indonesia: Journal of Structural Geology, v. 3, p. 467–475.
—— and Karig, D. E., 1980, Structural geology of Nias Island, Indonesia: Implications for subduction zone tectonics: American Journal of Science, v. 280, p. 193–223.

Moore, J. C., 1978, Orientation of underthrusting during latest Cretaceous and earliest Tertiary time, Kodiak Islands, Alaska: Geology, v. 6, p. 209–213.

Nagle, F., 1972, Chaotic sedimentation in north-central Dominican Republic: The Geological Society of America Memoir 132, p. 415–428.

Naylor, M. A., 1981, Debris flow (olistostromes) and slumping on a distal passive continental margin: The Palombrini limestone-shale sequence of the northern Apennines: Sedimentology, v. 28, p. 837–852.

——1982, The Casanova Complex of the northern Apennines: a melange formed on a distal passive continental margin: Journal of Structural Geology, v. 4, p. 1–18.

Nelson, K. D., 1982, A suggestion for the origin of mesoscopic fabric in accretionary melange, based on features observed in the Chrystalls Beach Complex, South Island, New Zealand: Geological Society of America Bulletin, v. 93, p. 625–634.

Ozkaya, I., 1981, Origin and tectonic setting of some melange units in Turkey: Journal of Geology, v. 90, p. 269–278.

Page, B. M., 1978, Franciscan melange compared with olistostromes of Taiwan and Italy: Tectonophysics, v. 47, p. 223–246.

——and Suppe, J., 1981, The Pliocene Lichi Melange of Taiwan: Its plate-tectonic and olistostromal origin: American Journal of Science, v. 281, p. 193–227.

Peive, A. A., 1980, The Structure and composition of the Rytgyl Belt of serpentinite melange (Koryak Highland): Moscow University Geology Bulletin, v. 35, no. 3, p. 107–110.

Phipps, S. P., 1984, Ophiolitic olistostromes in the basal Great Valley Sequence, Napa County, northern California Coast Ranges, *in* Raymond, L. A., ed., Melanges: Their nature, origin, and significance: Geological Society of America Special Paper 198 (this volume).

Pique, A., 1975, Differentiation of sedimentation regions in the northwestern Meseta, Morocco; the Devonian-Dinantian distribution: Comptes rendus de l'Academie des Sciences (Paris), Series D 281, v. 12, p. 767–770.

Platt, L. B., Loring, R. B., Papaspyros, A., and Stephens, G. C., 1972, The Hamburg Klippe reconsidered: American Journal of Science, v. 272, p. 305–318.

Raymond, L. A., 1984, Classification of Melanges, *in* Raymond, L. A., ed., Melanges: Their nature, origin, and significance: Geological Society of America Special Paper 198 (this volume).

Raznitzin, Y. N., 1978, Serpentinite melange and olistostromes in the southeastern part of the east Sakhalin Mountains: Geotectonics, v. 12, no. 2, p. 142–150.

Ries, R. C., Shackleton, R. M., Graham, R. H., and Fitches, W. R., 1983, Pan-African structures, ophiolites and melange in the Eastern Desert of Egypt: A traverse at 26°N: Journal of the Geological Society, v. 140, part 1, p. 75–95.

Rigo de Righi, M., 1956, Olistostromi neogenici in Sicilia: Bollettino Servizio Geologia Italia, v. 75, no. 3, p. 185–215.

Ryan, P. D., Sawal, J. K., and Rowlands, A. S., 1983, Ophiolitic melange separates ortho- and para-tectonic Caledonides in western Ireland: Nature, v. 302, no. 3, p. 50–52.

St. Julien, P., 1968, Les "Argile-a-blocs" du Sud-ouest des Appalaches du Quebec: Naturaliste Canadien, v. 95, no. 6, p. 1345–1356.

Saleeby, J., 1979, Kaweah serpentinite melange, southwest Sierra Nevada foothills, California: Geological Society of America Bulletin, v. 90, p. 29–46.

——1982, Polygenetic ophiolite belt of the California Sierra Nevada, geochronological and tectonostratigraphic development: Journal of Geophysical Research, v. 87, p. 1803–1824.

——1984, Tectonic significance of serpentinite mobility and ophiolitic melange, *in* Raymond, L. A., ed., Melanges: Their nature, origin, and significance: Geological Society of America Special Paper 198 (this volume).

Sarwar, G., and DeJong, K. A., 1984, Composition and origin of the Kanar melange, Southern Pakistan, *in* Raymond, L. A., ed., Melanges: Their nature, origin, and significance: Geological Society of America Special Paper 198 (this volume).

Schultz, A. P., 1981, Progressive development of melange fabrics in allochthons of the Pulaski Thrust, Southwest Virginia: Geological Society of America Abstracts with Programs, v. 13, p. 549.

Shackleton, R. M., Ries, A. C., Graham, R. H., and Fitches, W. R., 1980, Late Precambrian ophiolitic melange in the eastern desert of Egypt: Nature, v. 285, no. 12, p. 472–474.

Shah, S. K., and Sharma, M. L., 1977, A preliminary report on the fauna in radiolarites of ophiolite-melange zone around Mulbekh, Ladakh: Current Science, v. 46, no. 23, p. 817.

Sharp, R. P., 1940, A Cambrian slide breccia, Grand Canyon, Arizona: American Journal of Science, v. 238, p. 668–672.

Shcherba, I. G., 1978, Pliocene-Quaternary olistostromes of Crimea and mechanism of their formation: International Geology Review, v. 21, no. 10, p. 1140–1148.

Silver, E. A., and Beutner, E. C., 1980, Melanges: Geology, v. 8, p. 32–34.

Sporli, K. B., and Bell, A. G., 1976, Torlesse melange and coherent sequences, eastern Ruahine Range, North Island, New Zealand: New Zealand Journal of Geology and Geophysics, v. 19, no. 14, p. 427–447.

Swarbrick, R. E., and Naylor, M. A., 1980, The Kathikas melange, SW Cyprus: late Cretaceous Submarine debris flow: Sedimentology, v. 27, p. 63–78.

Tselikov, N. N., 1977, Tectonic slabs and melange in the Medes regions, Sakmarian Zone of the Urals: Geotectonics, v. 10, p. 256–260.

Vollmer, F. W., and Bosworth, W., 1984, Formation of melange in a foreland basin overthrust setting: Example from the Taconic orogen, *in* Raymond, L. A., ed., Melanges: Their nature, origin, and significance: Geological Society of America Special Paper 198 (this volume).

Williams, H., 1977, Ophiolitic melange and its significance in the Fleur de Lys Supergroup, northern Appalachians: Canadian Journal of Earth Sciences, v. 14, no. 5, p. 987–1003.

——and Talkington, R. W., 1977, Distribution and tectonic setting of ophiolites and ophiolitic melanges in the Appalachian orogen, *in* Coleman, R. G., and Irwin, W. P., eds., North American Ophiolites: Portland, State of Oregon Department of Geology and Mineral Industries Bulletin 95, p. 1–11.

Wood, D. S., 1974, Ophiolites, melanges, blueschists, and ignimbrites: early Caledonian subduction in Wales?, *in* Dott, R. H., and Shaver, R. H., eds., Modern and Ancient Geosynclinal Sedimentation: Tulsa, Society of Economic Paleontologists and Mineralogists, Special publication no. 19, p. 334–344.

Worley, P.L.H., and Raymond, L. A., 1978, Petrographic and structural reconnaissance of the Franciscan Complex (Dothan/Otter Point) in the Dillard Block, southwest Oregon: Geological Society of America Abstracts with Programs, v. 10, p. 155.

Yilmaz, P. O., and Maxwell, J. C., 1984, An example of an obduction melange: The Alakir Cay Unit, Antalya Complex, SW Turkey, *in* Raymond, L. A., ed., Melanges: Their nature, origin, and significance: Geological Society of America Special Paper 198 (this volume).

Zhang, Z., and Jin, M., 1979, Two kinds of Melange and their tectonic significance in Xiangcheng-Derong area, Southwestern Sichuan: Scientia Geologica Sinca, v. 7, p. 205–214.

MANUSCRIPT ACCEPTED BY THE SOCIETY JULY 12, 1984

Printed in U.S.A.

Geological Society of America
Special Paper 198
1984

Classification of melanges

Loren A. Raymond
Department of Geology
Appalachian State University
Boone, North Carolina 28608

ABSTRACT

Melanges and related rock bodies have been defined, described, and classified in a variety of ways. Classifications, divisible into eight types, reflect specific definitions of the term *melange*. Although melanges were originally considered to be tectonic in origin, currently the term is used widely in a descriptive rather than a genetic sense. Melanges are here defined as bodies of rock mappable at a scale of 1:24000 or smaller and characterized both by the lack of internal continuity of contacts or strata and by the inclusion of fragments and blocks of all sizes, both exotic and native, in a fragmented matrix of finer-grained material. Matrix composition and fabric are rejected as definitive criteria. Progressive fragmentation and mixing of original sedimentary, igneous, and/or metamorphic protoliths give rise to broken formations, dismembered formations, and melanges—three classes of unit that, together with formations, represent a continuum of rock bodies ranging from coherent stratified units at one extreme to chaotically mixed masses at the other. Melanges may originate through sedimentary processes, diapiric processes, tectonic processes, or combinations of these processes, and are classified accordingly.

INTRODUCTION

Melange (Fr. - mixture) is a term first introduced in a geological context by Greenly (1919). In mapping the rocks of Anglesey, Wales, Greenly invoked the term to denote rock bodies, underlying an entire region, that are characterized by phacoids of more durable rock enclosed in a schistose matrix. Greenly believed the matrix to be formed by "the shearing-down of more tractable material" (1919, p. 193), and he noted that this shearing process resulted in the general destruction of primary features. Greenly's use of the term *shearing-down* and his general discussion clearly indicate that he thought the Anglesey melanges were formed by structural (i.e., tectonic) processes.

In mapping the Anglesey terrane, Greenly (1919) distinguished variations in the character of the melange and distinguished between various types (i.e., Autoclastic General melanges and Autoclastic Phyllite-and-Grit melange). Wood (1974) contends that Greenly would not have prefixed the term *melange* had he intended a genetic meaning for the root term. Further, Shackleton (1954, 1969), Wood (1974), and Wood and Schuster (1978) suggest that the data show that the melanges of Anglesey, especially the Gwna melange, were formed by sedimentary rather than tectonic processes. Thus, regardless of Greenly's *beliefs* about the origin of the Angelsey melanges, the origin of the type melange is moot. Similar conflicts have evolved in the interpretation of other melanges (e.g., Blake and Jones, 1974; Cowan, 1974, 1978; Maxwell, 1974; Raymond, 1975). In fact, during the Penrose Conference on Melanges (1978), there was so little agreement about either the origin of melanges or the definition of the term that no consensus was reached; and Silver and Beutner (1980), in reporting the conference, were forced to offer their own definition which they indicate "grew from" the conference.

The definition of the term *melange* that I proposed earlier (Raymond, 1975) is essentially the one to which I still adhere. I here define a melange as a *body of rock mappable at a scale of 1:24000 or smaller and characterized both by the lack of internal continuity of contacts or strata and by the inclusion of fragments and blocks of all sizes, both exotic and native, embedded in a fragmented matrix of finer-grained material.* This is not the most widely used definition, but there is a growing consensus that whatever the definition of melange is to be, it should be based, as this one is, on observable criteria rather than on presumed origin. Yet the question of whether or not the term melange should be reserved for mixtures of *tectonic origin* remains a major controversy. A second, but related issue of debate is, what kinds of unit should be included under the name *melange?* Whether or not exotic blocks should be a definitive criterion for melange recognition is a third controversy, one in which the decidedly minority position is that exotic blocks are definitive.

The purposes of this paper are: (1) to review various uses of the term *melange,* as well as terms applied to related rock bodies; (2) to discuss classifications of melanges and related rock bodies *implied* by the various definitions and uses of the term *melange;* (3) to discuss proposed classifications and their problems; (4) to offer and defend the modified definition of the term *melange* presented above; and (5) to present a classification of the continuum of rock bodies that lie between formations and melanges. The origin of melanges will be discussed only where necessary to clarify the classification problems.

THE NATURE OF MELANGES AND RELATED ROCK BODIES

Greenly's (1919) original description of the Anglesey melanges indicates: (1) dominantly sedimentary protoliths; (2) a generally schistose matrix; (3) phacoids of more durable rock in a finer-grained matrix; (4) general destruction (i.e., the absence of) primary features; (5) melanges with no exotic elements (the Phyllite-and-Grit melange); and (6) melanges containing exotic elements (Autoclastic General melange). Yet Greenly's own work (1919) and that of subsequent workers (e.g., Shackleton, 1956, 1969; Wood, 1974) reveal conflicting data. For example, Greenly (1919, p. 306) reported that at some sites the melange is a "pseudoconglomerate" in which the matrix lacks a strong schistosity. Wood (1974) confirmed that observation noting that locally a rather crude cleavage is the only evidence of tectonism. Thus, the matrix, although generally schistose, is not ubiquitously so. Greenly (1919, p. 66, 199, 305, 306, Fig. 77) also described and figured primary features, including bedding, within the melange—creating ambiguity about the completeness of their general destruction. Recognizing these conflicts, it would be unfortunate to be rigidly bound to a definition of the term *melange* that is not generally consistent with the melanges of the type locality.

Subsequent to its resurrection by Bailey and McCallien (1950, 1953, 1961, 1963), Gansser (1955), and Hsu (1966, 1968, 1969), the term *melange* has been widely applied to rock bodies throughout the world. The characteristics of the many bodies called melange vary, but all bodies lie within a continuum of unit types bounded at one extreme by formations produced by "normal" depositional processes and at the other by chaotic (fragmented and mixed) masses of rock that lack continuity of internal stratigraphy or primary contacts and contain a bewildering array of unrelated rock types (exotic blocks) (Fig. 1). Few would argue against calling units of type α in Figure 1A formations or δ-type units in Figure 1B melanges (especially if the latter were known to be of tectonic origin). Here agreement ends.

In most units called melange, the dominant protoliths that give rise to the matrix (and many blocks) were either clastic sedimentary rock (sandstone, shale, or both) or ophiolitic rocks. In a few, the dominant lithology of the protolith was some other rock type, e.g., volcanic rock, carbonate rock, or evaporite. Clastic protoliths became matrix in the Gwna melange of Anglesey (Greenly, 1919; Wood, 1974); many Franciscan melanges including "melange unit #3" of the San Simeon area (Hsu, 1969; Cowan, 1978) and the Garzas melange of the Diablo Range (Cowan, 1974); the Taconic melanges of New York (Vollmer, 1980; Bosworth and Vollmer, 1981 a, b; Vollmer and Bosworth, this volume); the "Diabas-Hornstein Formation" of the Dinarides (Dimitrijevic and Dimitrijevic, 1973); parts of the Ankara melange of Turkey (e.g., Bailey and McCallien, 1953; Boccaletti and others, 1966); the Kanar Melange of Pakistan (Sarwar and DeJong, this volume); some Himalayan melanges (e.g., Gansser, 1974); the Lichi melange of Taiwan (Wang, 1976; Page and Suppe, 1981); and the Pohangina melange of New Zealand (Sporli and Bell, 1976) to name just a few. Melanges with matrix materials derived from ophiolitic rocks include the Tutkhun River melanges of the Lesser Caucasus (Gasonov and others, 1974); some melanges of the Xiangcheng-Derong area of China (Zhang and Jin, 1979); and the Kaweah serpentinite melange of California (Saleeby, 1979; Saleeby and others, 1978). Melanges derived from both main types, e.g., some "ophiolitic" melanges like the Coachman's melange of Newfoundland (Williams, 1977), the Pre-Cambrian ophiolitic melange of Egypt (Shackleton and others, 1980), and the Alakir Cay melanges (Yilmaz and Maxwell, this volume), have matrix protoliths that are clastic sedimentary rocks or sediments, but phacoid protoliths dominantly or abundantly of ophiolitic materials.

Clastic sedimentary formations and ophiolite complexes, as well as disrupted equivalents of these, have been recognized in orogenic zones around the world. Figure 1 depicts the continuum in structural states between coherent, undisturbed formations (Fig. 1A) or ophiolite complexes (Fig. 1B) and totally fragmented and mixed rock units containing exotic blocks, derived from these protoliths. A continuum of this type may be subdivided into two, three, four, or more parts. Most authors have utilized two or three subdivisions, although a fourfold subdivision seems equally reasonable (type VIII, Fig. 1C). At the left are stratigraphically coherent units, formations (Fig. 1A) or complexes (Fig. 1B) here designated as units of type α. With increasing brittle deformation (fracturing and/or faulting), ductile deformation (boudinage), or both, units of type α grade into disrupted units (type β) that are locally designated as "broken formations" (Fig. 2; also the Cambria Pines Slab = Broken Formation A of Hsu, 1969, photos 1 and 2) or faulted ophiolites. Considerable stratal continuity persists at this stage. Further deformation leads to a condition in which bedding or other primary connectors (e.g., dikes, flows, or ultramafic tectonite bands) are extensively fragmented, disrupted, and enmeshed in the more ductile matrix. The primary features are dismembered. Units in the latter condition, units of type γ (Fig. 3), have been called broken formations, melanges, and olistostromes (e.g., Abbate and others, 1970; Hsu, 1968; Hopson and

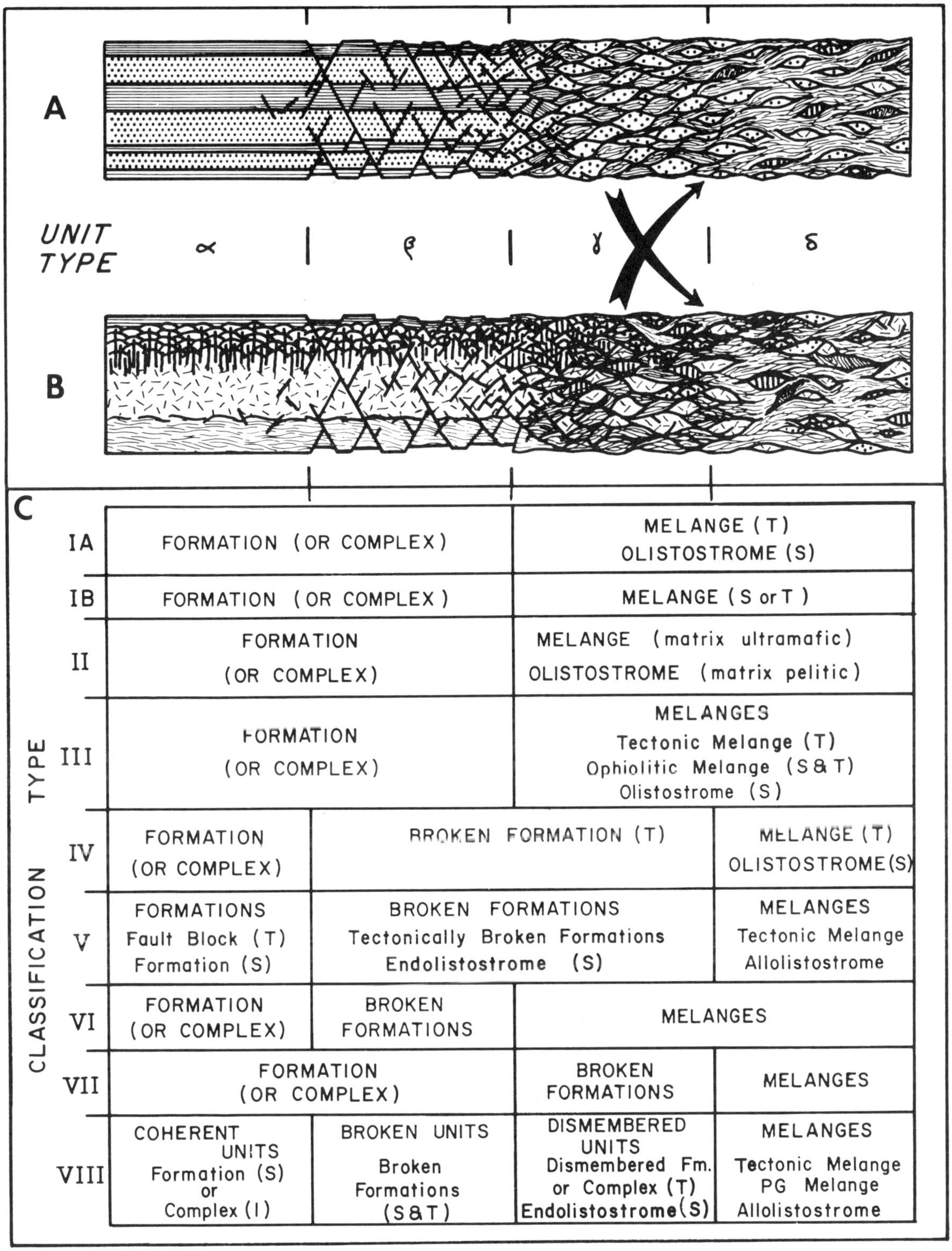

Figure 1. Character and classification of melanges and related rock bodies. A - Schematic diagram depicting progressive fragmentation and mixing (left to right) of interbedded sandstone/shale sequence. Continuum is divided into four types of unit - α, β, γ, and δ. B - Schematic diagram depicting ophiolite progressively fragmented and mixed (from left to right). Arrows show mixing of one protolithic rock type into fragmented unit derived from another protolith. C - Eight possible types of classification of the melanges and related rock bodies shown in A and B above. I = igneous, S = sedimentary, and T = tectonic genesis. PG = polygenetic.

Figure 2. Outcrop of type β unit. Fractured submarine fan facies B and C sandstone/shale sequence within the Franciscan Complex exposed along the West Fork of Hospital Creek, northeastern Diablo Range, California. Note hammer with 32 cm handle right of center for scale.

others, 1981; Elter and Trevisan, 1973), as well as other names such as wildflysch, chaos, and endolistostrome (e.g., Hsu, 1974; Suppe, 1973; Elter and Raggi, 1965). Where fragmentation exceeds the limits of the original protolithic unit and mixing of fragments from other units (exotic blocks) with phacoids from the original protolith (native blocks) occurs, a rock body results that lacks stratal continuity and is composed of blocks and fragments of all sizes, both exotic and native, in a finer-grained, commonly sheared matrix. Such units (type δ of Figs. 1A and 1B; Fig. 4) are widely designated as melanges, but also are called olistostromes, especially where the protoliths were dominantly sedimentary rock. Clearly, as emphasized by Hsu (1968), the development of a melange involves the two processes—fragmentation and mixing.

ROCK BODIES FORMED BY FRAGMENTATION AND MIXING

Numerous processes lead to fragmentation and mixing of grains, fragments, blocks, and slabs of native and exotic materials in a finer-grained matrix. Sedimentary processes commonly do so. Cavern collapse or marine erosion and redeposition of carbonate rocks may give rise to carbonate breccias (Trumpy, 1975; Johns, 1978). Glacial erosion and deposition, mudflows, subaerial landsliding (*sensu lato*), and submarine landsliding (*sensu lato*) give rise to a block-in-matrix texture. Terrigenous sedimentary rocks with such textures are called diamictites (Flint and others, 1960), have a general resemblance to melange rocks, and include the rocks comprising the sedimentary melanges (olistostromes) of some authors (Fig. 5).

Diamictites may be subdivided into genetic types on the basis of both diagnostic features and associated rock types. Thus,

Figure 3. Outcrop of type γ unit. Dismembered bedding in sandstone/shale unit of the Franciscan Complex, Broken Formation unit of Aalto (1982) at Enders Beach near Crescent City, California. Outcrop about two meters high.

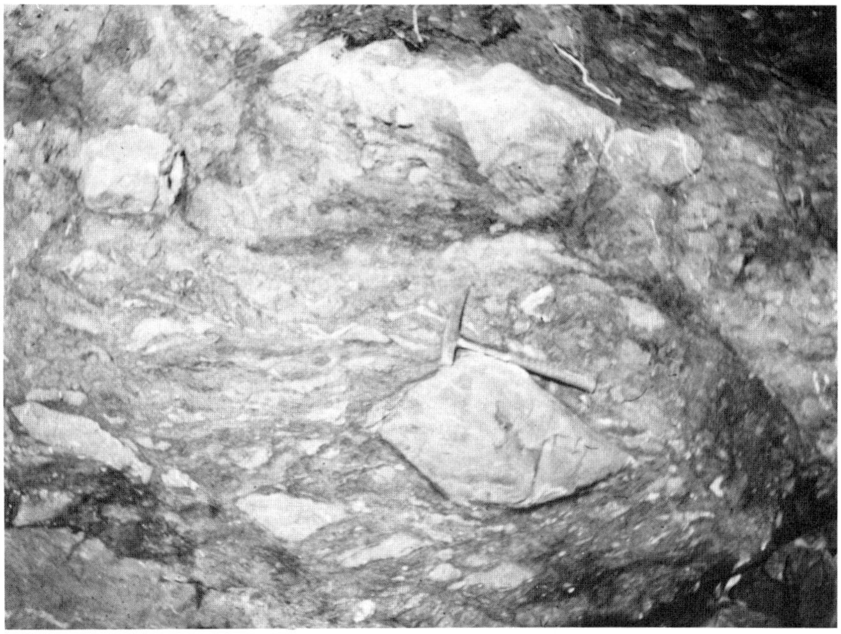

Figure 4. Outcrop of type δ unit. Blueschist-, greenstone-, and metagraywacke-bearing Franciscan Complex melange with sheared shale matrix. San Simeon area, California.

COMPARISON OF DIAMICTITE AND MELANGE ROCKS

	DIAMICTITES				MELANGE ROCKS
			Sedimentary		Tectonic
TEXTURE	Epiclastic with blocks, fragments, and/or grains in a finer grained matrix		Epiclastic to foliated with slabs, fragments, and/or grains in a finer grained matrix		Foliated with slabs, blocks, fragments, and/or grains in a finer grained matrix
CLAST SIZE*	Fragments to small blocks <1-15m	Fragments to small blocks <1-15m	Fragments to small slabs <1-1500m	Fragments to Large slabs <1->1500m	Fragments to Large Slabs <1->1500m
BLOCK/MATRIX	Low to Moderate	Generally Low	Generally High	Low to High Generally Low	Low to High
CONTACTS	Depositional		Depositional to Sheared		Sheared
ASSOCIATED ROCKS AND FEATURES	Rhythmites, Drop-stones, Striated clasts, Polished pavements Boreal Fossils	Crossbedded and channelled fluvial sand-stone, shale and conglomerate Non-marine Fossils	Colluvium, Alluvium, Fluvial Sedimentary Rocks, Continental Organic Debris	Marine Sedimentary Rocks (including turbidites); Marine Fossils	Any kind of rock (including gouge;breccia); slickensides
ROCK TYPES	TILLITE	TILLOID or PEBBLY MUDSTONE	TILLOID, BRECCIA, or MEGABRECCIA SF-TECTONITE	TILLOID, MEGABRECCIA, PEBBLY or BOULDERY MUDSTONE or SANDSTONE, SF-TECTONITE	BRECCIA, MEGABRECCIA, MYLONITE (sensu lato), SF-TECTONITE
ROCK UNIT	Formation or Member	Formation or Member	Formation, Member, Broken Formation, Landslide	Formation, Member, Broken Formation, Olistostrome	Broken Formation, Dismembered Formation, Tectonic Melange
ORIGIN	Glacial Deposition	Mudflow	Subaerial Landslide (sensu lato)	Submarine Landslide	Brittle or Ductile Deformation in a Fault or Shear Zone

*(After Hsu, 1969)

Figure 5. Comparison of diamictites and melange rocks. Central two columns show rocks with characteristics of both diamictites and sedimentary melanges.

on the basis of *descriptive criteria,* names, in some cases denoting genesis (e.g., tillite), are applied to the rocks. (The same process of subdivision is applied to melanges, below). In all cases, given an appropriate thickness, because the rock bodies are essentially rock-stratigraphic units with depositional contacts, which were deposited in a stratigraphic succession, they may be designated as formations. However, landslide deposits (either submarine or subaerial) may have characteristics of rock bodies of types β, γ, or δ (Figs. 1A and 1B); i.e., they may be fractured, fragmented, sheared, and/or mixed bodies of rock that also may be designated as broken formations, melanges, or olistostromes (Fig. 5).

The distinctions shown in Figure 5 are somewhat arbitrary, as conceivably diamictites could be defined as types of melange rocks. However, it seems advisable to avoid redundancy in terminology by restricting the term *melange* to *rock bodies* that may contain large to very large fragments or blocks and that do not fit the usual definition of a formation. Diamictites are *rock types* characteristic of depositional sedimentary units.

Tectonic processes also result in the genesis of block in matrix rock types (Fig. 5). Brittle and/or ductile deformation, especially of polylithologic rock associations or soft sedimentary rocks, result in rock units of types γ and δ (Fig. 1A and 1B). Such deformation occurs in accretionary complexes (subduction zones), as well as in all kinds of fault zones, resulting in various "breccia zones," "megabreccias," and melanges. That such rock bodies develop along high-angle strike-slip or transform faults, as well as along thrusts, e.g., in forelands or subduction zones (Blake and Jones, 1974; Moore and Karig, 1980; Schultz, 1981; Bosworth, 1982; Vollmer and Bosworth, this volume), has been demonstrated by Saleeby (1978, 1979, this volume) and Craw (1979). In addition, polydeformation of high-grade metamorphic terranes results in a superficially similar block-in-matrix geometry.

CONFLICTING DEFINITIONS AND CLASSIFICATIONS

Classification of melanges and related rock bodies depends on the definitions of the units involved. Only Raymond (1978) has published a specific classification, but the definitions of other workers imply classifications. As indicated above, for Greenly (1919, esp. p. 65 and 193–199) the melanges of Anglesey were tectonically deformed rock bodies derived from dominantly sedimentary protoliths, consisting characteristically of fragments of more durable rock "floating in a schistose matrix." Greenly described both native and exotic fragments.

Greenly (1919), by defining melanges as he did, implied a twofold classification of rock units, i.e., a division of rock units into formations and melanges, resulting in a classification of type IA (Fig. 1C). Bailey and McCallien (1950) and Gansser (1955), in following Greenly, adopted a similar subdivision. Cowan (1974), like Greenly, adopted a type IA classification, rejecting the threefold type IV classification implied by Hsu (1968) and adopted in modified form by Berkland and others (1972). Cowan (1974), using Hsu's (1968) wording, but with some omissions, supported a definition of melanges as:

> mappable bodies of deformed rocks characterized by the inclusion of tectonically mixed fragments or blocks . . . in a pervasively sheared, fine-grained, and commonly pelitic matrix.

Cowan (1974) adopted the modifier "tectonic" (i.e., tectonic melange) in order to emphasize his view that, by definition, melanges are bodies of rock with a tectonically generated "internal fabric dominated by penetrative mesoscopic shear fractures." Cowan was following Hsu (1968) in restricting the application of the term *melange* to rock bodies characterized by a matrix with a penetrative, mesoscopic shear fracture fabric, but both authors failed to account for the fact that Greenly (1919) had noted that some of the Anglesey melanges lack such a fabric. Because they assumed that such fabrics can be produced only by tectonism and that fragmentation and mixing were produced by the same process that created the fabric, both Hsu (1968) and Cowan (1974) chose to adopt a genetic definition (i.e., that melanges are tectonic) based on Greenly's controversial interpretation of the origin of the Anglesey melanges and their own similar interpretations of the Franciscan melanges of California.

Berkland and others (1972) and Raymond (1975) rejected the use of type IA classifications, as implied by their definition of the term *melange,* because of the problems involved with naming rock units on the basis of inferred genesis. Earlier Horne (1969) called attention to this problem in his study of the Dunnage melange of Newfoundland by noting that the petrogenesis of chaotic rock bodies, in general, is "poorly understood and conjectural." Horne's detailed work on the Dunnage melange showed it to be polygenetic, with an early sedimentary history of fragmentation and mixing overprinted by a deformation that produced a penetrative fabric. The type melange, the Gwna melange of Anglesey, apparently had a similar history (Shackleton, 1969; Wood, 1974). In a restudy of Hsu's melange #3 of the Franciscan Complex at San Simeon, Cowan (1978) was able to show that, as in the Gwna melange, fragmentation and mixing preceded deformation. Cowan found that two post-depositional deformations produced a mesoscopic shear fracture fabric and elliptically shaped boudins. The early processes of fragmentation and mixing were attributed by Cowan (1978) to sedimentary events. As a result, Cowan (1978) abandoned his earlier genetic definition (1974) in favor of a purely descriptive one that fits a classification of type IB. Hsu (1974), following an international field trip, had already done so.

Subdivision of the continuum of rock units (α - δ of Fig. 1A and 1B) into three units of the kind shown in classification type II is widely employed in Eurasia. In classifications of this type, a chaotic unit with a matrix derived from a sedimentary protolith is considered to be an olistostrome (see discussion of this term below), whereas a chaotic unit with a serpentinite matrix is, by definition, a melange (e.g., Knipper, 1971; Kurenkov, 1978; Zhang and Jin, 1979). Zhang and Jin (1979) suggest that such classifications have genetic significance. Both Gansser (1974) and Zhang and Jin (1979) propose that ophiolitic melanges represent zones of compressive stress developed at sites of plate convergence (also see comments on this topic by Phipps, Saleeby, and Sarwar and DeJong in this volume). In their view, the clastic ("olistostromic") melanges are formed through sedimentary processes. Within the context of such subdivisions, some authors argue that melanges are *only* generated by tectonic processes (e.g., Knipper, 1971).

Gansser (1974) introduced the term "ophiolitic melange" to designate hybrid melanges formed by both tectonic and sedimentary processes and composed, at least in part, of ophiolitic materials. In doing so, he sought to distinguish these hybrid types of melange both from those he considered essentially sedimentary in origin—the olistostromes—and those formed solely by tectonic processes. Ophiolitic melanges contain exotic blocks, by definition. The classification implied is shown as type III in Figure 1C.

Clearly Gansser (1974) implies a twofold subdivision of the continuum (Fig. 1A and 1B) into formations and melanges. Melanges are then subdivided into three types, in part on the basis of genesis—*tectonic melanges* formed by tectonic processes, *ophiolitic melanges* formed by both tectonic and sedimentary processes, and *olistostromes* formed by sedimentary processes. However, it is unclear whether or not Gansser believes that *all* melanges contain exotic blocks. Therefore, the assignment of units of type γ is ambiguous. That units of type γ are not distinguished in a type III classification makes such classifications similar to those of type IA and II. Nevertheless, type III classifications depart significantly from the latter in that olistostromes are considered to be a type of melange rather than a separate class of unit—a view proffered by Berkland and others (1972).

Two problems confront the terminology of Gansser (1974). First, it has not been demonstrated that submarine sliding *cannot* give rise to fragmentation and mixing in the production of ophiolitic melanges as assumed by Gansser (1974). In fact, some ophiolite-bearing melanges such as the Lichi melange are thought to have evolved as a result of such a process (Page and Suppe, 1981). Second, because Gansser (1974) was concerned only with ophiolitic melanges, a complete classification of units was not implied and complete definitions of units other than the ophiolitic melange were not given. As a consequence, classifications of type III fail to allow definitive assignment of units of type γ (Fig. 1A).

Threefold subdivision of the continuum of rock bodies shown in Figures 1A and 1B, i.e., a classification of type IV, was implied by the definitions of Hsu (1968). As noted above, Hsu followed Greenly (1919) in the latter's belief that melanges are tectonic in origin. Hsu (1968, p. 1065) defined melanges as:

> mappable bodies of deformed rocks characterized by the inclusion of tectonically mixed fragments or blocks, which may range up to several miles long, in a pervasively sheared, fine-grained, and commonly pelitic matrix. Each melange includes both exotic and native blocks and a matrix.

The requirement that each melange includes both native and exotic blocks, a new requirement added by Hsu, distinguishes his definition (and classification) from those offered by Greenly (1919) and Cowan (1974) in that an intermediate unit, the *broken formation,* having the characteristics of units like types β and γ (Fig. 1A, 1B), is injected between formations and melanges. Hsu (1968, p. 1065–1066) defines a broken formation as:

"A body of broken strata that contains no exotic elements . . . ." He further notes that "such a body, regardless of its broken state, functions as a rock stratigraphic unit."

Hsu (1968) recognized that fragmentation and mixing could result from sedimentary processes. However, in 1968 he argued that the term *melange* should be reserved for tectonic mixtures. Sedimentary mixtures, designated as olistostromes (following the usage of Flores, 1955), were not considered by Hsu to be melanges, even though some workers had previously applied both terms to the same rock bodies (Rigo de Righi and Cortesini, 1964).

Olistostromes (slide accumulations) were first defined by Flores (1955). In paraphrased form, Flores' definition is as follows:

> An olistostrome is a mappable sedimentary slide deposit, within a normal geologic sequence, which is characterized by lithologically or petrographically heterogeneous bodies of harder rock mixed and dispersed in a matrix of prevalently pelitic, heterogeneous material.

Flores specified further that olistostromes: (1) accumulate as semi-fluid bodies; (2) show no true bedding; and (3) include turbidites. Abbate and others (1970), in a review of olistostromes and olistoliths, sought to modify the definition by deleting the requirements that olistostromes be mappable, be enclosed in a normal geologic sequence, be accumulated as semi-fluid bodies, be lithologically or petrographically heterogeneous, or include turbidites. These modifications result in a definition, adopted here, that *an olistostrome is a sedimentary slide deposit characterized by bodies of harder rock mixed and dispersed in a matrix.* Such units have been widely recognized and described in both ancient and modern deposits; e.g., in the Apennines (Abbate and others, 1970; Elter and Trevisan, 1973; Maxwell, 1959; Naylor, 1981), in Cyprus (Swarbrick and Naylor, 1980), in Egypt (Shackleton and others, 1980), in the Dinarides (Dimitrijevic and Dimitrijevic, 1973), in Turkey (Gansser, 1950; Rigo de Righi and Cortesini, 1964; Boccaletti and others, 1966), in Pakistan (Sarwar and DeJong, 1981, this volume) and across the Himalayas (Gansser, 1959), in southwestern Sichuan, China (Zhang and Jin, 1979), in Taiwan (Wang, 1976; Page and Suppe, 1981; Page, 1981), in Peru (DeJong, 1974), in the Franciscan subduction Complex of California (Suppe, 1973; Gucwa, 1975; Cowan, 1978), in the Great Valley arc trench sequence of California (Emerson and Rich, 1960; Moiseyev, 1970; Phipps, this volume), in the Grand Canyon of Arizona (Sharp, 1940), in Newfoundland (Horne, 1969; Kay, 1976; Jacobi, this volume), in Wales (Wood, 1974; Wood and Schuster, 1978), and along active and passive continental margins (Embley, 1976; Jacobi, 1981, this volume).

Both the definition and the descriptions of olistostromes reveal that substantial similarities exist between rock bodies called melanges and those designated as olistostromes. The distinction, where it is made, is commonly a genetic one. Yet, as has been discussed above, many melanges are recognized to be olistostromes. Are there olistostromes that are not melanges? From two perspectives the response to this query must be yes. First, some olistostromes are too small to be mapped. Therefore, as melanges are recognized generally as mappable bodies of rock (scale $\leqslant$1:24000), those olistostromes that are too small to be mapped cannot be melanges. In addition, Hsu (1968) and some other workers (e.g., Berkland and others, 1972; Raymond, 1975, 1978) require that melanges contain exotic blocks. Some olistostromes do not. Elter and Raggi (1965) distinguished the latter as *endolistostromes*, defining those units with materials from different depositional basins (exotic blocks) as *allolistostromes.* Raymond (1978) used these definitions in a classification of type V (Fig. 1C) in which endolistostromes are sedimentary equivalents of Hsu's (1968) broken formations and allolistostromes are sedimentary melanges (with exotic blocks). Of course, if one adopts classifications of types IA, III, or IV, or if one defines a melange as having a deformed matrix, then some or all olistostromes are excluded from the category of melange.

The threefold, type IV classification of Hsu (1968) was adopted, with modifications, by some authors. Berkland and others (1972) redefined the term *melange,* but adopted Hsu's definition of broken formation. Raymond (1973), Christensen (1973), Crawford (1975), Jones and others (1978), and Blake and others (1981) used a type IV classification for rock units within the Franciscan Complex of California. Yet neither Hsu's exact definitions nor his implied classification has been widely accepted, in spite of the fact that his 1968 paper is widely cited and credited with popularizing the melange concept.

Several criticisms have been raised against various aspects of Hsu's (1968) definitions and implied classification. The problem of using a genetic definition, i.e., that melanges are essentially tectonic bodies, was discussed above. Gansser (1959) and Abbate and others (1970) pointed out the difficulties in applying such a terminology and Horne (1969), Berkland and others (1972), and Raymond (1975) argued against its use. Berkland and others (1972) and later Raymond (1975) essentially argued that chaotic bodies of unknown origin need to be assigned a designation free of genetic implications and that to define all melanges as tectonically formed bodies precludes raising the question of genesis. Following an international excursion in 1973, Hsu (1974) agreed.

Hsu (1968) indicated in his definition that the matrix of melanges is "pervasively sheared, fine-grained and commonly pelitic." Available data call for modification of these strictures. As noted above, the type melange (Gwna melange) of Anglesey does not everywhere exhibit a pervasively sheared fabric (Greenly, 1919; Wood, 1974). Similarly, a variation from foliated to non-foliated matrix was noted in the "Williams Chaos," a Franciscan melange mapped by Suppe (1973). In some units called melange, a penetrative shear fracture fabric reflecting pervasive shearing is entirely lacking. Consequently, such a restriction seems undesirable. Similarly, albeit it is true many a melange with a matrix that is pelitic has been described, it is equally true many matrices are serpentinite or other materials, so that the specification of a commonly pelitic matrix serves no useful purpose. Berkland and others (1972) also suggested that the specification that melange matrix be fine-grained should be eliminated, as some melanges have a matrix of coarse sand or other material. "Finer-grained" (than the blocks) is a suitable substitute phrase.

Probably the most controversial aspect of Hsu's (1968) definition of melange was his inclusion of the requirement that melanges contain both native and exotic blocks. Berkland and others (1972), Crawford (1975), and Raymond (1975, 1978) adopted this requirement, but most other workers did not. Cowan (1974) argued that the essence of the melange was the fabric, which he attributed to a deformation, noting that "the deformational style in question is not limited to terranes containing demonstrably exotic blocks." DeJong (1974) abandoned the requirement of exotic blocks to avoid applying two different designations to stratigraphically equivalent parts of the Titicaca olistostrome of Peru, which locally contains exotic blocks (Pirin area), but in some areas lacks these fragments (Puno area). A similar interregional variation in block content was described by Hopson and others (1981) in the "Hawasina Group" of Oman. Additional objections to the stricture that melanges must contain exotic blocks were raised by Beutner (1975) and by some participants at the Penrose Conference on Melanges. The arguments raised are that the process of designating tectonic blocks as exotic versus native is interpretive rather than descriptive and that the distinction may be meaningless as far as petrogenesis is concerned. Similarly, Knipper (1971) argues that the composition of blocks is accidental and does not reflect the melange site or conditions of formation. Beutner (1975) reiterated Cowan's (1974) view, that the fabric of the melange is the more significant feature. Beutner (1975) correctly points out that Greenly (1919) designated his Phyllite-and-Grit unit, a unit without exotic elements, as a melange.

By eliminating the exotic block requirement, one eliminates the distinction between melanges (unit type δ - Fig. 1A, 1B) and broken formations (unit type $\beta + \gamma$ - Fig. 1A, 1B) in Hsu's type of classification (type IV). This and the recognition that all melanges are not tectonic in origin (e.g., Hsu, 1974; Raymond, 1975; Maxwell, 1974) lead many workers back to a classification of type I, but one in which both sedimentary and tectonic processes are recognized as agents of fragmentation and mixing. Such a classification is here designated as a type IB classification (Fig. 1C).

At present (1982), classification type IB is seemingly the type most commonly accepted by North American geologists, as reflected by the wide use of a descriptive definition of the term *melange* like that proffered by Silver and Beutner (1980) following the Penrose Conference on Melanges. These authors define melange as follows (1980, p. 32).

> 'Melange' is a general term describing a mappable (at 1:25000 or smaller scale), internally fragmented and mixed rock body containing a variety of blocks, commonly in a pervasively deformed matrix.

Similar definitions were adopted in studies of the melanges of the Fleur de Lys Supergroup of Newfoundland by Williams (1977), of Franciscan melanges by Cowan (1978) and Page (1978), of the Uyak melange of Alaska by Moore and Wheeler (1978), of the Oyo Complex of Nias Island, Indonesia, by Moore and Karig (1980), of the Piney Branch Complex of Virginia by Drake and Morgan (1981), of the Hawasina Group of Oman by Hopson and others (1981), and of the Taconic melanges of New York by Bosworth and Vollmer (1981b) and Vollmer and Bosworth (this volume).

Another tripartite subdivision of the continuum (Fig. 1A and 1B), a type V classification (Fig. 1C), was proposed by Raymond (1978). Although parts of this classification were noted in the AGI Glossary of Geology (Bates and Jackson, 1980), few workers have adopted it. Raymond combined ideas from Greenly (1919), Hsu (1968), and Elter and Raggi (1965) in a classification founded first on observable criteria, the presence or absence of disrupted bedding and exotic blocks, and second on genetic criteria. Units without disrupted bedding were designated as formations, further subdivided into units with sedimentary contacts (formations) and units with faulted contacts (fault blocks). Broken formations were designated as units with disrupted bedding, but without exotic blocks. Sedimentary broken formations were called *endolistostromes,* whereas tectonic units were designated *tectonically broken formations.* The term *melange* denoted bodies with both disrupted bedding and exotic blocks. Sedimentary melanges were assigned the appellation *allolistostrome* and tectonic units the name *tectonic melange.*

Following presentation of this classification at the 1977 Cordilleran Section meeting of the Geological Society of America, I received several comments and criticisms. Notably, objections were raised regarding the use of the term *fault block* to designate a fault-bounded formation and the use of *exotic blocks* as a criterion for the definition of a melange. The term *fault block* is dropped. *Exotic blocks* are discussed below.

In contemplating the conflicting uses of the term *melange,* the controversies over the origins of chaotic rock bodies, and the general nature of the rock bodies involved (Figs. 1A and 1B), it seems clear to me that the most important questions relating to chaotic rock bodies are: (1) how do such rock bodies form, and (2) what is the significance of such rock bodies in the geologic record? The answers to these questions are not yet resolved. Yet attempts to solve them are continually clouded by misunderstandings about definitions and resulting classifications of the rock bodies involved. These problems raise two additional questions, specifically: (3) how many distinct types of rock bodies are generally recognizable in those terranes containing chaotic units, and (4) should recognizable types of rock bodies be lumped or split in a classification?

As noted above, the continuum represented by Figures 1A and 1B may be split into two, three, four, or more parts. I have here designated four types of unit—α, β, γ, and δ—the former being well-stratified and/or coherent and the latter being quite disrupted. Communications with colleagues suggest that confusion has resulted from variations in combining these four types of units into usable definitions (and a classification). Some workers combine units of types α and β under the name *formation* and γ and δ under the name *melange* (classification type I). Others designate units of type α as formations (or complexes), units of type β as broken formations, and units of type $\gamma + \delta$ as melanges. (This type of classification, type VI, was not discussed above because I found no published account of such a usage; yet some workers have argued orally for such a subdivision.) Conceivably, other workers might adopt a classification of type VII, although I know of no published account of such a usage. Still others designate unit type α as a formation, $\beta + \gamma$ as a broken formation, and δ as a melange (classification type V). Therefore, I suggest that the four types of unit described here (α, β, γ, and δ) have been recognized and described and should be distinguished by name for the sake of clarity in communication; i.e., the continuum should be split. Towards that end, a new classification based upon a fourfold subdivision of the continuum in Figure 1, is offered here, as is a defense of the terminology and the classification.

A NEW CLASSIFICATION

A new classification of chaotic rock bodies and related units is presented in Figure 6. Figure 6 is an expanded version of classification VIII, Figure 1C, a classification based upon a fourfold subdivision of the continuum of rock bodies shown in Figures 1A and 1B. Like its predecessor, the type V classification of Raymond (1978), this type VIII classification is based primarily on descriptive criteria and secondarily on genesis. As in the type V classification, the descriptive criteria are degree of stratal continuity and presence or absence of exotic blocks. The rationale for retaining these criteria follows.

Several major descriptive criteria reappear in the literature in definitions of the term *melange* and those of related units. These include mappability, stratal continuity, matrix composition, matrix fabric, and block composition. That the unit should be mappable at a reasonable scale (e.g., 1:24000) is widely accepted (Hsu, 1968; Berkland and others, 1972; Silver and Beutner, 1980) and will be accepted here without further discussion.

The lack of stratal continuity generally gives rise to the recognition of a unit that differs from normal formations and is a feature widely recognized as characteristic of chaotic units, especially melanges (Rigo de Righi and Cortesini, 1964; Hsu, 1968,

	COHERENT UNITS	BROKEN UNITS	DISMEMBERED UNITS	MELANGES
	stratigraphic units with internal stratal continuity	stratigraphic units with locally broken internal stratal continuity	rock bodies without internal stratal continuity or exotic blocks	rock bodies without internal stratal continuity, but with exotic blocks
Diapiric and Tectonic units		Broken Formations or Complexes (S or T)	Dismembered Formation or Complex	Diapiric Melange Tectonic Melange Polygenetic Melange
Igneous or Sedimentary units	Formation (S) or Complex (I)		Endolistostrome	Allolistostrome

Figure 6. Classification of melanges and related rock units. This type of classification, type VIII of Figure 1, divides the continuum of rock units into four types. S = sedimentary genesis, fracturing, fragmentation, and/or mixing. T = tectonic genesis, fracturing, fragmentation, and/or mixing. I = formation through primarily igneous processes.

1969; Cowan, 1974; Raymond, 1978). Units of type α (Fig. 1) clearly possess stratal continuity, whereas units of type δ lack it. However, the degree of stratal disruption involved in the definition of a broken formation is unclear. Hsu (1968) wrote: "A body of broken strata that contains no exotic elements, such as Greenly's 'Phyllite-and-Grit Melange' is called a *broken formation,* because such a body, regardless of its broken state, functions as a rock-stratigraphic unit." Hsu envisioned such units as tectonically broken. A tectonically fragmented unit of type γ will not have normal depositional (or primary) contacts and may, in fact, have contacts that transect depositional contacts within the sequence in which it occurs. Thus, it will not function as a rock-*stratigraphic* unit. In contrast, a unit of type β will, in general, retain normal, albeit discontinuous, stratigraphic contacts and will therefore function as a rock-*stratigraphic* unit. Consequently, a distinction between units of types β and γ is desirable. I here have assigned the term *broken formation* to units of type β that, although broken, retain internal stratal continuity and thus, in spite of their broken state, function as rock-stratigraphic units. Units of type γ, i.e., units that lack internal stratal continuity, that may or may not function as normal rock-stratigraphic units, and that lack exotic elements are called *dismembered formations.* Tectonically dismembered formations or complexes are distinguished from endolistostromes (sedimentary units) where the genesis of the fragmentation has been recognized.

Matrix composition is a descriptive criterion used by some authors to characterize or classify melanges (e.g., Zhang and Jin, 1979). As noted, the two dominant matrix types are pelitic and ultramafic in composition. Knipper (1971) and many other Eurasian authors (see discussion above) define melanges as chaotic units with serpentinite matrices (unit type γ + δ, Fig. 1B) and olistostromes as chaotic units with pelitic matrices (unit type γ + δ, Fig. 1A). As noted by Zhang and Jin (1979), such a criterion is easily applied in the field. Matrix composition as a criterion is indeed easy to apply, but has one major drawback. Defining a melange as a unit with an ophiolitic or serpentinite matrix and an olistostrome as a unit with pelitic matrix excludes the type melange, the Gwna melange, from the class of units assigned that designation. In addition, many classic melanges, including most in the Franciscan Complex, would be eliminated from the melange category.

Zhang and Jin (1979) suggest that a type III classification, i.e., one based on matrix composition, may be genetic. They propose that ophiolitic melanges are formed under compressive stresses in the lithosphere (at convergent plate margins), whereas olistostromes develop in aqueous environments under horizontal stresses. However, not all ophiolitic chaotic bodies of type γ or δ (Figure 1B) are tectonic in origin (Abbate and others, 1970; Moiseyev, 1970; Bonatti and others, 1973; Phipps, this volume; Sarwar and DeJong, this volume; Yilmaz and Maxwell, this volume), nor are all units of types γ and δ (of Figure 1A) thought to be sedimentary in origin (e.g., Blake and Jones, 1974; Byrne, 1981; Bosworth and Vollmer, 1981a, b; Vollmer and Bosworth, this volume; Schultz, 1981). Thus, such a genetic distinction must be rejected. Matrix composition is not used in the classification proposed in Figure 6.

Matrix fabric, like matrix composition, is an observable and therefore descriptive criterion upon which classification could be based. Hsu (1968, 1974) specified that melange matrix should be pervasively sheared and Cowan (1974) and Beutner (1975) argued that such a fabric is a distinctive and characteristic feature of melanges. However, as noted above, neither the type melange (Gwna) nor other melanges are everywhere characterized by matrix that is pervasively sheared. Apparently both Cowan (1978) and Beutner (Silver and Beutner, 1980) dropped this stricture and it is not adopted here.

Block content has been used by some authors to define and characterize melanges. Hsu (1968), in a controversial departure from Greenly's usage, specified that melanges must contain exotic blocks. Berkland and others (1972) and Raymond (1975, 1978) adopted this usage. Gansser (1974) distinguished ophiolitic melanges from other types because of the genetic significance he attached to such bodies. Cowan (1974) suggested that, where necessary, prefixes such as *exotic* or *graywacke-chert-greenstone* could be affixed to the root term *tectonic melange* in order to distinguish particular lithologic variations, but he rejected the specification proposed by Hsu that melanges contain exotic blocks.

Few authors object to the idea of using lithologic prefixes such as *ophiolitic* to distinguish one melange from another. However, the exotic block concept, adopted here, is highly controversial and demands justification. *Exotic block,* as used herein, *is a term that denotes a fragment of rock greater than 64mm in diameter that was not interbedded or intimately associated with the dominant formation(s) that give rise to the matrix ± blocks of a chaotic unit* (i.e., it was a part of another formation or other unit, perhaps with a different history, that has become mixed with the parent rock of a melange). Major arguments against the use of exotic blocks either as a criterion in the definition of the term *melange* or as a feature with which melanges are distinguished from other chaotic units include the following.

(1) In practice, it is difficult to distinguish whether or not specific blocks are exotic (Cowan, 1974; Beutner, 1975).

(2) Recognition of exotic blocks is an interpretive rather than descriptive process (Beutner, 1975).

(3) The composition of exotic elements does not reflect the origin of the melange (Knipper, 1971).

(4) Units exposed over a wide region may contain exotic blocks in some areas and not in others (DeJong, 1974).

(5) In a unit such as the Franciscan Complex, only blueschists can be considered to be exotic, as only they are "far removed" from their site of origin (Beutner, 1975).

(6) Greenly (1919) used the term *melanges* for units with exotic blocks (the Autoclastic General melange) and for units that lack such blocks (the Phyllite-and-Grit melange). Thus, requiring melanges to contain exotic blocks is not consistent with the exact usage of the term *melange* employed originally by Greenly.

There is no doubt that it is difficult in some cases to distinguish between exotic and native blocks. For example, recognizing a wacke derived from a sequence different from that which gave rise to the matrix and dominant wacke blocks in a unit such as one of type δ, Figure 1A would indeed be difficult. Petrographic and stratigraphic analyses may clearly separate exotic from native blocks (e.g., see how Cowan, 1974, distinguished sandstones of different composition). In other cases, the recognition that certain blocks such as glaucophane schist, eclogite, and peridotite were not interbedded with dominant lithologies and, in fact had quite different histories, is not ambiguous at all. In both of the above cases, the designation of a block as exotic is based upon observable, not purely interpretive, criteria. Occasional difficulty in applying a criterion is not, in my view, justification for rejecting a useful and meaningful classification based on that criterion. Criteria in many classifications in geology, in some cases, are difficult to apply. For example, note the problems of distinguishing, in the field, between an andesite and a basalt using many igneous rock classifications.

Interpretation *is* required in certain cases. For example, in the case of a greenstone block within a dominantly clastic unit of type γ or δ, it may be exceedingly difficult to find evidence to resolve the question of whether or not the original basalt was interbedded with the clastic rocks or was later incorporated in the unit as a result of a mixing process. If mafic flows were originally interbedded with clastic rocks, fragmentation and dismemberment would give rise to a unit of type γ. If the block was later incorporated, the unit would be of type δ. If available evidence is ambiguous, one worker might interpret the rock body as a melange (according to Figure 6), whereas another would interpret the body as a dismembered formation. Inevitably, use of a classification of type VIII produces ambiguous cases. So too does the use of some rock and fossil classifications *where evidence is incomplete,* but we do not discard the classifications on that basis. The argument that the exotic block criterion should be rejected on this basis is a "red herring."

Knipper (1971) argues that the inclusion of exotic blocks is accidental and does not reflect the origin of the melange. To the contrary, I argue that inclusion of exotic elements reflects *extra-formational mixing*—and, as mixing is one of the two fundamental melange-forming processes, is indeed indicative of genesis. Both exotic and native phacoids reveal the protoliths that gave rise to a given melange. The presence of exotic elements further indicates fragmentation and mixing in more than a single, local or restricted unit.

De Jong (1974) rejected the exotic block criterion in his work on the Titicaca melange of Peru because rock bodies in two areas, interpreted to be stratigraphically equivalent, differed in block content. Exotic blocks are present in the Pirin area, but are absent in the Puno area. Thus, to avoid applying two different designations to equivalent units, De Jong designated both types of unit *melange.* The stratigraphic record is replete with cases of time-equivalent units that differ in lithology and, therefore, environmental significance. I argue that variations in block content from area to area within a given chaotic unit, e.g., as described in the Hawasina Group of Oman by Hopson and others (1981), similarly are meaningful to interpretations of the history of the unit. In a tectonic melange, the incorporation of exotic elements indicates that fragmentation has exceeded the limits of a given formation, group, or complex and mixing has been extensive enough to involve rocks formed in different temporal or physical environments. Similarly, an allolistostrome has resulted from sliding of a stratigraphically and/or structurally extensive or complex source area. If chaotic rock units differ from region to region, they should be given different names, just as facies equivalents (e.g., shale and limestone) are given different names and assigned to different formations.

Beautner's (1975) argument, that the exotic block concept should be rejected because only blueschists are "truly exotic" in the Franciscan, is a specious "Strawman." Apparently his criteria for "truly exotic" are:

(1) the block must be far from its site of metamorphism; and

(2) the blocks should not be time-equivalent with the melange unit.

I concur with Beutner (1975) that the strictures that exotic blocks be metamorphic rocks and that they be "far removed" from their site of metamorphism are stultifying, but it was he who suggested such criteria. With regard to his second criterion, radiometric dates on Franciscan blueschists (e.g., Lanphere and others, 1978) indicate that many are time-equivalent with other Franciscan rocks. Therefore, by Beutner's own criterion, many blueschists cannot be exotic. Further, Beutner (1975) considers the time-equivalent, structurally juxtaposed, arc trench gap Great Valley sequence rocks and their basement "to be 'normal' components of a melange derived from such a provenance." This statement implies the undocumented suggestion that Franciscan (subduction zone) melanges were derived from the Great Valley (arc trench gap) sequence. It further suggests, without explanation, that there is such a thing as a "normal" component of a melange and that any largely time-equivalent rocks within the "provenance" of the melange are normal components. It is unclear what the "provenance" of a melange is, but if it were considered to be any area that could provide landslide materials to a marine depositional basin plus any structural unit that could potentially be fragmented to produce blocks for a tectonic melange, given Beutner's time-equivalent stricture, exotic blocks would be defined only as blocks with ages different from those of the main components of the melange. This, too, is a stultifying stricture. If exotic blocks were defined as Beutner (1975) defines them, I, too, would reject the use of exotic blocks as a criterion for the definition of the term *melange*. Rejection of Beutner's "truly exotic" block concept has little bearing on the exotic block concepts of Hsu (1968) and Raymond (1978). The broad definition presented here is, I believe, quite useful.

A final argument raised against the use of exotic blocks as a criterion for defining the term *melange* is that Greenly (1919) employed no such criterion (Beutner, 1975). Indeed he did not. Yet as noted above, Greenly's use of the term was not entirely consistent with the character of the melange to which he applied it. Thus, some changes are required. I suggest that Hsu's (1968) exotic block concept is a useful and genetically significant change that should be retained.

SUMMARY AND CONCLUSIONS

In summary, a melange is here defined as *a body of rock mappable at a scale of 1:24000 or smaller and characterized both by the lack of internal continuity of contacts or strata and by the inclusion of fragments and blocks of all sizes, both exotic and native, embedded in a fragmented matrix of finer-grained material.* The criteria of matrix composition and fabric are not employed in the definition of the term *melange.* Further, as both sedimentary and tectonic processes may play a role in melange formation, no genetic significance is implied by the term. Melanges may be subdivided into tectonic melanges, diapiric melanges, allolistostromes, and polygenetic melanges where the primary cause of fragmentation and mixing is known to be tectonic, diapiric, sedimentary, or a combination of processes, respectively.

Dismembered formations or complexes are bodies of rock mappable at a scale of 1:24000 or smaller that both lack continuity of internal contacts or strata and are characterized by the inclusion of native blocks and fragments of all sizes in a matrix of finer-grained material. Both sedimentary and tectonic processes may give rise to such units. Units of the former type are designated endolistostromes and of the latter type, tectonically dismembered formations.

Formations broken by faults, but which retain substantial continuity of contacts and internal stratigraphic units, are called broken formations. No distinction is made between units deformed by gravity-induced fracturing and those with tectonically created fractures. Unbroken formations and complexes retain their normal definitions in the classification of units proposed here.

The proposed classification retains as many of Greenly's original criteria as is possible, but modifies the definition of melange as required by the known features of the Gwna melange and other similar units. It is hoped that use of the proposed classification of rock bodies, which is based on the four types of unit widely recognized, will reduce the communication gap and allow us to deal with the important questions of melange genesis and significance.

ACKNOWLEDGMENTS

I wish to thank Richard Abbott, William Bosworth, and William Brown for critically reading an earlier version of this manuscript and Carole Muirhead, Tom Terranova, and Susan Wilson for technical assistance. Research was supported by the Division of Earth Sciences, National Science Foundation, NSF Grant EAR 76-06062 and by Appalachian State University.

REFERENCES CITED

Aalto, K. R., 1982, The Franciscan Complex of northernmost California: Sedimentation and tectonics: *in* Leggett, J. K., ed., Trench-Forearc geology: Sedimentation and tectonics on modern and ancient active plate margins: Geological Society of London Special Publication 10, p. 419–432.

Abbate, E., Bortolotti, V., and Passerini, P., 1970, Olistostromes and olistoliths: Sedimentary Geology, v. 4, p. 521–557.

Bailey, E. B., and McCallien, W. J., 1950, The Ankara Melange and the Anatolian Thrust: Nature, v. 166, p. 938–940.

—— 1953, Serpentine lavas, the Ankara Melange and the Anatolian Thrust: Transactions Royal Society of Edinburgh, v. 62/2, p. 403–443.

—— 1961, Structure of the Northern Apennines: Nature, v. 191, p. 1136–1137.

—— 1963, Liguria Nappe: Northern Apennines: Transactions Royal Society of Edinburgh, v. 65, p. 315–333.

Bates, R. L., and Jackson, J. A., eds., 1980, Glossary of geology (2nd ed.): American Geological Institute, Falls Church, Virginia, 749 p.

Berkland, J. O., Raymond, L. A., Kramer, J. C., Moores, E. M., and O'Day, M., 1972, What is Franciscan?: American Association of Petroleum Geologists Bulletin, v. 56, p. 2295–2302.

Beutner, E. C., 1975, Tectonite and melange—A distinction: Comment: Geology, v. 3, p. 358–359.

Blake, M. C., Jr., and Jones, D. L., 1974, Origin of Franciscan melanges in Northern California: *in* Dott, R. H., Jr., and Shaver, R. H., eds., Modern and Ancient Geosynclinal Sedimentation: Society of Economic Paleontologists and Mineralogists Special Publication no. 19, p. 345–357.

—— Jayko, A. S., and Howell, D. G., 1981, Sedimentology and structure of Franciscan assemblage, Yolla Bolly terrane, Northern California: American Association of Petroleum Geologists Bulletin, v. 65, p. 902.

Boccaletti, M., Bortolotti, V., and Sagri, M., 1966, Ricerche sulle ofioliti delle Catene Alpine: 1 - Osservazioni sull'Ankara Melange nella zona di Ankara: Bollettin della Societa Geologica Italiana, v. 85, p. 485–508.

Bonatti, E., Honnorez, J., and Gartner, S., Jr., 1973, Sedimentary serpentinites from the Mid-Atlantic Ridge: Journal of Sedimentary Petrology, v. 43, p. 728–735.

Bosworth, W., 1982, Evolution and structural significance of master shear zones within the parautochthonous flysch of eastern New York: Vermont Geology, v. 2, p. 6–13.

—— and Vollmer, F. W., 1981a, Formation of melange in a foreland basin overthrust setting: Example from the Taconic Orogen: Geological Society of America Abstracts with Programs, v. 13, p. 413.

—— 1981b, Structures of the medial Ordovician flysch of eastern New York: Deformation of synorogenic deposits in an overthrust environment: Journal of Geology, v. 89, p. 551–568.

Byrne, T., 1981, Structural geology of the Ghost Rocks Formation Melange terranes, Kodiak Islands, Alaska: Geological Society of America Abstracts with Programs, v. 13, p. 421.

—— 1984, Early deformation in melange terranes of the Ghost Rocks Formation, Kodiak Islands, Alaska, *in* Raymond, L. A., ed., Melanges: Their nature, origin, and significance: Geological Society of America Special Paper 198 (this volume).

Christensen, W. P., 1973, Structure of Franciscan terrane of the Occidental Guerneville area, California: Geological Society of America Bulletin, v. 84, p. 1283–1288.

Cowan, D. S., 1974, Deformation and metamorphism of the Fransiscan subduction zone complex northwest of Pacheco Pass, California: Geological Society of America Bulletin, v. 85, p. 1623–1634.

—— 1978, Origin of blueschist-bearing chaotic rocks in the Franciscan Complex, San Simeon, California: Geological Society of America Bulletin, v. 89, p. 1415–1423.

Craw, D., 1979, Melanges and associated rocks Livingstone Mountains, Southland, New Zealand: New Zealand Journal of Geology and Geophysics, v. 22, p. 443–454.

Crawford, K. E., 1975, The Geology of the Franciscan tectonic assemblage near Mount Hamilton, California [unpubl. Ph.D. dissertation]: Univ. California, Los Angeles, 137 p.

De Jong, K. A., 1974, Melange (olistostrome) near Lago Titicaca, Peru: American Association of Petroleum Geologists Bulletin, v. 58, p. 729–741.

Dimitrijevic, M. D., and Dimitrijevic, M. N., 1973, Olistostrome melange in the Yugoslavian Dinarides and Late Mesozoic plate tectonics: Journal of Geology, v. 81, p. 328–340.

Drake, A. A., and Morgan, B. A., 1981, The Piney Branch Complex—A metamorphosed fragment of the central Appalachian ophiolite in northern Virginia: American Journal of Science, v. 281, p. 484–508.

Elter, P., and Raggi, G., 1965, Contributo alla conoscenza dell'Appennino ligure: 1. Osservazioni preliminari sulla posizione della ofioliti nella zona di Zignago (La Spezia); 2. Considerazioni sul problema degli olistostromi: Bollettino della Societa Geologica Italiana, v. 84, p. 303–322.

—— and Trevisan, L., 1973, Olistostromes in the tectonic evolution of the northern Apennines: *in* De Jong, K. A., and Scholton, R., eds., Gravity and Tectonics: John Wiley and Sons, New York, p. 175–188.

Embley, R. W., 1976, New evidence for occurrence of debris flow deposits in the deep sea: Geology, v. 4, p. 371–374.

Emerson, D. O., and Rich, E. I., 1966, Field trip Sacramento Valley and Northern Coast Ranges: *in* Bailey, E. H., ed., Geology of Northern California: California Division of Mines and Geology Bulletin 190, p. 473–485.

Flint, R. F., Sanders, J. E., and Rodgers, J., 1960, Diamictite, a substitute term for symmictite: Geological Society of America Bulletin, v. 71, p. 1809.

Flores, G., 1955, Discussion: Proceedings of the 4th World Petroleum Conference, Sec. I/A/2, p. 121–122.

Gansser, A., 1955, New aspects of the geology in central Iran: Proceedings of the 4th World Petroleum Conference, Sec. I/A/5, p. 279–298.

—— 1959, Ausseralpine ophiolithprobleme: Ecologae Geologicae Helvetiae, v. 52, p. 659–680.

—— 1974, The ophiolitic melange, a world-wide problem on Tethyan examples: Ecologae Geologicae Helvetiae, v. 67, p. 469–507.

Gasanov, T. A., Gadzhiyev, T. G., and Ali-zade, S. A., 1980, Structure of the serpentinite melange in Tutkhun River basin (Lesser Caucasus): International Geology Review, v. 22, p. 797–804.

Greenly, E., 1919, The geology of Anglesey: Great Britain Geological Survey Memoir, v. 1, p. 980.

Gucwa, P. R., 1975, Middle to Late Cretaceous sedimentary melange, Franciscan Complex, northern California: Geology, v. 3, p. 105–108.

Hopson, C. A., Coleman, R. G., Pallister, J. S., and Bailey, E. H., 1981, Geologic section through the Samail Ophiolite and associated rocks along a Muscat-Ibra transect, southeastern Oman Mountains: Journal of Geophysical Research, v. 86, p. 2527–2544.

Horne, G. S., 1969, Early Ordovician chaotic deposits in the central volcanic belt of northeastern Newfoundland: Geological Society of America Bulletin, v. 80, p. 2451–2464.

Hsu, K. J., 1966, Melange concept and its application to an interpretation of the California Coast Range geology: Geological Society of America Abstracts for 1966, p. 99–100.

—— 1968, The principles of melanges and their bearing on the Franciscan-Knoxville paradox: Geological Society of America Bulletin, v. 79, p. 1063–1074.

—— 1969, Preliminary report and geologic guide to Franciscan melanges of the Morro Bay-San Simeon area, California: California Division of Mines and Geology Special Publication, 35, 46 p.

—— 1974, Melanges and their distinction from olistostromes: *in* Dott, R. H., Jr., and Shaver, R. H., eds., Modern and Ancient Geosynclinal Sedimentation: Society of Economic Paleontologists and Mineralogists Special Publication No. 19, p. 321–333.

Jacobi, R. D., 1981, Modern submarine sediment slides and their implications for "melanges": Geological Society of America Abstracts with Programs, v. 13, p. 479.

——1984, Modern submarine sediment slides and their implications for melange and the Dunnage Formation in north-central Newfoundland, *in* Raymond, L. A., ed., Melanges: Their nature, origin, and significance: Geological Society of America Special Paper 198 (this volume).

Johns, D. R., 1978, Mesozoic carbonate rudites, megabreccias and associated deposits from central Greece: Sedimentology, v. 25, p. 561–573.

Jones, D. L., Blake, M. C., Jr., Bailey, E. H., and McLaughlin, R. J., 1978, Distribution and character of upper Mesozoic subduction complexes along the west coast of North America: Tectonophysics, v. 47, p. 207–222.

Kay, M., 1976, Dunnage melange and subduction of the Protacadic Ocean northeast Newfoundland: Geological Society of America Special Paper 175, 49 p.

Knipper, A. L., 1971, Development of serpentinite melange in the lesser Caucasus: Geotectonics, v. 6, p. 384–390.

Kurenkov, S. A., 1978, The serpentinite melange and olistostrome complexes of the Alai Range (southern Tien Shan): Geotectonics, v. 12, p. 376–382.

Lanphere, M. A., Blake, M. C., Jr., and Irwin, W. P., 1978, Early Cretaceous metamorphic age of the South Fork Mountain Schist in the Northern Coast Ranges of California: American Journal of Science, v. 278, p. 798–815.

Maxwell, J. C., 1959, Turbidite, tectonic and gravity transport, Northern Apennine Mountains, Italy: American Association of Petroleum Geologists Bulletin, v. 43, p. 2701–2719.

——1974, Anatomy of an orogen: Geological Society of America Bulletin, v. 85, p. 1195–1204.

Moiseyev, A. N., 1970, Late serpentinite movements in the California Coast Ranges: New evidence and its implications: Geological Society of America Bulletin, v. 81, p. 1721–1732.

Moore, G. G., and Karig, D. E., 1980, Structural geology of Nias Island, Indonesia: Implications for subduction zone tectonics: American Journal of Science, v. 280, p. 193–223.

Moore, J. C., and Wheeler, R. L., 1978, Structural fabric of a melange Kodiak Islands, Alaska: American Journal of Science, v. 278, p. 739–765.

Naylor, M. A., 1981, Debris flow (olistostromes) and slumping on a distal passive continental margin: The Palombrini limestone-shale sequence of the northern Apennines: Sedimentology, v. 28, p. 837–852.

Page, B. M., 1978, Franciscan melanges compared with olistostromes of Taiwan and Italy: Tectonophysics, v. 47, p. 223–246.

——1981, Ophiolite- and arc-derived conglomerate/breccia blocks in the Kenting melange, Taiwan: Geological Society of America Abstracts with Programs, v. 13, p. 524.

——and Suppe, J., 1981, The Pliocene Lichi melange of Taiwan: Its plate-tectonic and olistostromal origin: American Journal of Science, v. 281, p. 193–227.

Phipps, S. P., 1984, Ophiolitic olistostromes in the basal Great Valley Sequence, Napa County, northern California Coast Ranges, *in* Raymond, L. A., ed., Melanges: Their nature, origin, and significance: Geological Society of America Special Paper 198 (this volume).

Raymond, L. A., 1975, Tectonite and melange—A distinction: Geology, v. 3, p. 7–9.

——1978, A classification of melanges and broken formations: Geological Society of America, Abstracts with Programs, v. 10, p. 143.

Rigo de Righi, M., and Cortesini, A., 1964, Gravity tectonics in foothills structure belt of southeast Turkey: American Association of Petroleum Geologists Bulletin, v. 48, p. 1911–1937.

Saleeby, J., 1979, Kaweah serpentinite melange, southwest Sierra Nevada foothills, California: Geological Society of America Bulletin, v. 90, p. 29–46.

——1984, Tectonic significance of serpentinite mobility and ophiolitic melange, *in* Raymond, L. A., ed., Melanges: Their nature, origin, and significance: Geological Society of America Special Paper 198 (this volume).

——Goodin, S. E., Sharp, W. D., and Busby, C. J., 1978, Early Mesozoic paleotectonic-paleogeographic reconstruction of southern Sierra Nevada region: *in* Howell, D. G., and McDougall, K. A., eds., Mesozoic Paleogeography of the Western United States: The Pacific Section Society of Economic Paleontologists and Mineralogists, Los Angeles, p. 311–336.

Sarwar, G., and De Jong, K. A., 1981, Kanar melange, South Central Pakistan: Geological Society of America Abstracts with Programs, v. 13, p. 545.

——1984, Composition and origin of the Kanar melange, Southern Pakistan, *in* Raymond, L. A., ed., Melanges: Their nature, origin, and significance: Geological Society of America Special Paper 198 (this volume).

Schultz, A. P., 1981, Progressive development of melange fabrics in allochthons of the Pulaski Thrust, southwest Virginia: Geological Society of America Abstracts with Programs, v. 13, p. 549.

Shackleton, R. M., 1954, The structure and succession of Anglesey and the Lleyn peninsula: British Association for the Advancement of Science, v. 11, p. 106–108.

——1956, Notes on the structure and relations of the Precambrian and Ordovician rocks of southwestern Lleyn (Caernarvonshire): Liverpool and Manchester Geological Journal, v. 1, p. 400–409.

——1969, The Precambrian of North Wales: *in* Wood, A., ed., The Precambrian and lower Paleozoic rocks of Wales: Cardiff, Univ. Wales Press, p. 1–22.

——Ries, A. C., Graham, R. H., and Fitches, W. R., 1980, Late Precambrian ophiolitic melange in the eastern desert of Egypt: Nature, v. 285, p. 472–474.

Sharp, R. R., 1940, A Cambrian slide breccia, Grand Canyon Arizona: American Journal of Science, v. 238, p. 668–672.

Silver, E. A., and Beutner, E. C., 1980, Melanges: Geology, v. 8, p. 32–34.

Sporli, K. B., and Bell, A. B., 1976, Torlesse melange and coherent sequences, eastern Ruahine range, North Island, New Zealand: New Zealand Journal of Geology and Geophysics, v. 19, p. 427–447.

Suppe, J., 1973, Geology of the Leech Lake Mountain-Ball Mountain Range, California: University of California Publications in Geological Sciences, v. 107, p. 82.

Swarbrick, R. E., and Naylor, M. A., 1980, The Kathikas melange, Southwest Cyprus: Late Cretaceous submarine debris flows: Sedimentology, v. 27, p. 63–78.

Trumpy, R., 1975, Age and location of Mesozoic scarp breccias in the Swiss Alps: IX International Congress of Sedimentology, p. 313–320.

Vollmer, F. W., 1980, Progressive deformation, fold rotation and melange formation in middle Ordovician Flysch near Albany, New York: Geological Society of America Abstracts with Programs, v. 12, p. 542–543.

——and Bosworth, W., 1984, Formation of melange in a foreland basin overthrust setting: Example from the Taconic orogen, *in* Raymond, L. A., ed., Melanges: Their nature, origin, and significance: Geological Society of America Special Paper 198 (this volume).

Wang, Chao-Siang, 1976, The Lichi Formation of the Coastal Range and arc-continent collision in eastern Taiwan: Bulletin of the Geological Survey of Taiwan, v. 25, p. 73–86.

Williams, H., 1977, Ophiolitic melange and its significance in the Fleur de Lys Supergroup, northern Appalachians: Canadian Journal of Earth Sciences, v. 14, p. 987–1003.

Wood, D. S., 1974, Ophiolites, melanges, blue schists, and ignimbrites: early Caledonian subduction in Wales?: *in* Dott, R. H., Jr., and Shaver, R. H., eds., Modern and Ancient Geosynclinal Sedimentation: Society of Economic Paleontologists and Mineralogists Special Publication, No. 19, p. 334–343.

——and Schuster, D. C., 1978, The nature of melanges; criteria for recognition of their origin with reference to the "type" melange in Wales: Geological Society of America Abstracts with Programs, v. 12, p. 519.

Yilmaz, P. O., and Maxwell, J. C., 1984, An example of an obduction melange: The Alakir Cay Unit, Antalya Complex, Southwest Turkey, *in* Raymond, L. A., ed., Melanges: Their nature, origin, and significance: Geological Society of America Special Paper 198 (this volume).

Zhang, Z., and Jin, M., 1979, Two kinds of melange and their tectonic significance in Xiangcheng-Derong Area, southwestern Sichuan: Scientia Geologica Sinica, v. 7, p. 205–214.

Manuscript Accepted by the Society July 12, 1984

Printed in U.S.A.

Geological Society of America
Special Paper 198
1984

Early deformation in melange terranes of the Ghost Rocks Formation, Kodiak Islands, Alaska

Tim Byrne*
Earth Sciences Board
University of California
Santa Cruz, California 95064

ABSTRACT

The Ghost Rocks Formation represents the youngest phase of a Late Cretaceous to early Tertiary episode of accretion during which about 80 percent of Kodiak Island was added to the continental margin. The formation consists of both structurally coherent terranes and terranes of complexly deformed sandstone-shale melange. Two mesoscopically ductile and regionally occurring deformations and three younger, relatively brittle deformations have affected the melange. D_1 is characterized by a foliation, S_1, which dips northwest and a lineation, L_1, which plunges gently southwest. S_1 is defined by disrupted lithologic layering, pinch and swell structures, the alignment of inclusions, and an anastomosing cleavage in the shales. L_1 is defined by pinch and swell axes, the long axes of inclusions, mullions, and folds of phyllosilicates in the shales. D_2 is characterized by a foliation, S_2, and a lineation, L_2, which respectively dip and plunge parallel to the D_1 fabrics. S_2 is a closely spaced, roughly planar, slaty-like cleavage that commonly transposes, but only locally folds S_1. L_2 is defined by D_2 fold axes and the intersection of S_1 and S_2 which locally is a pencil cleavage. D_1 significantly and substantially altered the original stratigraphy in the melange, and the distinctive disruptive style of the melange is probably a direct consequence of this deformation. The most important structural element that contributed to this deformation, and therefore to S_1 and L_1, was a three-dimensional web structure of cataclastic shear zones. This structural element occurs in essentially every sandstone in the melange and suggests, but does not prove, that D_1 was tectonically induced. Other common D_1 structural features are calcite-filled veins, swirls and folds of detrital or diagenetic phyllosilicates, and planar shear zones in the shales. Asymmetric D_1 structures, especially disrupted tuff horizons, suggest D_1 occurred during northwest directed layer-parallel shear and underthrusting, although other noncoaxial strain histories are also possible. Layer-parallel simple shear, and coaxial plane, constrictional, and flattening strains appear to be ruled out, however, because they do not explain the general, mesoscopic strains observed in the melange.

INTRODUCTION

Melanges are mappable, but internally chaotic, rock units found in orogenic belts throughout the world (Silver and Beutner, 1980). Despite their universal occurrence, however, they are still poorly understood and their origin and tectonic significance is controversial (see other papers in this volume and Cowan, in press). Many melange terranes are intensely deformed and are associated with blocks of high-pressure and low-temperature metamorphism. Consequently, these terranes are commonly interpreted to be the exposed remnants of ancient subduction zones. In contrast, some melanges have been interpreted as gravitationally mixed rock bodies that accumulated at either passive or convergent margins. Documenting a specific deformational environment for melanges is usually dependent, however, on distin-

*Present address: Department of Geological Sciences, Brown University, Providence, Rhode Island 02912.

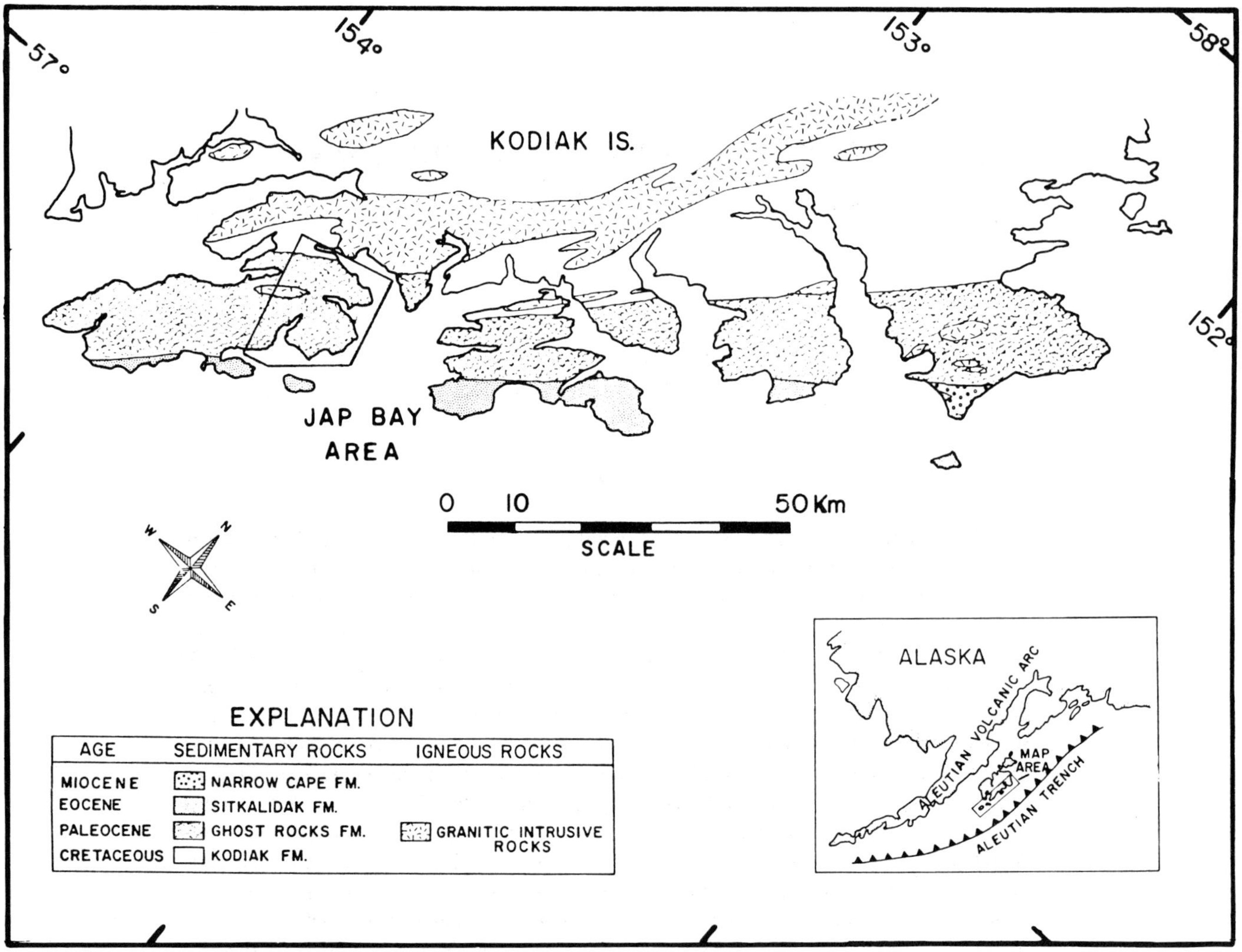

Figure 1. Geologic map of southeastern Kodiak Island. Compiled from G. Moore (1967), C. Moore, and A. Allwardt (1980), and Moore and others (1983).

guishing tectonically and gravitationally induced structures; these structures, and the associated processes, are probably gradational in nature. As a consequence, what are believed to be "obvious" interpretations can lead to what are really arbitrary conclusions. These problems are particularly acute in sandstone-shale melanges because of the inherent ambiguities in determining which soft-sediment structures are of purely tectonic origin. Two equally important and controversial problems in melanges are: What specific strain histories are responsible for the distinctive fabrics and structures, and what role do melanges play in the subduction of oceanic crust and the accretion of deep-sea sediments at convergent margins? These questions are especially relevant if we are to understand the geological consequences of plate convergence and plate reorganizations, and to interpret ancient plate boundaries accurately. The Ghost Rocks Formation is an excellent unit in which these questions can be addressed because it is young, only mildly metamorphosed, contains a variety of deformational structures, and rests in an unambiguous tectonic setting.

The continental margin of southwest Alaska has experienced repeated episodes of subduction-related magmatism and accretion since at least the early Mesozoic. Magmatism has been concentrated on the Alaska Peninsula, and is continuing today, while accretion of deep-sea rocks progressively broadened the continental shelf to the southeast. The Kodiak Islands constitute an emergent portion of this enlarged continental shelf and expose northeast-trending belts of rocks that dip northwest and young progressively toward the present day Aleutian trench (Figure 1; Moore and Connelly, 1979; Moore, 1969). The Ghost Rocks Formation (Armentrout, 1979a; Moore, 1969) comprises the second youngest sequence of deep-sea deposits in the Kodiak Islands and is in fault contact with older rocks to the northwest and younger rocks to the southeast. Geological mapping and recent paleomagnetic data suggest that the Kodiak margin was probably far south of its present position (Stone and others, 1981; Plumley and others, 1982, 1983; Moore and others, 1983) at the time when the Ghost Rocks Formation was deposited and deformed.

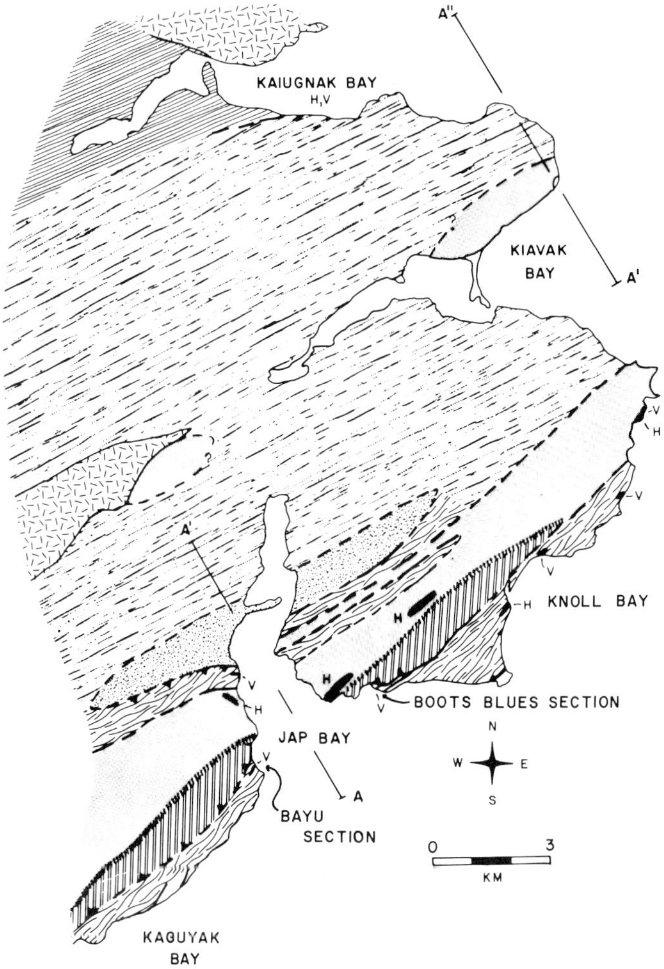

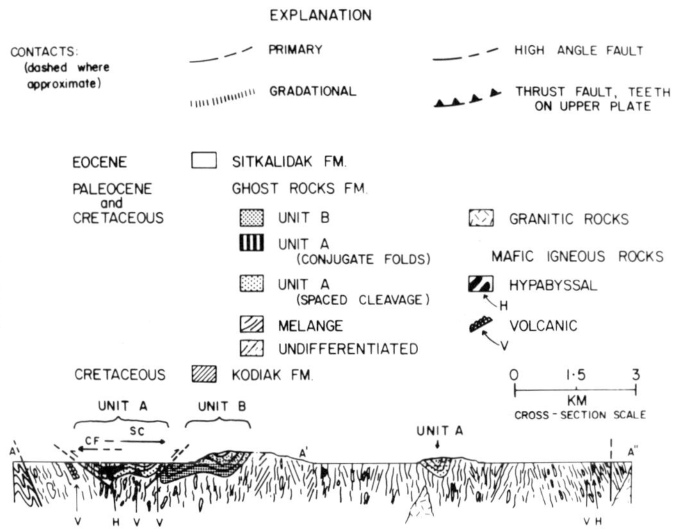

Figure 2. Geologic map and cross section of the Jap Bay area. Generalized from coastline (excellent exposure) and inland (poor exposure) traverses. In cross section, SC and CF, respectively, show extent of spaced cleavage and conjugate fold belts.

Geochemical trends (Reed and Lanphere, 1974) in the Alaska-Aleutian Range batholith series and plate reconstructions (Moore and others, 1983; Wallace and Engebretson, 1984) suggest however that plate convergence was toward the north or northwest (present coordinates) during the time of deposition and accretion of the Ghost Rocks Formation.

The Ghost Rocks Formation crops out in a belt 160 km long and 15 km wide along the southeast side of the Kodiak Islands (Figure 1). The formation consists of complexly deformed sandstone and shale with interbedded volcanic rocks and local hypabyssal intrusions. The sedimentary rocks are dominantly thin to thick bedded (1-5 cm to tens of meters), consisting primarily of massive sandstones and interbedded shale. Pebbly mudstone and conglomerate also occur locally. Although tectonic deformation has severely dismembered almost all original stratigraphy, sedimentary structures in the sandstones and shales indicate turbidite deposition; and the occurrence of massive sandstone, pebbly mudstone and conglomerate suggest local deposition in channelized environments. The igneous rocks of the Ghost Rocks Formation are both interbedded with and intrude the sedimentary rocks and include pillow lava, pillow breccia, tuff, and dikes and sills of andesite and basalt (Moore and others, 1983; Byrne, 1982a; Reid and Gill, 1980; Hill, 1978; Reid, in preparation). Fossiliferous pelagic limestones of Paleocene and late Cretaceous age typically overlie or are interbedded with the volcanic rocks (Moore and others, 1983; Byrne, 1982a).

Silicic plutonic rocks also intrude the Ghost Rocks Formation. These consist of quartz dioritic to tonalitic plutons that probably represent satellitic intrusions of the Kodiak batholith. The pluton exposed northwest of Jap Bay (Figure 2) is similar to other plutonic rocks of the Ghost Rocks Formation and yields ages that range from 60 to 63 m.y. (Byrne, 1982a; Armentrout, personal communication, 1978).

The conclusions presented here are derived from a detailed study in the Ghost Rocks Formation of the Jap Bay area (Figure 2; Jap Bay was apparently named for a Japanese family that lived in the area in the 1920's). In this area, detailed mapping revealed the presence of two dominant structural styles. These structural styles are broadly differentiated as melange and coherent terranes, which occur as mappable belts that strike northeast, parallel to the structural grain of Kodiak Island (Figure 2). The coherent terranes are mildly deformed sedimentary deposits that I interpret to be analogous to the slope basin deposits recognized in many modern convergent margins (Byrne, 1982a). The melange terranes are more complexly deformed than the coherent terranes and, because of their unique deformation history, may have been deposited in or near a tectonically active trench environment. In this paper, I present the results of detailed study of the mesoscopic scale features that characterize the melange and suggest possible

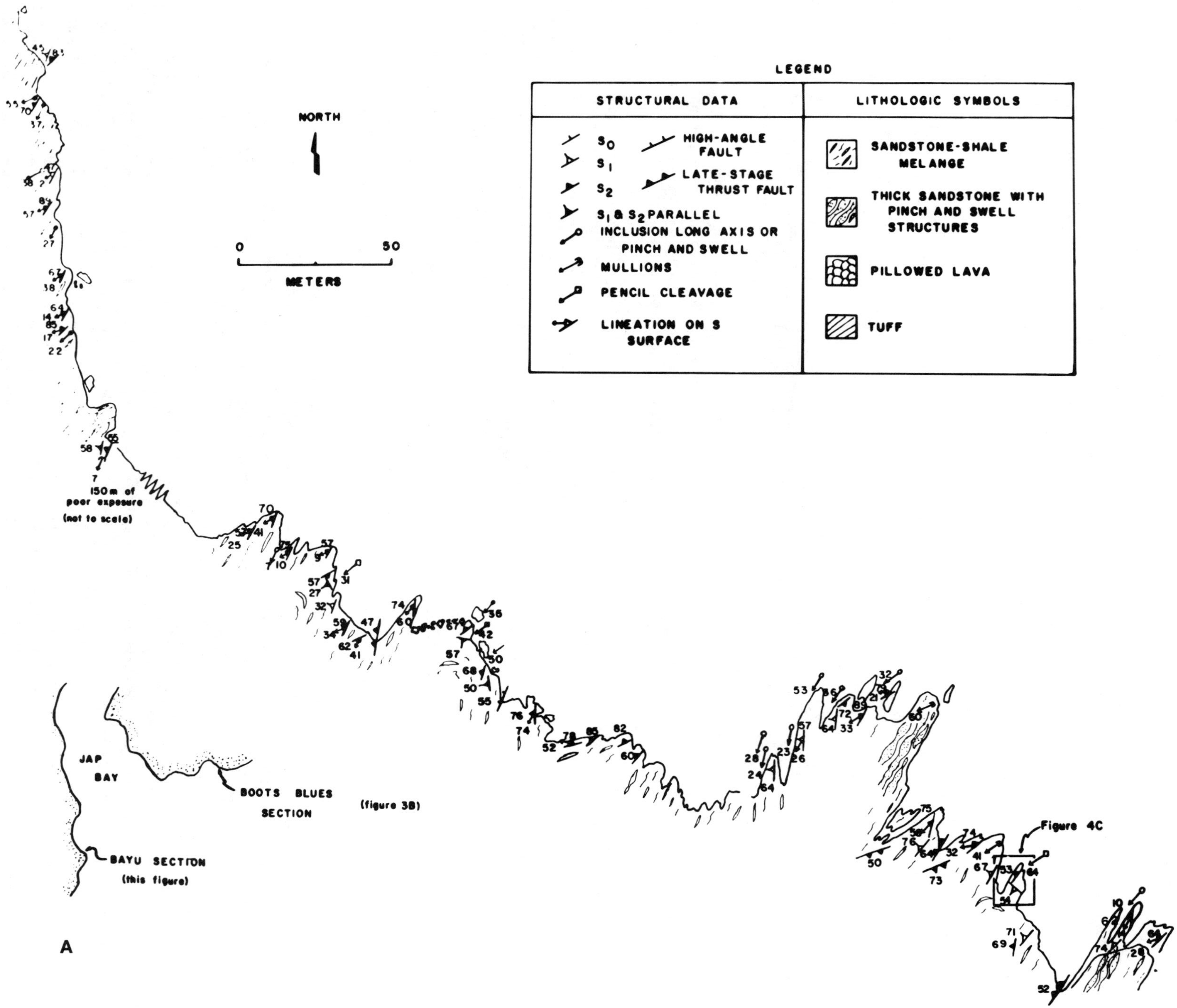
LEGEND
STRUCTURAL DATA
S_0
S_1
S_2
S_1 & S_2 PARALLEL
INCLUSION LONG AXIS OR PINCH AND SWELL
MULLIONS
PENCIL CLEAVAGE
LINEATION ON S SURFACE
HIGH-ANGLE FAULT
LATE-STAGE THRUST FAULT
LITHOLOGIC SYMBOLS
SANDSTONE-SHALE MELANGE
THICK SANDSTONE WITH PINCH AND SWELL STRUCTURES
PILLOWED LAVA
TUFF
NORTH
0
50
METERS
150m of poor exposure (not to scale)
Figure 4C
JAP BAY
BOOTS BLUES SECTION
(figure 3B)
BAYU SECTION
(this figure)
A

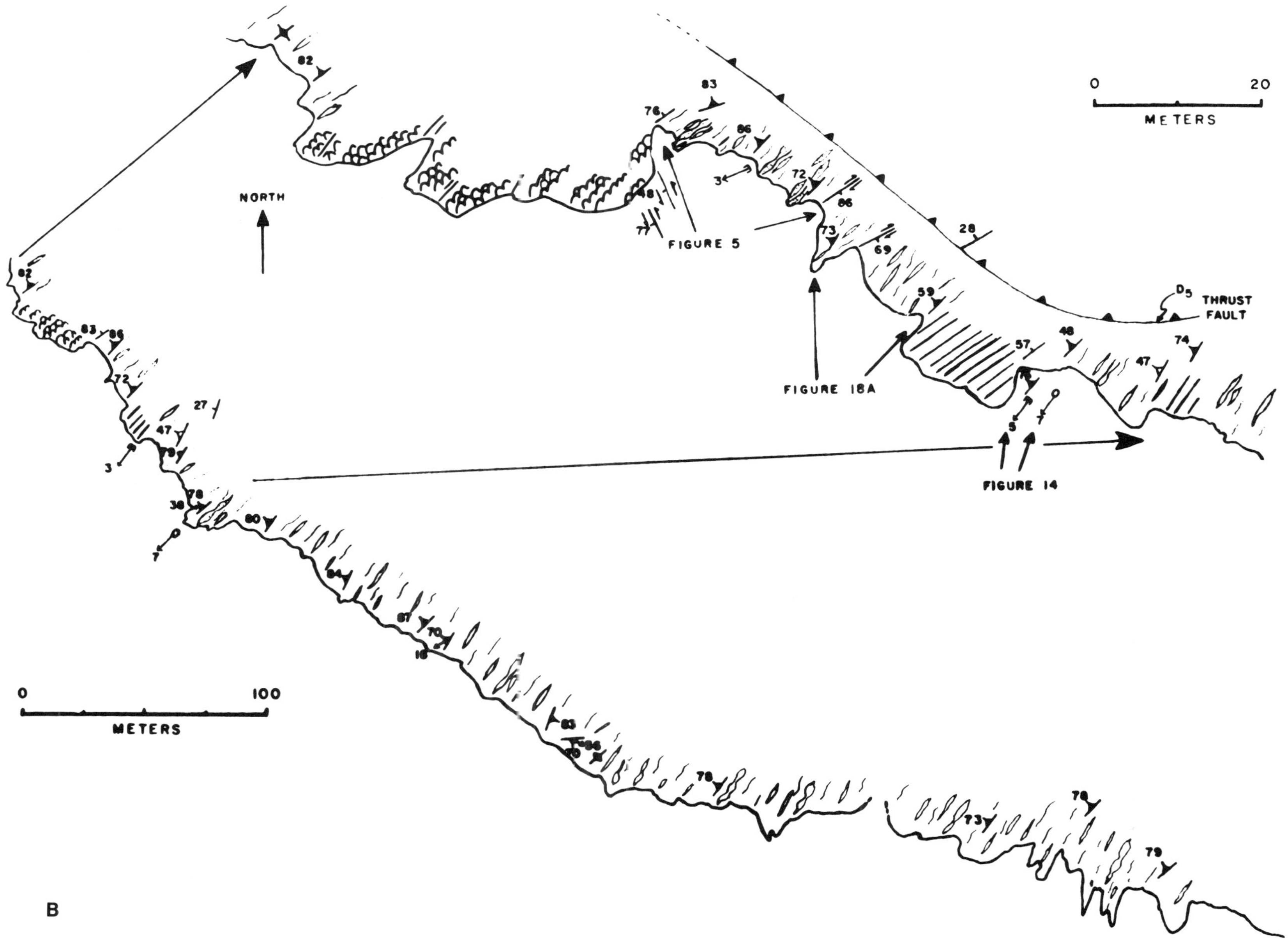

Figure 3. Structural and geologic strip maps across the Bayu (A) and Boots Blues (B) melange zones.

deformation histories for its development. Thorough studies of the microscopic structures and associated strains are in progress.

MELANGE TERRANES

Areas designated as melange (Figure 2) are characterized by strongly foliated and discontinuous sandstone-shale units in which characteristic structures and individual lithologic blocks are not resolvable at any convenient mapping scale. Sandstone and shale are the dominant rock types, but conglomerate, pebbly mudstone, limestone, pillow lava and tuffaceous units also occur, both interlayered in the matrix and as inclusions. On a mesoscopic scale, the foliation occurs only in the pelitic rocks and distinctively anastomoses around the more competent sandstone and greenstone blocks, resulting in a "phacoids in a matrix" appearance at most outcrops. Characteristically, the sandstones and foliated shales display a well-developed linear fabric discussed in more detail below.

No inclusions of exotic metamorphic rocks have been found. The volcanic layers and inclusions, although not typical trench lithologies, are similar to other volcanic rocks clearly interbedded with sedimentary rocks in less deformed sequences in the Jap Bay area and along strike (see Moore and others, 1983 and Byrne, 1982a). Thus, the melange terranes do not contain lithologically or metamorphically exotic inclusions. The melanges are, instead, characterized by their extreme stratigraphic disruption (see discussion by Raymond, this volume and Cowan, in press). Overall, the melange terranes are lithologically indistinguishable from the oldest coherent terrane (Unit A, Byrne, 1982a).

The melange in the Jap Bay area contains fossils of planktonic foraminifera of Upper Cretaceous age (R. Poore, written communication, 1980). The fossils occur in highly deformed pods and layers of limestone that are intercalated with basaltic pillow lavas. Locally, shale and sandstone are also interlayered with the volcanic rocks and limestones. These relationships and the common occurrence of interpillow pelagic limestones interbedded with volcanic rocks in less-deformed areas of the Ghost Rocks Formation (Moore and others, 1983) suggest that at least some of the melange lithologies in the Jap Bay area are of Cretaceous age. The age of deformation of the melange is therefore Cretaceous or earliest Paleocene (pre-62 m.y. B.P.), as the intruding plutons are only mildly deformed (Byrne, 1982a; Moore and others, 1983).

Diagnostic metamorphic minerals and mineral assemblages are rarely present in the melange. Pumpellyite has been recognized in veins and in the altered groundmass of the volcanic rocks, but has not been observed in the sedimentary rocks. However, tuff layers, originally interbedded with the sedimentary rocks of the melange (discussed below), currently contain the metamorphic minerals prehnite and quartz. The low grade of metamorphism indicated by these minerals is also suggested, in the sedimentary rocks, by the absence of dynamic recrystallization in deformed quartz and calcite, by the alteration of plagioclase to albite + calcite ± epidote, and by the absence of K-feldspar. Detrital grains of K-feldspar are present in the less-deformed coherent terranes. Fine-grained crystals of stilpnomelane, epidote, chlorite and quartz commonly occur as replacement minerals on the matrix of the sedimentary rocks.

My detailed structural studies of the melange were completed along two cross sections exposed at the entrance of Jap Bay (Figure 3). Each section was mapped using the pace and compass technique at a scale of approximately 1:300. Although not depicted by the resulting maps, the melange is excellently exposed along the sea cliffs in this area and structural features commonly can be observed in three dimensions. For convenient reference, I informally designate the section along the southwest side of Jap Bay as the "Bayu" section (for USCGS benchmark located along the section), and the section exposed along the northeast side of Jap Bay as the "Boots Blues" section (for the water-filled boots that were consistently required to land a boat in this area). Traverses were also made along the sea cliffs of Knoll and Kaguyak Bays to the northeast and southwest. All of these sections are lithologically similar and contain a distinctive sequence of volcanic rocks that can be traced almost the entire length of the melange in this area (approximately 20 km). Structurally, however, the sections across Jap Bay are surprisingly different.

STRUCTURAL GEOLOGY

The melange terranes are characterized by four dominant structural elements (see Figures 4, 5, 6 and 7).

S_1 is a foliation defined by lithologic layering, the alignment of inclusions and a weak, hackly cleavage in the shales.

L_1 is a linear element formed by elongate inclusions and pinch and swell structures in S_1, a wavy mullion-like feature that occurs on the surface of sandstone units, and a lineation on S_1 in the shales.

S_2 is a steep NW-dipping, planar, pressure solution cleavage that is best developed in the shales. This cleavage locally folds and commonly transposes S_1.

L_2 is a linear element formed in part by the intersection of S_1 and S_2, and is locally a pencil cleavage.

The two linear fabrics, L_1 and L_2, are parallel and plunge gently to steeply southwest. The deformations related to these structural fabrics are designated D_1 and D_2, respectively.

Primary bedding features, S_0, were recognized in only a few isolated inclusions. In these inclusions, S_0 was oriented either parallel to S_1 or it had been rotated 30 to 40 degrees away from S_1 by antithetic normal faults. Thus, I interpret S_1, and the associated across strike variations, as a feature inherited from S_0 (i.e., ghost-or pseudo-stratigraphy; see Figure 5, 14 and 18A). As discussed below, neither S_0 nor S_1 appears to be isoclinally folded on a regional scale.

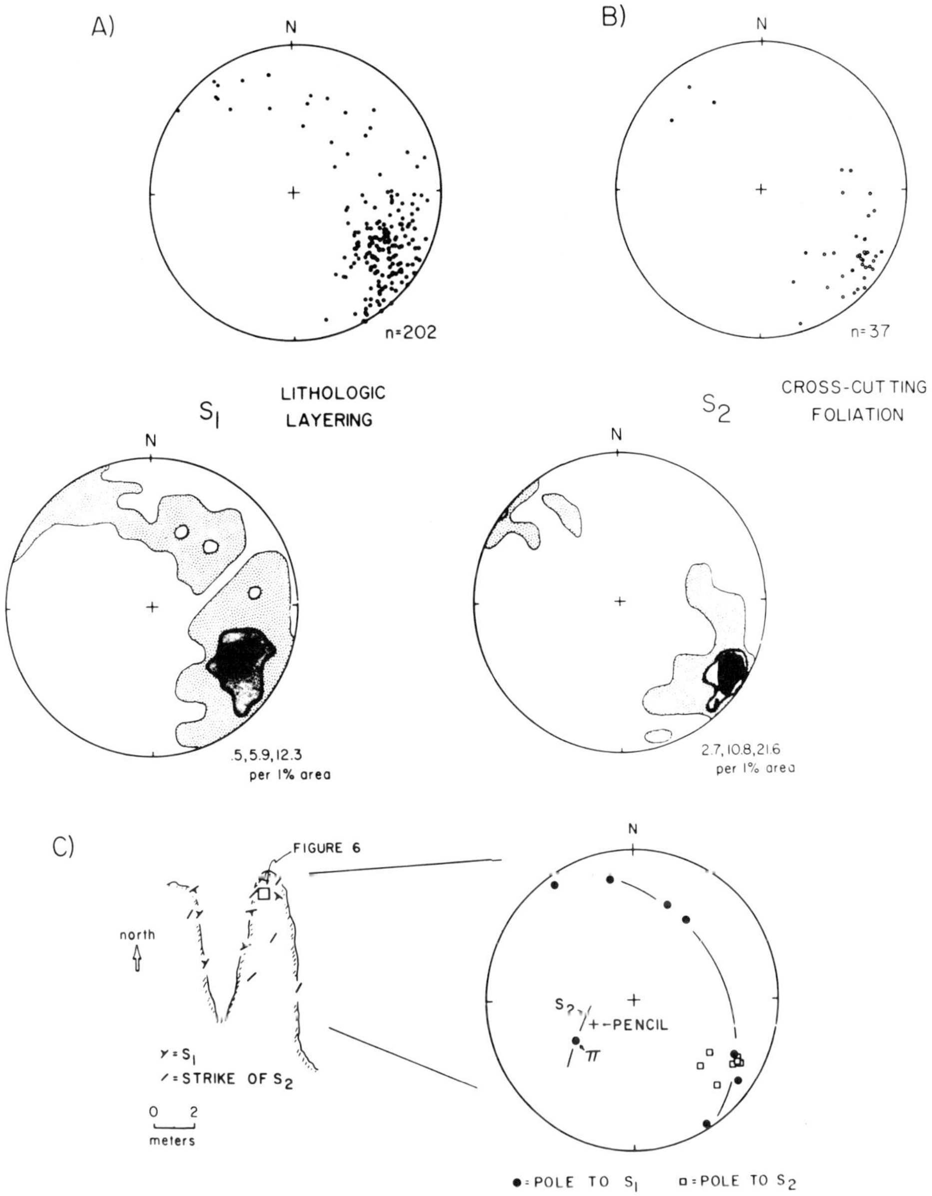

Figure 4. Stereographic plots of poles to S_1 (A) and poles to S_2 (B) and detailed structural map of area of D_2 fold (C). Location of map is shown in Figure 3A. In Figure 4C the stereographic projection shows poles to S_1 and S_2 in the area of the D_2 fold. The pi-axis for the D_2 fold, the average pencil cleavage, and the intersection of S_1 and S_2 are also shown. The pencil cleavage occurs only in the hinge of the D_2 fold.

The rocks in the melange have experienced several later, less penetrative brittle deformations as well (summarized in Figure 8). The oldest is represented by generally east-trending, right-lateral strike-slip faults and north-trending left-lateral strike-slip faults. These fault systems appear to have been contemporaneous and indicate subhorizontal, N45W directed compression. These conjugate faults are cross-cut by moderately dipping thrust faults (shown as D_5 in Figures 3B and 8 but as D_2 of the coherent terrane deformations in Byrne 1982a). The thrust faults dominantly dip to the northwest, but conjugate sets are also present. Slickenside lineations on the fault surfaces indicate underthrusting of the melange toward N25W. The youngest brittle deformation is recorded by east-west trending conjugate normal faults. These faults are relatively rare in the melange terranes, but the few observed faults have the same relative age as, and are parallel to, the normal faults that are very common in the youngest coherent terrane (Unit B). Although these younger deformations have affected the melange, in this paper I discuss in detail only the regionally penetrative and mesoscopically ductile deformations, D_1 and D_2. A comparison and correlation of deformations recognized in the coherent and melange terranes is in preparation (see also Byrne, 1982b).

Figure 5. Photograph of Boots Blues melange zone. Note general continuity (pseudostratigraphy) of some completely disrupted sandstone layers. Well-developed mullions are present on the lower surface of the thick sandstone in the center of the photograph. S_1 and S_2 are parallel in this photograph.

Figure 6. Photograph showing cross-cutting relationships between S_1 and S_2 in the hinge area of the D_2 fold shown in Figure 4C. Outcrop surface shown in photograph is approximately normal to S_1/S_2 intersection.

Figure 7. Photograph showing very well-developed lineation in melange matrix. The lineation is formed in part by small D_1 folds and swirls of the shale, by elongate sandstone inclusions, and by the intersection of S_1 and S_2. S_1 and S_2 intersect at a very low angle in this sample.

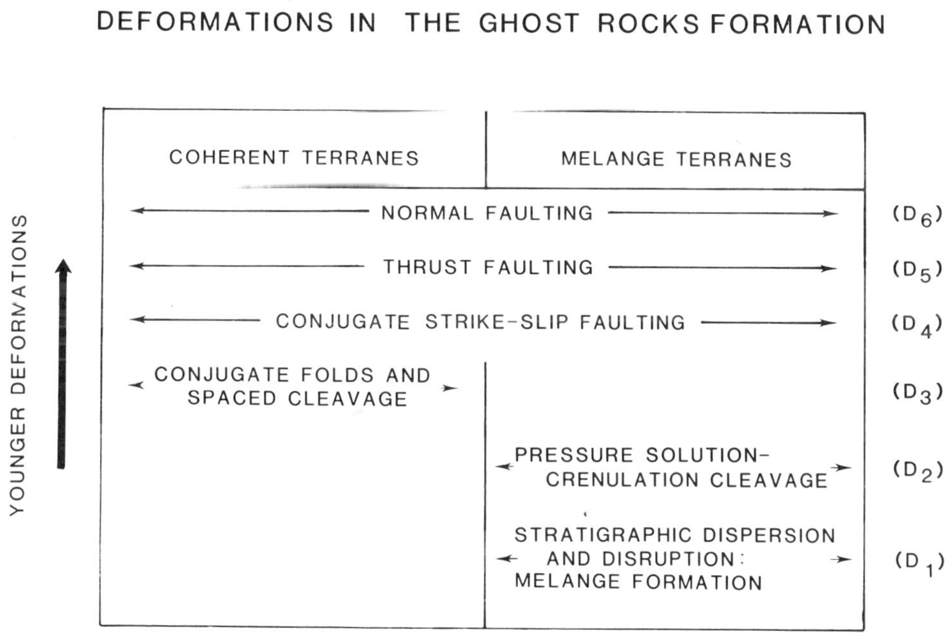

Figure 8. Summary of deformations recognized in the Ghost Rocks Formation. D_1 and D_2 are discussed in detail in this paper. D_3 and D_5 are discussed in Byrne (1982a). D_1 through D_4 are believed to be Late Cretaceous or early Tertiary (pre-62 m.y.) in age (Byrne, 1982b).

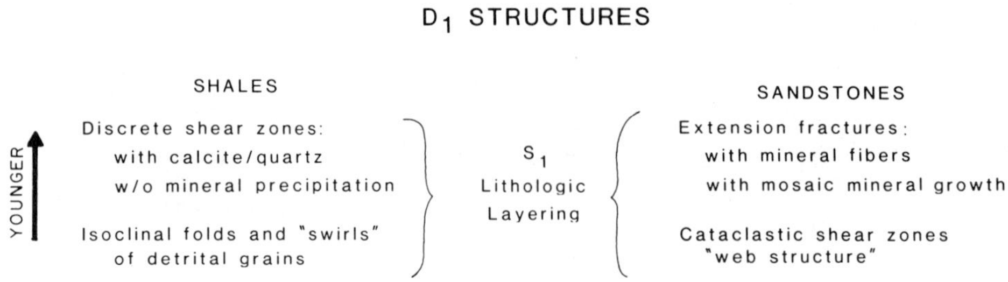

Figure 9. Summary of D_1 structural features recognized in the shale matrix and sandstone inclusions. All D_1 structures together define S_1 and L_1. In general, the structures get younger towards the top of the diagram.

FIRST DEFORMATION (D_1)

The deformation responsible for the development of S_1 substantially and significantly altered the original sedimentary bedding and stratigraphy in the melange. Most exposures of the melange consist of subrounded to subangular elongate ellipsoidal sandstone inclusions dispersed in a matrix of dark gray foliated shale (e.g., see Figures 5 and 6). Tabular and continuous (up to at least several meters) layers of sandstone and tuff display pinch and swell structures and grade into the completely isolated and dispersed ellipsoids. For descriptive purposes I have chosen the nongenetic terms "inclusions" and "matrix" to separate objects of known limited dimensions from the material that engulfs and surrounds them. Typically, but not always (e.g., the D_1 folds of phyllosilicates), the inclusions and matrix have different compositions.

Pinch and swell structures and trains of inclusions together generally define S_1 and L_1 in the melange. In detail, however, S_1 and L_1 are represented by a variety of textures and structures (Figure 9). In the following sections, I discuss these various D_1 features in the context of their relationships to the inclusions, the matrix, and the greenstone layers. D_1 structural features associated with the inclusions include: 1) the inclusions' external shape and texture; 2) an internal three-dimensional "web" structure of shear zones; and 3) calcite- and quartz-filled extension fractures, which are the latest D_1 structural elements in the inclusions. D_1 structural features found in the shale matrix are: 1) a fabric formed by the alignment of detrital phyllosilicates; 2) isoclinal folds and swirls of this fabric; 3) a microscopic set of shear zones; and 4) a mesoscopic set of shear zones, similar to 3), which also commonly have zones of calcite associated with them. These D_1 structural features are illustrated diagrammatically in Figure 10.

The discussion of the mesoscopic character of the deformed layers of greenstone and tuff units has been separated from that of the inclusions and matrix, mainly because of their variable response to D_1. The greenstones (pillowed lava, sills and dikes) behaved more rigidly than the sandstones, whereas the tuffs behaved in a manner intermediate between the sandstones and shales. At a more regional scale, these volcanic horizons also provide an important marker horizon within the melange terrane.

INCLUSIONS

Pinch and Swell Structures and the Shapes of Inclusions

Mesoscopic three-dimensional pinch and swell structures in the sandstones occur in two orthogonal orientations and resemble classic chocolate tablet boudinage (Ramsay, 1967). The "tablets" in the Ghost Rocks Formation melange, however, are elongate ellipsoids (Figure 10). They are not square or circular, and the layers do not appear to have extended uniformly in all directions (compare to Cowan, 1982). In general, the necks and long axes of the swell structures plunge gently to the southwest, trending approximately parallel to the strike of the sandstone layering, which dips steeply to the northwest. Because the pinch and swell structures may have formed from literal "pinching" and "swelling" of the sandstone layers, the amount of extension in any direction, if any, is not known. In general, greater extension may have occurred down-dip (to the northwest) than along strike (to the southwest), as the layers are more discontinuous in that direction. The pinch and swell structures appear to grade into the isolated sandstone inclusions at all scales.

At most outcrops, the melange inclusions are irregularly shaped, commonly elongate, oblate to prolate ellipsoids. Typically, the inclusions vary in size from being only resolvable with a hand lens to possibly tens of meters in their shortest dimension. To quantify their shapes, I exhumed and carefully measured 22 inclusions (Figure 11). The inclusions are always encrusted by a thin veneer of compacted, finely fissle shale (e.g., Figure 12). The veneer generally mimics the shape of the inclusions, but masks and smooths many of the complex, irregular details of the internal fabric of the sandstone. The external texture of the shale veneers varies from being glossy and scalloped (scaly) to dull, raspy, and slickensided. A triangular plot of the aspect ratios of the inclusion axes shows clearly that the inclusions are elongated ellipsoids with intermediate and short axes of slightly different lengths (Figure 11). The long axes of the inclusions plunge gently to moderately southwest (Figure 11).

The internal configuration of the inclusions is best observed on cut and polished slabs or polished beach exposures, because of the shale veneer that encrusts the inclusions. The inclusions are generally circular to lensoidal in sections perpendicular to their

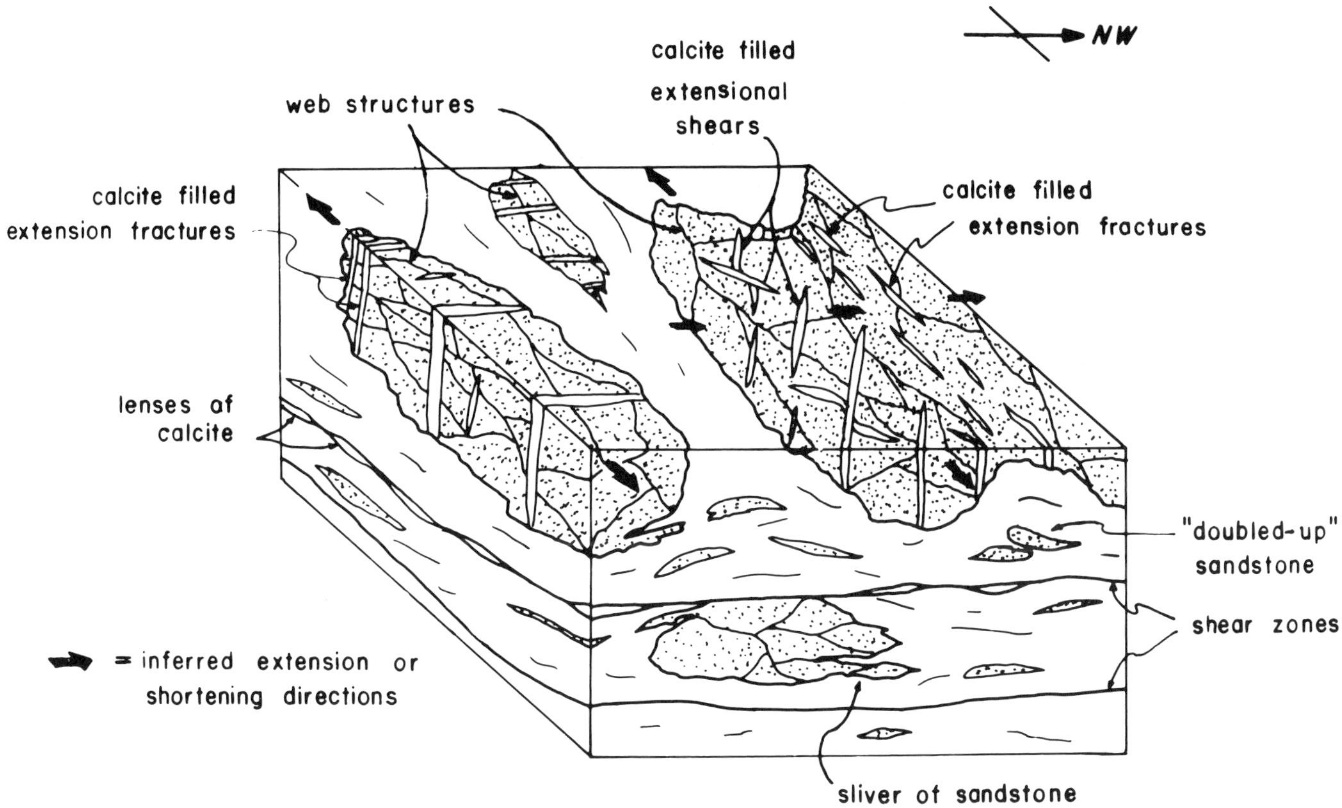

Figure 10. Schematic block diagram illustrating the distribution and relationship of structural features that define S_1 and L_1. S_1 is approximately parallel to the upper surface of the diagram, and L_1 parallels the southwest-trending edge. Stippled areas are sandstone and clear areas are shale, except where noted. Sandstone in upper right represents pinch and swell structure and sandstone at left represents isolated inclusion.

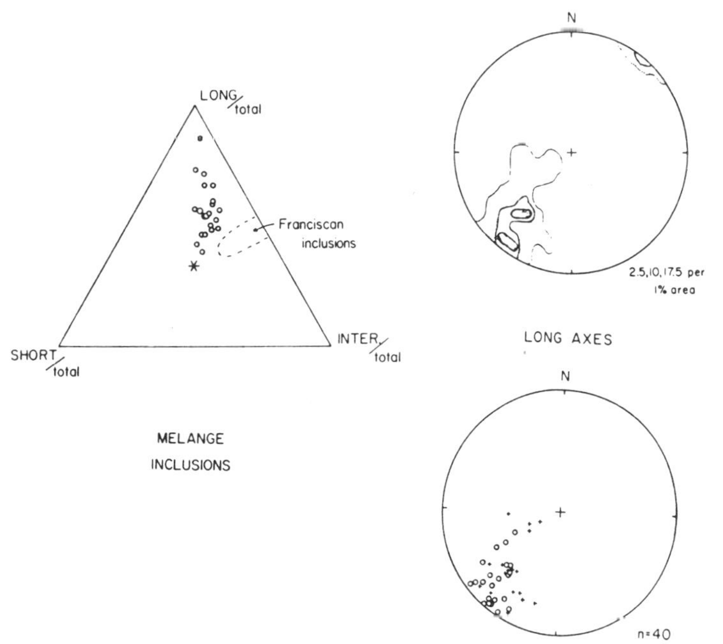

Figure 11. Triangular diagram showing the aspect ratios of sandstone inclusions in the melange. Only one-sixth of this diagram is applicable to defining inclusions shapes. Stereographic plot shows the southwest plunge of the inclusion long axes. General field for Franciscan inclusions is from Cowan (1982).

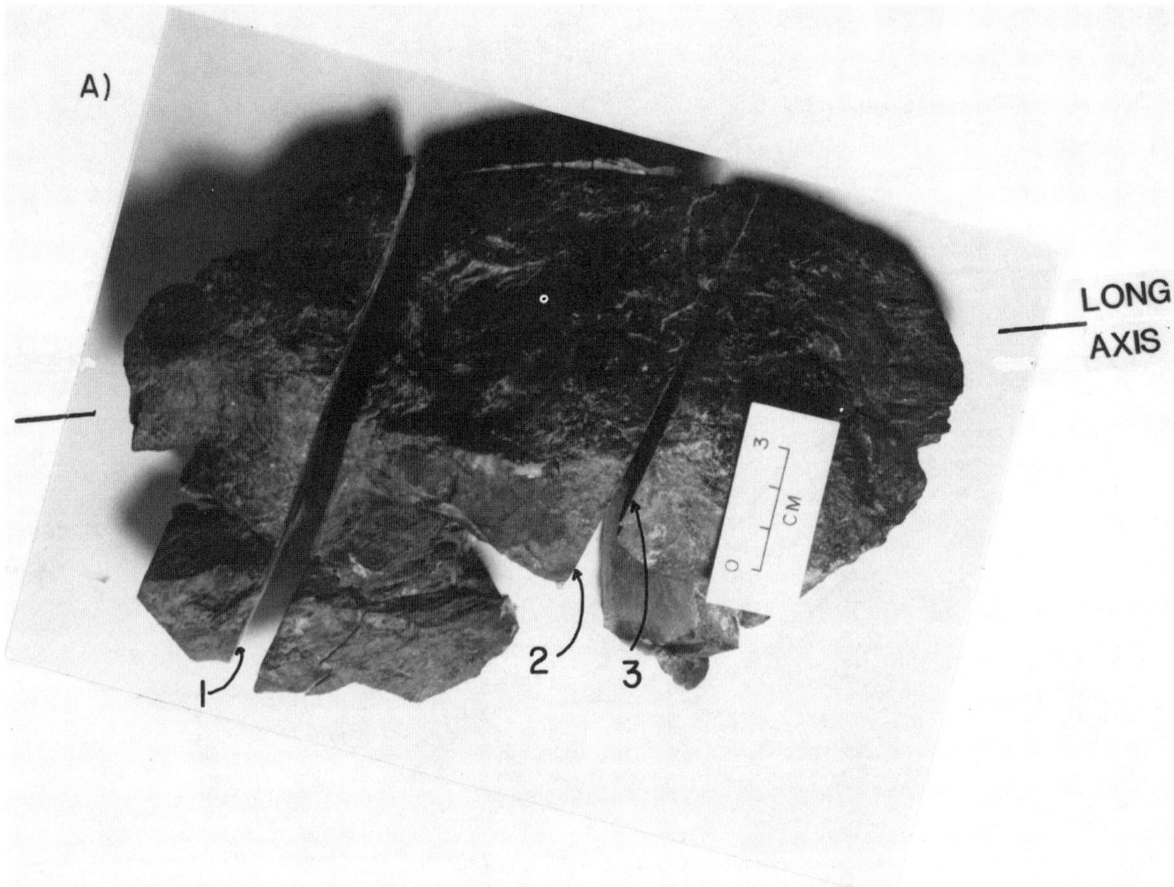
A)
LONG AXIS
1
2
3
0
3
CM

B)
0
3
CM
1
2
3

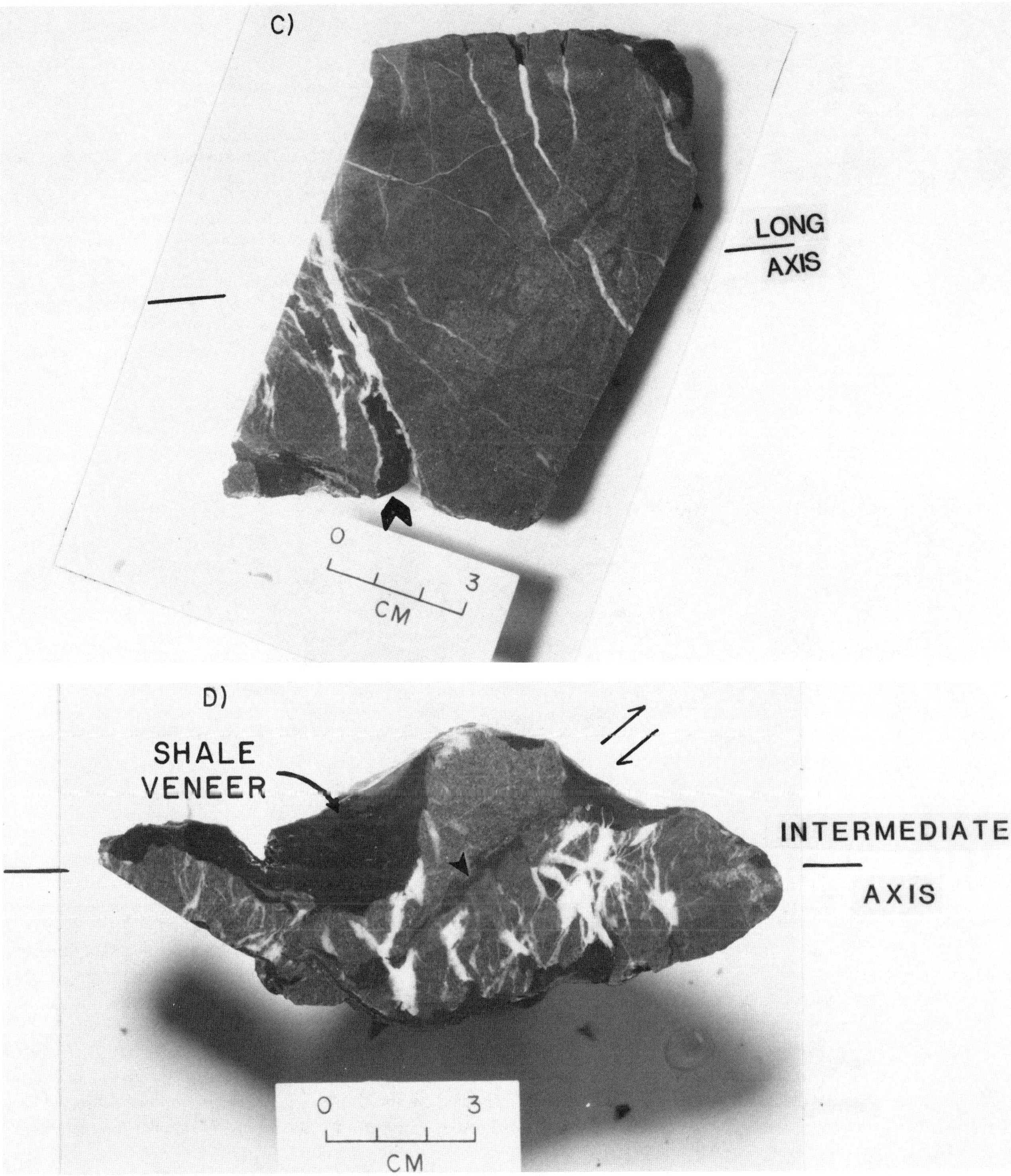

Figure 12. Photographs of the external and internal features of an exhumed sandstone inclusion. (A) shows the external shape and texture and the shale veneer that coats the inclusion. The long axis of the inclusion is shown. The orientation of the surfaces in (B) and (D) are also shown. (B) shows the internal textures and structures of the inclusions. Note that the "ductile" thrust (shown in D also) in slab 1 is not present in the other slabbed surfaces. (C) shows a slabbed and wetted surface oriented normal to the short axis of the inclusion. Photograph is in the same orientation as in A. Note the black tongue of shale that intrudes and widens the calcites vein just above the 1 cm mark (arrow). (D) shows the "ductile" thrust zone. The thrust zone has apparently been intruded by shale (see arrow) which is locally cross-cut by calcite veins. Note the shale veneer that encrusts the sandstone.

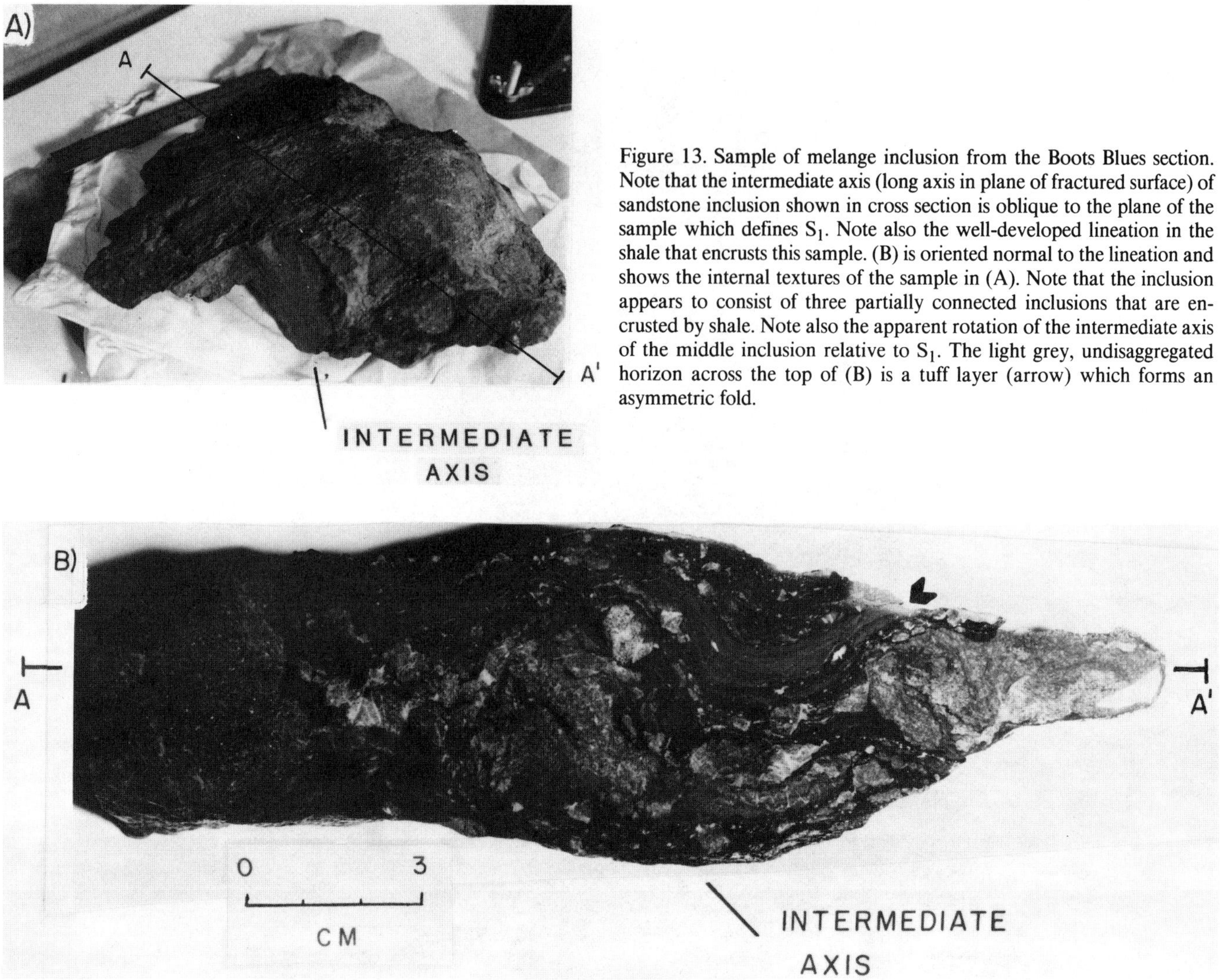

Figure 13. Sample of melange inclusion from the Boots Blues section. Note that the intermediate axis (long axis in plane of fractured surface) of sandstone inclusion shown in cross section is oblique to the plane of the sample which defines S_1. Note also the well-developed lineation in the shale that encrusts this sample. (B) is oriented normal to the lineation and shows the internal textures of the sample in (A). Note that the inclusion appears to consist of three partially connected inclusions that are encrusted by shale. Note also the apparent rotation of the intermediate axis of the middle inclusion relative to S_1. The light grey, undisaggregated horizon across the top of (B) is a tuff layer (arrow) which forms an asymmetric fold.

long axis, although in detail they have highly irregular boundaries. Bulbous protrusions of sand are often swirled or lobate and occur on the upper and lower surfaces of the inclusions. Some inclusions have angular to subrounded boundaries and appear to have been brittle when deformed. Others contain calcite veins that are fractured and cross-cut by thin injections of shale (see Figures 12C and 20). Many inclusions form distinct lensoid shapes in cross section and display "tails" or "wisps" of sandstone that taper out in the plane of the foliation. These inclusions often pinch out over a distance much less than the exposed width of the inclusions. In any exposure, there is a gradation between extension fractures filled with calcite or injections of shale to extreme pinch and swell structures. Mesoscopically, the sandstones in the melange display both brittle and ductile behavior.

One sandstone inclusion, exhumed and slabbed, appears to have been shortened, instead of extended, parallel to its intermediate axis (Figure 12). The external shape and texture of this inclusion are typical of others in the melange, including the shale veneer (1 mm to 1 cm thick) that coats the outside of the inclusion. The slabbed surface, oriented approximately normal to the long axis of this inclusion (Figure 12B and D), reveals, however, that the inclusion consists of two overlapping sandstone layers instead of a single pinched bed. These two sandstones apparently represent a single thrusted sandstone layer; the thrust zone is marked by a thin zone of shale. A second slabbed surface, oriented parallel to the first, shows a thickened sandstone layer but no well-delineated thrust zone (Figure 12B). The asymmetry represented by the thrust fault indicates northwest-directed underthrusting and is consistent with other asymmetric structures in the melange (discussed below).

One example of a mixed and extremely disrupted sandstone inclusion is shown in Figure 13. Note the striking difference between the disruption observed in hand sample and that observed on a slabbed surface of the same sample (Figures 13A and B). The sandstone is very disaggregated and displays a wide range of textures and structures, although there is a crude continuity that parallels S_1. The sandstone is laced by calcite and quartz fractures and a gray material. The traces and areas of this grayish,

Figure 14. Photograph of folded (F_1) sandstone inclusion (arrow) in Boots Blue section (see Figure 3B for location). Note the asymmetry of the fold. An isoclinal fold (synformal hinge only) also occurs in the right edge of the photograph. The massive, finely foliated units (marked with a "T") are tuff horizons that partially define S_1. Other horizons are sandstone and shale. S_2 dips more steeply to the left (northwest) than S_1 in this area.

fine-grained material are zones of grain size reduction and intense cataclasis. This is a good example of stratigraphic disruption through cataclastic flow and dispersion. The tuff layer is highly foliated, but does not display the mesoscopic scale disruption exhibited by the sandstone. A poorly developed asymmetry is suggested by the rotated intermediate axes of the center inclusion and by the folded tuff layer.

A few sandstone inclusions in the melange are folded or display swirl-like patterns. The folded inclusions are always rootless and have tight to isoclinal limbs (Figure 14). In most cases, pinched and swelled and highly disrupted sandstone layers appear to have been folded, whereas in a few folds primary bedding features define the fold. I observed only six folds of primary bedding in the melange and these all had axes that plunged to the southwest. Three of the folds were asymmetric and verged to the southeast. A few inclusions appear to have been swirled or rolled about their long axes. Sections cut normal to the long axes of these inclusions show that the inclusions are composed of several lens-shaped pieces of sandstone that are concentrically shape-oriented around the long axis. Serial sections reveal that the sand-

Figure 15. Photograph of wet, beach-polished surface of massive sandstone in the melange. The thin cross-cutting grey to black zones are areas of cataclasis and shearing and form a three-dimensional web structure. These features are characteristic of essentially all sandstones in the melange.

stone lenses pinch and swell and coalesce into larger pieces in the direction of the long axis of the intrusion. A matrix of indurated shale supports these sandstone pieces and forms a veneer that disguises the internal complexities of the inclusion. I observed no asymmetry in the swirl-like patterns in these inclusions. Both the folds and the swirled inclusions are riddled by web structure (discussed below) and grey areas of cataclasis.

About 10 isolated inclusions of vesicular volcanic rocks occur in a zone on the southeast sides of both the Boots Blues and Bayu sections. These inclusions are widely scattered within this zone, only mildly deformed, often display chilled margins, and appear to be similar in composition to other volcanic rocks in the Ghost Rocks Formation—suggesting that they are probably locally derived sedimentary clasts. Most of these inclusions are irregular spheroids, but a few are oblate in shape and oriented parallel to nearby sandstone inclusions.

The inclusions in the Ghost Rocks Formation are similar in shape to inclusions exhumed from several other melange terranes, suggesting that the process of inclusion formation may be common to other melanges. Schamel (1974; see also Silver and Beutner, 1980) has documented the formation of elongate sandstone inclusions during nappe emplacement in the Appenines. He was able to document a gradation between undeformed sandstone beds and disaggregated and distended inclusions of sandstone in a shale matrix. The long axes of the inclusions were oriented perpendicular to the direction of nappe transport. Bachman (1978) has reported elongate inclusions from the coastal belt of the Franciscan Complex also, but no kinematic significance was attributed to them. Cowan (1978), however, has described oblate inclusions from the Franciscan Complex near San Simeon, California, and suggested the inclusions formed during coaxial flattening and uniform layer parallel extension.

In interpreting the kinematic significance of the inclusion shapes and the associated pinch and swell structure in the melange, it is fundamental to know how they formed. Two models that represent the end-members of possibilities are: 1) the inclusions were originally spherical in shape and their present shapes represent an estimate of finite strain; or 2) they originally formed a continuous tabular bed of sandstone that experienced complete boudinage via pinch and swell and layer parallel extension. In a simple case, these models differ radically in their implications for melange kinematics. The first model suggests a pebbly mudstone or olistostromal origin, whereas the second allows the possibility of post-depositional disruption of melange. Field evidence can be used to argue for both of these possibilities; some inclusions occur isolated in shale matrix and appear "flattened" and "elongated," whereas many other inclusions are still partially connected to sandstone layers. Consequently, a more relative question, albeit more difficult to answer, is, through what strain path have the inclusions formed? For example, layer perpendicular flattening, layer parallel shear, or some combination of the two? These models will be discussed in a later section.

Internal Structure—"Web" Structure

The sandstone inclusions in the melange are penetratively laced with thin gray to black anastomosing surfaces here interpreted as cataclastic shear zones (Figures 10 and 15, and also 12, 17, and 20). These structures appear to be present in all sandstones in the melange, although they are best observed on wet and

Figure 16. Example of well-developed mullions on the lower surface of a sandstone inclusion. Mullions form, in part, by movement along webs of cataclastic shear zones. Ruler is 15 cm long. Location shown in Figure 6.

polished surfaces exposed in the marine intertidal zones. Single gray to black surfaces are typically no more than a millimeter in thickness, but zones composed of several anastomosing surfaces are up to 1 cm in width. In most sandstones, they are so concentrated that individual surfaces can be traced for only a few centimeters before they are cross-cut by younger shear surfaces that are in turn cross-cut by still younger shear surfaces. A distinct age progression in shear surfaces of different orientations was not recognized, however, and cross-cutting relationships were often contradictory, suggesting contemporaneity of these features. Also, the web structures appear to be annealed (perhaps during D_2), because they are not zones of weakness and preferential fracturing, making it difficult to determine their orientation in three dimensions.

Mesoscopically and structurally, these surfaces are similar to features Cowan (1982) has recognized in sandstone clasts in the Franciscan Complex. Cowan justifiably called the features "web structure" because of their criss-crossing, irregular habit in three dimensions. Cowan also suggested the features were in part "veins," as they contained more clay and silt than the surrounding sandstone. I have not seen any vein-like structures associated with the web structure in the Ghost Rocks Formation, and rarely is there an increase in clay and silt within them. Mesoscopic scale web structures have been recognized in deformed rocks of the Olympic Peninsula (Cowan, personal communication, 1978), in Mesozoic melange terranes of eastern Oregon (Dickinson, personal communication, 1980), and they are a conspicuous feature of deformed sandstones of the Sitka Greywacke of southeast Alaska (Decker, personal communication, 1978). Moore and Allwardt (1980) have described web-like structures in the sandstones collected from melange zones within the Sitkalidak Formation. They attribute formation of these structures to cataclasis and comminution of the surrounding sandstone during formation of the melange. They also thought pressure solution may have been important. J. C. Moore (personal communication, 1978) has also observed web structures of cataclasis in the sandstones of the Uyak Complex on Kodiak Island. I have recently seen web-like structures in the sandstones of the coastal belt of the Franciscan and Wright (personal communication, 1981) and I have observed similar features in the melange zones of the Normanskill Formation of New York. Fergusson (1984) has recently described web structure of cataclastic origin in the Gundahl Complex in Australia. Thus, these three-dimensional web structures appear to be common and perhaps characteristic features of many melange zones. As discussed below, I believe these structures reflect the tectonic deformation and stratigraphic disruption that typifies melange zones.

In the sandstones of the Ghost Rocks Formation melange terranes, the web structures in part accommodated the change in shape indicated externally by bulbous protrusions, pinch and swell structures, boudinage, and the irregular distortion that is obvious at most outcrops. This is especially clear on polished surfaces perpendicular to the mullion-like structures present on many sandstone layers and inclusions (e.g., Figure 16). On cut surfaces, individual web structures or shear zones can be traced into the small linear depression that separates two mullion ridges. Movement along individual shear zones is partly responsible for the formation of the mullions. Moreover, the linear fabric of the

Figure 17. Example of pinched and swelled sandstone that also displays mullions on its upper and lower surfaces (A). (B) is photograph of polished slab oriented normal to mullions. This is the same surface shown in (A). The arrow points to a common point in the two photographs. Note the intense internal disruption of the sandstone layer. (C) is photograph of polished slab oriented parallel to mullions and normal to the layering.

Figure 18. Example of outcrop scale stratigraphic disruption (A). (B) and (C) are larger-scale photographs of the sandstone in (A). Note particularly the small sliver of pinched and swelled sandstone that is decorated by mullions on its upper and lower surfaces (C). Location of (A) is shown in Figure 3B.

mullions indicates that the line formed by the intersecting shear zones parallels the mullion structures.

Some sandstone layers and inclusions appear only mildly deformed externally but have an unusually dense system of web structures internally. For example, one sandstone layer displays only subtle pinching and swelling externally (Figure 17), whereas internally there is a dense network of cross-cutting, anastomosing shear surfaces that outline elongate areas of relatively undeformed grain clusters (Figures 17B and C). These relatively undeformed areas have cigar-like shapes that have irregular, but generally circular, cross sections (Figures 17B and C). Externally, the same structures form a lineated and wavy topography on the upper and lower surface of the sandstone. This lineation is parallel to other L_1 features in the melange. The contrasting amounts of deformation suggested by the internal and external structures are puzzling and suggest a complex strain history. In a relatively simple case, one involving layer parallel extension and perpendicular flattening, the web structure may have accommodated extreme thinning of the sandstone layer. However, this strain history should have resulted in undeformed regions that were elliptical, and not circular, in cross section (see Vollmer and Bosworth, this volume, and Reches, 1978). Alternative strain histories are discussed in a later section.

A progressive relationship between the web structure within sandstone inclusions and the formation of isolated sandstone inclusions is suggested by their geometric similarities; both form elongate ellipsoids. More directly, detailed observations of partially disrupted sandstone beds indicated disaggregation of the sandstone along anastomosing shear surface. Figures 18 and 19 show good examples of this disaggregation. At an outcrop scale the sandstone layers vary from being semi-continuous to consisting of an aggregate of lens-shaped pieces that are shape-oriented roughly parallel to the layering. Close inspection of these pieces shows that they are often interconnected, although highly irregular in external form (Figures 18B, C, and 19). Delicate mullion structures decorate the surface of the pinched and swelled sandstone in Figure 18C even though it is still connected to the main sand body. Internally this tadpole-shaped sandstone is laced with the gray to black surfaces distinctive of web structure. Moreover, where the sandstone is connected to the main sandstone layer, a thin cataclastic shear zone parallels the pinched and swelled sandstone. The sandstone was apparently slivered off the main sandstone layer during movement along the shear zone. As the slivered sandstone was engulfed by the shale, it continued to change shape, forming pinch and swell and mullion structures. These structures parallel other D_1 structures found throughout the melange.

A similar style of disruption is inferred for the other lens-shaped appendage exposed just to the left of the one shown in detail (Figure 18B). This sliver is also still connected to the main sandstone body, but has been slivered off in a direction opposite to that of the sliver next to the pencil. Together, these slivers and the overall structure of the sandstone layer suggest that the sandstone is being disaggregated along many irregular shear surfaces that occur at a low angle to the sandstone layering. The opposing slivers shown in Figure 18 (and less clearly in Figure 19) suggest a clockwise asymmetry (viewed to the northeast) to the deformation. The geometry of these shear surfaces is identical to the well-developed web structure described earlier. Both share a common line of intersection that parallels the mullions and pinch and swell structures. In these examples, however, the undeformed regions outlined by the web structure are being separated and engulfed by shale, probably leading to the complete isolation of elongate ellipsoids of sandstone immersed in shale matrix.

Microscopically, the shear zones or "web structures" in the Ghost Rocks Formation are characterized by an extreme reduction in grain size caused by cataclasis and comminution. The boundaries between undeformed regions and individual shear zones are usually marked by abrupt to gradational changes in grain size. In sandstones containing a low concentration of web structures, the undeformed regions between shear zones display many primary sedimentary textures. Original detrital shapes are preserved even in the lithic fragments. Sandstone inclusions containing higher concentrations of web structures are more deformed throughout. The regions between shear zones display very few primary textures, and undulatory extinction angles are ubiquitous. These sandstones are composed almost totally of plagioclase and quartz grains that are intensely undulatory, appear broken, and are separated from one another by much smaller grains of quartz and feldspar and less commonly by calcite or fine-grained phyllosilicates and epidote. The larger, siliceous grains are commonly cross-cut by straight, through-going fractures that, in some cases, have offset the grain boundaries. The sharp, angular boundaries of some broken grains can often be matched with the angular boundaries of nearby grains. These matching broken grains are usually also nearly optically continuous even though they are separated from one another. I observed very little evidence for dynamic recrystallization (e.g., subgrains, or core-mantle textures). These zones of grain size reduction developed through breakage and granulation of the detrital grains without concomitant dynamic recrystallization.

Figure 19. Second example of outcrop scale stratigraphic disruption. Note the interconnecting fingers and ellipses of sandstones. Location shown in Figure 5.

The web structure of shear zones is microscopically and mesoscopically similar to small faults Johnson and Aydin (1982)

a

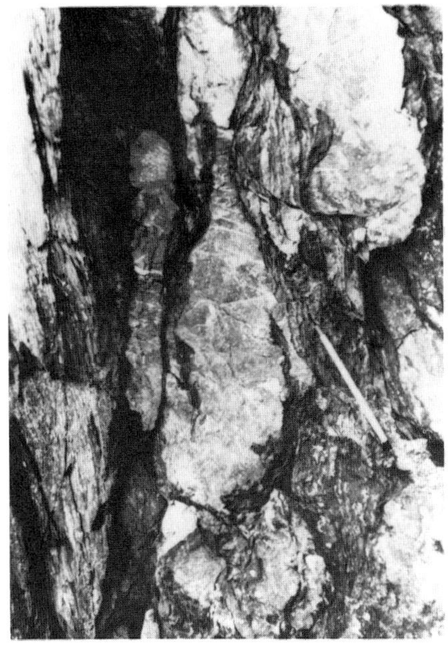
b

c

Figure 18. Example of outcrop scale stratigraphic disruption (A). (B) and (C) are larger-scale photographs of the sandstone in (A). Note particularly the small sliver of pinched and swelled sandstone that is decorated by mullions on its upper and lower surfaces (C). Location of (A) is shown in Figure 3B.

mullions indicates that the line formed by the intersecting shear zones parallels the mullion structures.

Some sandstone layers and inclusions appear only mildly deformed externally but have an unusually dense system of web structures internally. For example, one sandstone layer displays only subtle pinching and swelling externally (Figure 17), whereas internally there is a dense network of cross-cutting, anastomosing shear surfaces that outline elongate areas of relatively undeformed grain clusters (Figures 17B and C). These relatively undeformed areas have cigar-like shapes that have irregular, but generally circular, cross sections (Figures 17B and C). Externally, the same structures form a lineated and wavy topography on the upper and lower surface of the sandstone. This lineation is parallel to other L_1 features in the melange. The contrasting amounts of deformation suggested by the internal and external structures are puzzling and suggest a complex strain history. In a relatively simple case, one involving layer parallel extension and perpendicular flattening, the web structure may have accommodated extreme thinning of the sandstone layer. However, this strain history should have resulted in undeformed regions that were elliptical, and not circular, in cross section (see Vollmer and Bosworth, this volume, and Reches, 1978). Alternative strain histories are discussed in a later section.

A progressive relationship between the web structure within sandstone inclusions and the formation of isolated sandstone inclusions is suggested by their geometric similarities; both form elongate ellipsoids. More directly, detailed observations of partially disrupted sandstone beds indicated disaggregation of the sandstone along anastomosing shear surface. Figures 18 and 19 show good examples of this disaggregation. At an outcrop scale the sandstone layers vary from being semi-continuous to consisting of an aggregate of lens-shaped pieces that are shape-oriented roughly parallel to the layering. Close inspection of these pieces shows that they are often interconnected, although highly irregular in external form (Figures 18B, C, and 19). Delicate mullion structures decorate the surface of the pinched and swelled sandstone in Figure 18C even though it is still connected to the main sand body. Internally this tadpole-shaped sandstone is laced with the gray to black surfaces distinctive of web structure. Moreover, where the sandstone is connected to the main sandstone layer, a thin cataclastic shear zone parallels the pinched and swelled sandstone. The sandstone was apparently slivered off the main sandstone layer during movement along the shear zone. As the slivered sandstone was engulfed by the shale, it continued to change shape, forming pinch and swell and mullion structures. These structures parallel other D_1 structures found throughout the melange.

A similar style of disruption is inferred for the other lens-shaped appendage exposed just to the left of the one shown in detail (Figure 18B). This sliver is also still connected to the main sandstone body, but has been slivered off in a direction opposite to that of the sliver next to the pencil. Together, these slivers and the overall structure of the sandstone layer suggest that the sandstone is being disaggregated along many irregular shear surfaces that occur at a low angle to the sandstone layering. The opposing slivers shown in Figure 18 (and less clearly in Figure 19) suggest a clockwise asymmetry (viewed to the northeast) to the deformation. The geometry of these shear surfaces is identical to the well-developed web structure described earlier. Both share a common line of intersection that parallels the mullions and pinch and swell structures. In these examples, however, the undeformed regions outlined by the web structure are being separated and engulfed by shale, probably leading to the complete isolation of elongate ellipsoids of sandstone immersed in shale matrix.

Microscopically, the shear zones or "web structures" in the Ghost Rocks Formation are characterized by an extreme reduction in grain size caused by cataclasis and comminution. The boundaries between undeformed regions and individual shear zones are usually marked by abrupt to gradational changes in grain size. In sandstones containing a low concentration of web structures, the undeformed regions between shear zones display many primary sedimentary textures. Original detrital shapes are preserved even in the lithic fragments. Sandstone inclusions containing higher concentrations of web structures are more deformed throughout. The regions between shear zones display very few primary textures, and undulatory extinction angles are ubiquitous. These sandstones are composed almost totally of plagioclase and quartz grains that are intensely undulatory, appear broken, and are separated from one another by much smaller grains of quartz and feldspar and less commonly by calcite or fine-grained phyllosilicates and epidote. The larger, siliceous grains are commonly cross-cut by straight, through-going fractures that, in some cases, have offset the grain boundaries. The sharp, angular boundaries of some broken grains can often be matched with the angular boundaries of nearby grains. These matching broken grains are usually also nearly optically continuous even though they are separated from one another. I observed very little evidence for dynamic recrystallization (e.g., subgrains, or core-mantle textures). These zones of grain size reduction developed through breakage and granulation of the detrital grains without concomitant dynamic recrystallization.

Figure 19. Second example of outcrop scale stratigraphic disruption. Note the interconnecting fingers and ellipses of sandstones. Location shown in Figure 5.

The web structure of shear zones is microscopically and mesoscopically similar to small faults Johnson and Aydin (1982) have described in sandstones in the San Rafael Desert of southeastern Utah (also see Aydin and Johnson, 1978; Aydin, 1978). Johnson and Aydin (1982) recognized a progression from deformation bands to zones of deformation bands to discrete planes of slip. They argue that this progression reflects strain-hardening of the sandstone. Crushing and consolidation within the band cause the material there to become stiffer than the surrounding sandstone. Consequently, other zones of deformation bands develop. Strain-hardening is believed to be caused by the increase in grain-to-grain contacts (and therefore friction) brought on by the reduction of grain size and porosity within the zones of deformation. In the shear zones of the Ghost Rocks Formation, reduction of grain size through cataclasis is a typical feature of the web structure. The reduction of porosity in these zones is more difficult to prove, however, because the space between framework grains in undeformed regions is occupied by a clayey to silty detrital matrix. The primary porosity of the sandstones, therefore, may have been relatively low. It is probably significant, however, that the weaker matrix material is completely disaggregated within the shear zones and the zones are dominated by comminuted plagioclase and quartz grains. Thus, the characteristic feature of these zones is the intense grain size reduction which, in itself, may have caused the shear zones to strain harden.

Some relatively small sandstone inclusions (less than 15 to 20 cm in greatest dimension) are so riddled by web structure that individual shear zones are no longer recognizable. Microscopically, almost the entire sandstone consists of intensely cataclastically deformed and comminuted quartz and plagioclase. There are very few detrital grains remaining and no grains are free of undulatory extinction. The perimeter of these smaller inclusions is wavy, irregular and bulbous or diapir-like. Individual grains or grain clusters appear to be partially removed from the inclusions and engulfed by the shale matrix. There are essentially no microscopic or mesoscopic scale faults related to this deformation. The inclusions have changed shape and flowed ductilely at a mesoscopic scale, but were deformed brittlely and penetratively at a microscopic scale. These sandstones have cataclastically flowed.

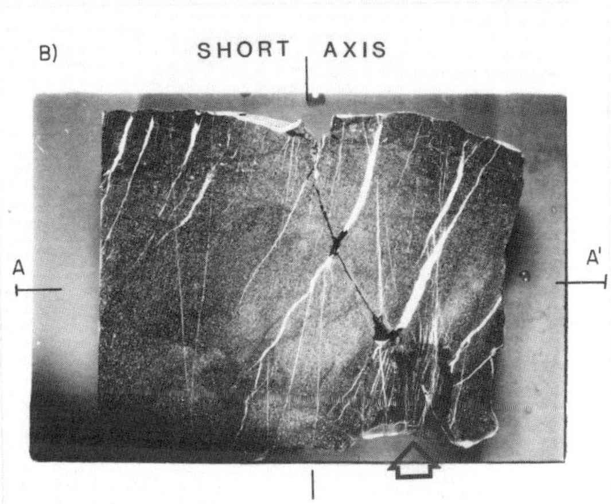

Figure 20. (A) shows an example of the well-developed extension fractures that are commonly present in the sandstone inclusions. Photograph is parallel to S_1. (B) is a photograph of the slabbed and wetted surface oriented normal to the intermediate axis of the sandstone inclusion. The fractures are filled with calcite and oriented at a high to moderate angle to the long axes of the inclusion. The "V" in the top of the sample coincides with the hole in the sample in A. Note the shale intrusion in the lower right corner (arrow).

Calcite Fractures

Veins and irregular shaped pods of calcite and minor quartz are commonly associated with the sandstones in the melange. At most exposures, these conspicuous features are oriented perpendicular to S_1 and occur in the pinched areas of pinch and swell structures. These fractures occur perpendicular to both the long and intermediate axes of the pinch and swell and inclusion structures. Thus, the sandstones in the melange appear to have experienced two directions of principal extension, one perpendicular to, the other parallel to, the linear fabric in the melange (i.e., down-dip and parallel to strike). The geometry and distribution of the calcite fractures indicate that they represent a later, mesoscopically brittle phase of the pinch and swell deformation. Consequently, the qualitative displacements they record can be used to infer the displacement histories related to the earlier pinch and swell structures, for which there are no strain markers. This, of course, provides only a first approximation of the earlier displacements because the fractures record only a small increment of the total strain history.

Calcite-filled fractures are also commonly present within the swelled part of the pinched and swelled sandstone (Figures 10 and 21) and within the isolated inclusions (Figures 10, 12, and 20). These fractures, however, are often masked externally by the thick shale veneer that coats the sandstones. I sampled and orthogonally slabbed approximately 30 inclusions so that the internal fractures could be carefully studied and documented. In all of these inclusions, the calcite fractures indicate an extension direction approximately parallel to the long axes of the inclusions, which plunge gently to the southwest (e.g., Figures 10, 12, and 20). This direction of extension is 90° to the down-dip extension

Figure 21. Photograph of surface slabbed parallel to the plane of pinch and swell structures in continuous sandstones. All calcite fractures are oriented perpendicular to the plane of the photograph. Note different orientations of fractures in swell area and in pinch area. See text for discussion.

direction inferred from the most common set of pinch and swell structures and the direction of most disruption discussed above. A few inclusions show extension directions parallel to both their long and intermediate axes. In these latter inclusions, the two fracture sets are domainal, and cross-cutting relationships cannot be confirmed; they appear to have been contemporaneous. The curvature of some veins (see the surface of the sandstone in Figure 20A) and the slight angular discordance between the short axis and the trace of the veins in Figures 20B and 12B are not understood at present. However, thorough three-dimensional mesoscopic and microscopic studies and strain analyses of twinned calcite within the veins are in progress.

An orthogonally slabbed, discontinuous layer of sandstone that displays pinch and swell structures is shown in Figure 21. The linear fabric of the pinch and swell structures is oriented subhorizontally and parallels the other linear features in the melange. S_1 and S_2 are parallel in the location where this sample was collected. Limited exposure prevented observation of the complete three-dimensional shape of this sandstone layer. The slabbed surface, oriented parallel to the long and intermediate axes of the pinch and swell structures (S_1 and S_2), reveals two sets of fractures (Figure 21). One set occurs in the pinch area of the pinch and swell structure and is oriented perpendicular to the intermediate axis, consistent with extension normal to the pinch and swell structure. A second set of fractures occupies the remaining part of the inclusion (the "swell" part of the pinch and swell structure) and contains a conjugate pair of fractures. Offset relationships between the conjugate fractures indicate that both the maximum extension and shortening directions are in the plane of the layering. The shortening direction is not perpendicular to either S_1 or S_2. Instead, the shortening direction parallels the intermediate axis of the swell which is also parallel to the extension direction inferred from the fractures in the nearby pinch area. Thus, the sandstone layer appears to have extended in the pinch area in the same direction as it shortened in the swell area. The development of these two fracture sets in restricted zones of the same sandstone layer suggests their contemporaneity, although petrographic and cathodluminescence studies, in progress, should confirm or disprove this suggestion.

Even if the veins did not develop contemporaneously, the 90° rotation in principal strains is surprising and difficult to explain. The shortening direction inferred for the swell area, however, is consistent with the shortening direction inferred from the doubled-up sandstone discussed above (see Figure 12); both directions are approximately parallel to the intermediate axis. Also, the direction of extension in the swell area of the pinched and swelled sandstone is parallel to the extension directions determined from the calcite fractures within isolated inclusions (discussed above). Thus shortening within the plane of S_1 and parallel to the intermediate axes of inclusions and extension parallel to the long axes seem to be real phenomenan in these rocks. In total, the calcite fractures in this sandstone layer, and perhaps the pinch and swell structures in general, record complex displacement histories and suggest that the displacements commonly inferred from mesoscopic pinch and swell structures are misleading in their simplicity.

If the strain fields suggested by the orientation of the calcite fractures can be applied to the formation of the pinch and swell structures prior to the development of the fractures, then the pinch and swell structures may represent areas of literal "pinching and swelling." Movement of sandstone from the pinches and into the swells may have been accommodated by slippage along the "web structure" of ubiquitous shear zones that penetrate the sandstones. An interpretation of literal pinching and swelling is supported also by the doubled-up sandstone (Figure 12).

SHALE MATRIX

At an outcrop scale in most areas in the melange, it is impossible to distinguish S_1 and S_2 in the shale because they are, or are nearly, parallel. Where S_1 and S_2 occur at high angles, however, S_1 is defined by a weak, and variably spaced anastomosing cleavage (Figure 6).

Microscopically, the S_1 cleavage in the shale is defined by large, crudely oriented phyllosilicates of probable detrital or diagenetic origin. These phyllosilicates are ubiquitous in the shale matrix of the melange. The S_1 phyllosilicates do not form trains or stringers associated with opaque residues and other pressure solution phenomena, in contrast to S_2. Moreover, they are not more concentrated or more strongly oriented near the competent sandstone inclusions. Typically, the S_1 phyllosilicates completely envelop and often grade into the phyllosilicates that form the detrital matrix of the sandstone inclusions.

Thin-sections oriented normal to L_1 (and therefore S_1 and S_2) show S_1 phyllosilicates that change orientation radically and unpredictably in a single section. Tight to isoclinal folds, complex swirls and microscale faults appear to be responsible for this disruption. Phyllosilicates in concentric to spiral-like patterns are locally present. These swirl-like patterns are geometrically similar to the mesoscopic patterns observed in some sandstone inclusions. There is no obvious asymmetry to these patterns. In sections of other orientations, S_1 folds are less well-developed and no concentric or spiral-like patterns were observed. These S_1 folds and swirls in part form the prominent lineation present at most outcrops of the melange matrix (Figure 7).

Thin-section to outcrop scale shear zones occur in the shale matrix throughout the melange. These zones typically parallel, or cross-cut at a small angle (less than 10°), the S_1 fabric. The shear zones occur in two gradational modes (Figures 10 and 22): as sharp discrete zones of very fine-grained phyllosilicates and minor opaques, and as slightly irregular zones associated with calcite and quartz precipitation. The former zones are best recognized in slabbed and polished surfaces or in thin-sections oriented perpendicular to S_1 (Figure 22A). These thin zones usually are less than .5 mm in thickness and are spaced about 1 to 5 mm apart. Their traces are very straight and continuous, except where folded by S_2. The shear zones associated with calcite precipitation are easy to observe in the field or thin-section (Figure 22B). These zones can be much thicker (1-5 mm) and more widely spaced (possibly up to a meter) than the zones without calcite. Quartz is locally associated with the calcite and may be a replacement mineral.

Typically, the calcite and quartz are laminated parallel to the shear zone and occur in lens-shaped discontinuous pods that parallel the shear zones. These laminated pods were probably deposited in the voids that formed as asperities on the wall of the fracture moved past one another. Within pods, neither mineral appears to be shape-oriented relative to the shear zone. In some cases, the entire shear zone seems to be filled with calcite because shear surfaces several square meters in area are often covered by calcite.

These laminated veins are probably late D_1 shear zones, as they truncate the inclusions and S_1, but are folded by D_2. Also, both types of zones have sharp regular traces and there is no evidence from field and thin-section observations that indicates shortening perpendicular to the zones. Individual zones probably represent relatively small amounts of displacements; but because they occur throughout the melange, they may, in total, record significant amounts of layer-parallel displacement.

GREENSTONE AND TUFF HORIZONS

Continuous layers and isolated inclusions of coarse-grained tuff and porphyritic pillowed lava flows occur locally throughout the melange (Figure 3). Pillow lavas dominate the upper 30 meters of the melange exposed at Boots Blues and Knoll Bay. The pillow lavas are dark green to rusty red in color, highly fractured, and deformed. Continuous layers of coarse-grained tuff occur directly below the pillow lavas of Knoll Bay and about 40 meters below the lavas at Boots Blues. Similarly appearing tuffs, but no lavas, are present at the highest levels of the Bayu melange. In the Bayu section, the tuffs consist of seven to eight 3-15 cm thick horizons separated by several centimeters of shale. At Boots Blues, the tuffs are amalgamated into two much thicker horizons, as at Knoll Bay. In total, these tuffs and lava flows delineate marker horizons that can be traced for at least 20 km along strike within the melange. As these volcanic horizons are not present elsewhere in the melange, they were apparently not imbricated or folded on a regional scale during either D_1 or D_2.

The tuffs in the Bayu section provide conspicuous S_0 horizons that have been disrupted by D_1, but not by D_2. The tuffs dip moderately to the northwest and are sandwiched between several centimeters (up to 10 cm) of unusually well-indurated and non-fissile shale. In these tuff and shale layers, S_2 is a weak, wavy 1-10 cm spaced, fracture-like cleavage that cross-cuts the layers and dips steeply to the northwest. In the melange away from the tuff and associated zones of indurated shale, S_2 is a well-developed and closely spaced (few mm) cleavage that has apparently transposed S_1 because the two foliations are parallel. Apparently, the indurated horizons, cored by tuff, formed a rigid tabular unit that was not substantially reoriented by the imposition of S_2. The smaller, less continuous sandstone inclusions above and below the tuff layers, in contrast, appear to have been completely rotated or flattened into parallelism (transposed) with S_2.

The tuff layers in the Bayu section provide valuable insights into the style and symmetry of D_1 because they have been little affected by D_2 (Figure 23). Although the layers are regular and little disrupted on a regional scale, at mesoscopic and microscopic scales, the individual tuff layers are intensely disrupted and show

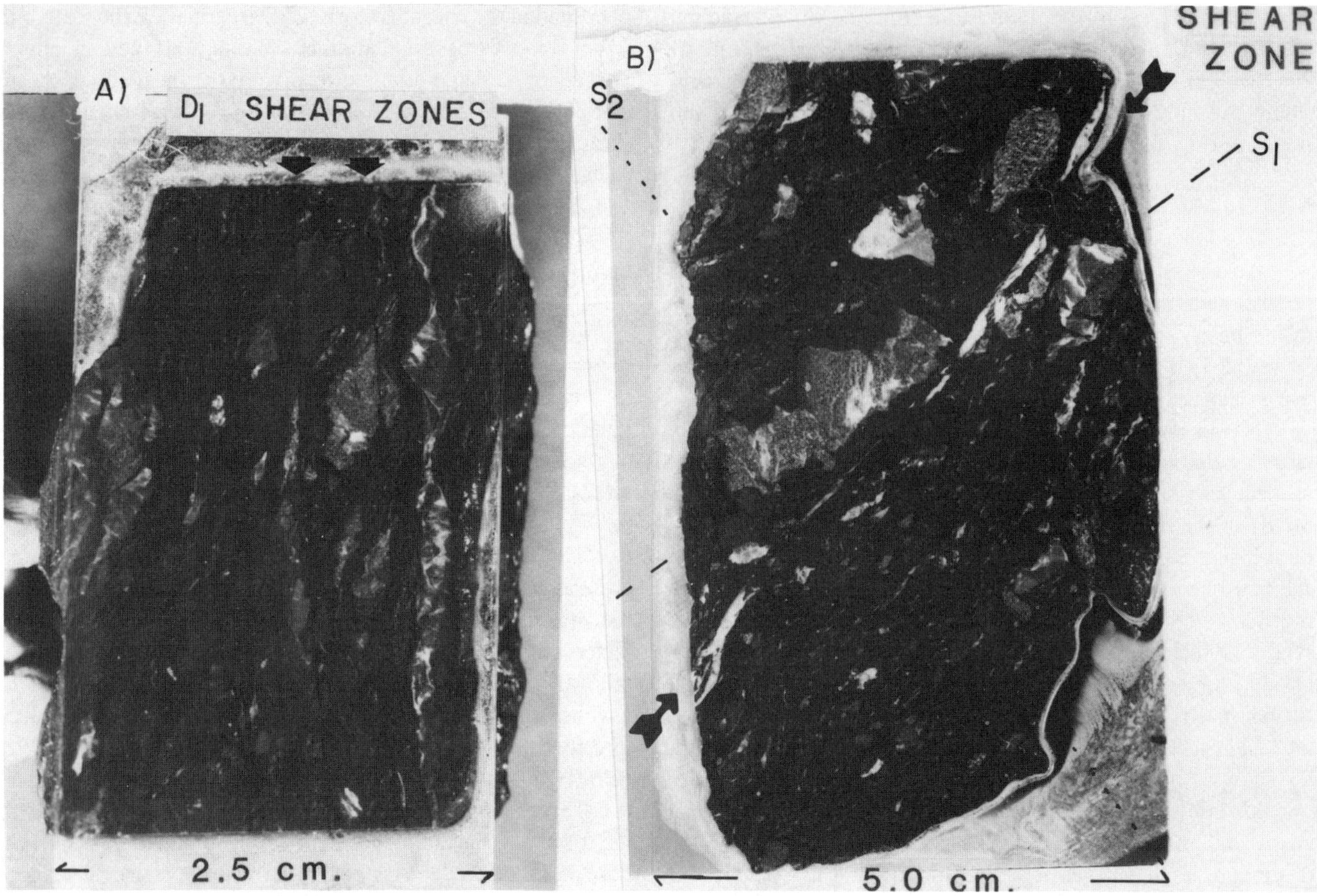

Figure 22. Two photographs of slabbed surfaces (with thin-section glass glued on) of melange shale matrix. Surface in (A) is oriented normal to S_1 and parallel to L_1. Thin dark zones that truncate the inclusions are one type of D_1 shear zones (see arrows). (B) is oriented normal to L_1. Note change in scale. Note also the mesoscopic ductile and brittle styles of deformation and the distinct shear zone (arrows), partially filled with calcite, which cross-cuts S_1 at a low angle and is itself cross-cut by S_2.

complex intermixing with the shale. The tuff and shales are both so faulted and ductilely swirled that isolated pieces of one are completely engulfed by the other (Figure 23). The tuff layers, however, are always contained within the zones of nonfissile and well-indurated shale. The upper and lower boundaries of the indurated shale show no indication of being as complexly deformed as the tuffs they surround. Induration apparently postdates the deformation of the tuff and shale layers and predates the development of S_2 because S_2 is so poorly developed within these areas. The close association of the indurated shale zones with the tuff horizons suggests that induration may be related to the diagenetic or metamorphic alteration of the tuffs.

On planes parallel to the tuff layering, the irregular boundaries between tuff and shale result in a well-developed lineation that plunges gently southwest parallel to other D_1 lineations in the melange. The lineation is formed by the interdigitation of elongate swirls and wisps of tuff in shale. On planes oriented normal to this lineation, the faulted and swirled tuff displays a distinct and consistent asymmetry (Figure 23).

Disruption of the tuff layers during D_1 is indicated because: 1) the zones are cross-cut by S_2, 2) the tuff lineation is parallel to D_1 lineations, and 3) the approximately layer-parallel disruption is characteristic of D_1 elsewhere. The disruption of the tuff layers probably was not synsedimentary, although the tuffs and shales may have been only partially lithified during deformation because: 1) the lineation parallels lineations of D_1 that are tectonic (e.g., extensional veins and cataclasic zones), and 2) the matrix and phenocrysts of the tuff horizons are often broken, fractured, and highly undulatory in thin-section. If this interpretation is correct, it is probably significant that the disruption of the tuff layers is consistently asymmetric and suggests movement of the lower plate to the northwest. Thus, the structural features in the tuff horizons provide an important insight into the asymmetry of D_1 kinematics.

Figure 23. Photographs of asymmetrically swirled and disrupted tuff horizons in Bayu section. Photographs are oriented normal to layering and L_1. Sense of shear is inferred to be upper part to the right.

RELATIVE AGES AND DISCUSSION OF THE ORIGIN OF D_1 STRUCTURES

The relative ages of the separate D_1 structures are difficult to determine because of their restricted spatial development. Most calcite/quartz extension fractures cross-cut the web structures in the sandstones, suggesting that they are the youngest structural features. Locally, however, the opposite cross-cutting relationships are present. The shear zones in the shale matrix appear to post-date most of the stratigraphic disaggregation, because they truncate the inclusions, and therefore they also probably post-date the web structures. The shear zones also truncate the swirls and folds of phyllosilicates in the shale matrix. In general, therefore, the earliest structural features in the melange are swirls and folds of phyllosilicates in the shale matrix and the three-dimensional web structure of shear zones in the sandstones. With these reservations, the age progression shown in Figure 9 generally reflects the development of structures through time.

D_1 structural features may be summarized as four general observations that provide insight into the environment of formation and genesis of the melange: First (1), the inclusions often grade into partially disaggregated sandstone layers. This observation suggests that at least some of the inclusions are not sedimentary clasts. Second (2), cataclasis is an important deformation process in the formation of the inclusions. Cataclasis and grain breakage have been reported in gravitationally driven slide blocks (see Crittenden and others, 1980; Frost and Martin, 1982; and many others) as well as in unlithified sediments drilled by DSDP (Lundberg and Moore, 1981). Thus, cataclasis, by itself, does not prove tectonic processes or the lithified state of the sediments. However, in the Ghost Rocks Formation melange terranes, the web structure of cataclastic shear zones apparently occurs in every sandstone. Even relatively small inclusions (less than a few cm in each dimension), that are engulfed by shale, are cataclastically deformed. This regionally pervasive cataclasis strongly suggests that confining pressures in the shales and sandstones were substantially higher than pressures normally associated with gravitationally driven submarine slides or slumps. It seems unreasonable that this style of deformation would pervade a gravity-driven slide that must have extended for 20 km or more along strike. If the melange consists of several smaller slides, why are there no interbedded sequences that are not cataclastically deformed? The third and fourth observations are: (3) the presence of calcite-filled veins and (4) the apparent progressive relationship, based on similar geometries, between the veins and the older pinch and swell structures. The veins are significant because they, like the cataclastic zones, suggest that the confining pressures, and probably pore pressures (Beach, 1977, 1980; and Etheridge, 1983), were higher than pressures normally expected for surficial, gravity-driven slides. The recognition of a progressive deformation between the web structure and the veins also supports this interpretation because the web structures are interpreted as tectonic structures. Apparently then, both the cataclastic web structures and the calcite veins were tectonically induced, although regional-scale and relatively deep-seated sliding cannot be completely ruled out.

If a tectonic interpretation is accepted, then it is important to ask what amount of the stratigraphic disruption in the melange can be related to the D_1 deformation? The calcite veins, although conspicuous features, appear to record only a small amount of the overall strain and are only indirectly related to the stratigraphic disruption. In contrast, the web structures appear to be an integral part of the disruption, and the amount of strain they record may be very large. Alternatively, a significant amount of the disruption may have occurred prior to the development of cataclastic shear zones. For example, a nongrain-breaking deformation, as described by Cowan (1982), may have occurred and resulted in stratigraphic disruption and disaggregation. The structural features related to this deformation may be obscured by younger deformations. In any case, it is important that the cataclastic shear surfaces intersect and anastomose to form elongate ellipsoids that are similar in shape to the completely isolated sandstone inclusions. This similarity and the preservation of partially disaggregated sandstone layers suggest a progressive deformation from anastomosing shear surfaces to the complete isolation of ellipsoidal sandstone inclusions in a shale matrix. This deformation apparently ended with the development of the geometrically similar calcite veins. With these reservations and assumptions about progressive deformation, I interpret D_1 to be a tectonic deformation that could have fully produced the stratigraphic disruption that characterizes the melange. However, older nongrain-breaking tectonically or gravitationally induced deformations also may have occurred.

An equally important question to ask about the stratigraphic disruption in the melange is what overall strain history does this deformation represent? This can be approached by first asking what are the orientations of finite principal strains in the melange? To a first approximation, the absence of folding in the tuff layers and the nearly layer-parallel disruption of the sandstone horizons suggest that at least one, and maybe two, principal strains were parallel to layering. This is also suggested by the orientation of calcite veins (Figure 10). One principal strain direction is indicated by veins in the inclusions and swell areas. This strain is extensional and plunges gently southwest, parallel to L_1. A second principal strain is suggested by the extensional veins in the pinch area between swell structures. This extensional strain plunges steeply northwest, down the dip of S_1. This principal strain, however, appears to have a complex history, because at least two inclusions show evidence of shortening strains that are parallel to the extensional strains. Unfortunately, it is not possible at present to prove the relative ages of these strains or to tell which principal strain is the maximum.

The shape of the inclusions and pinch and swell structures also gives an indication of the strains in the melange. Based on experimental and theoretical studies of the development of boudinage during layer-perpendicular flattening (a relatively simple case), the spacing between boudin necks (dominant wavelength) is believed to be a function mainly of the layer thickness prior to boudinage (Smith, 1977). Therefore, the amount of strain in any direction is not recorded by the shape of the boudins, as this is a function of layer thickness, but by the amount of space between boudins. Classic chocolate tablet boudinage is the expected result of coaxial shortening normal to layering, whatever the relative magnitude of the extensional strains (see discussions by Sanderson, 1974, and Ramsay, 1967, for more complex cases). Under what circumstances then is it possible to form elongate boudins? In the Ghost Rocks Formation, the calcite veins indicate an extensional strain parallel to the long axis of the inclusions (boudins?), suggesting that the inclusions have been elongated. The magnitude of the calcite strains, however, does not appear to be as large as the strain indicated by the aspect ratios of the inclusions (e.g., see Figure 20). Thus, if the inclusions have been elongated or stretched, then some other mechanism, perhaps the web structures, must have accommodated the extensional strains. An alternative to this model of "stretching" the inclusions is given below.

If we can believe that the strains recorded by the calcite veins and inclusion shapes, although poorly understood, are representative of the strains in the inclusions of the melange, then the strain histories appear to be inconsistent with either a coaxial plane or flattening strain history. Neither strain history can explain the parallel shortening and extension directions, noted above, or the development of elongate boudins. Moreover, a model of approximately layer-parallel simple shear (a type of noncoaxial plane strain) is also inconsistent with the strains in the melange because both principal strain axes in the plane of layering appear to have changed length.

A few other mesoscopic scale structures in the melange can also be used to test or are suggestive of specific strain histories. For example, the seven to eight tuff horizons exposed in the upper part of the Bayu section (Figure 23), which are little affected by D_2, appear not to have been folded or otherwise disrupted on a regional scale by D_1. Yet at a mesoscopic scale, the tuffs have been substantially disrupted by D_1. These observations suggest that the stratigraphic disruption associated with D_1 occurred approximately parallel to the lithologic layering (presumably bedding). Moreover, the asymmetrical style of the disruption suggests that the strains imposed on the tuffs during D_1 were also asymmetric. This asymmetry is suggestive of layer-parallel shearing. The disrupted sandstone layer with asymmetrically slivered appendages, exposed at Boot Blues (Figure 18), the doubled-up sandstone inclusion (Figure 12) at Bayu, the inclined intermediate axis of one inclusion (Figure 13), and the few asymmetric folds (e.g., Figure 14) are also consistent with shearing approximately parallel to the layering. The sense of shearing in these five examples is the same; the lower plate has moved to the northwest.

The strains recorded by the calcite fractures are also partially consistent with a model of layer parallel shear. Specifically, the shortening observed parallel to the intermediate axis of the swell structure (Figures 10 and 21) and the doubled-up sandstone may be the result of local internal-buckling within the shear zone (Cobbold and others, 1971; Platt, 1983).

A model of heterogeneous layer-parallel shear may also account for the elongate inclusions because they would not be the result of layer-perpendicular flattening. Instead, they may have formed as ellipsoids of sandstone were slivered and detached from each other, forming independent, often elongate inclusions (Figure 24). Their shapes may be more a function of the process of disaggregation than of the associated strains. This style of disruption is similar to that of many fault zones in which lens-shaped pieces of rocks are disaggregated by anastomosing shear surfaces (Tchalenko, 1970; Logan and others, 1979). In most examples of "brittle" fault zones, however, the rocks are relatively homogeneous throughout. In the Ghost Rocks Formation, the sand and shale sequences formed relatively inhomogeneous sequences. As these sequences experienced layer-parallel shearing, the shales deformed by folding, swirling and layer-parallel displacements; they flowed, mesoscopically and microscopically (submicroscopic studies of the phyllosilicates may reveal evidence for cataclastic processes in the shales also). The sandstone, in contrast, deformed through the development of a web structure of cataclastic shear zones. The sandstone fragments may have gradually become separated and disaggregated, like blocks in brittle fault zones. In the melange, however, the developing fault blocks were suspended in a matrix of ductile shale.

Although a model of layer-parallel shear explains many of the features of D_1, it also may have some inconsistencies. Most important, if the shear strains were similar to those of simple shear (plane strain), then, as discussed above, the model is not consistent with the southwest-plunging extension inferred from the calcite veins. Assuming the transport direction, as inferred from the tuff layers and the other asymmetric structures, was to the northwest, then in a simple shear case, the southwest direction would be the intermediate axis of the strain ellipsoid and should not record a change in length. Consequently, if D_1 developed during layer-parallel shear, then the strains were more complex than the plane strains of simple shear. In fact, if the shortening strains that parallel layering are attributed to local perturbations that caused ramping or internal-buckling, then the strains in the melange may have been in the field of flattening (two principal strains are extensional). Flattening strains in shear zones are not very common, but recent field studies in Arizona (Bykerk-Kauffman, 1983) have documented flattening strains in areas transitional to mylonitic shear zones. Thus, if the melange terranes record layer-parallel shear, then the strains may have been more like those in the Arizona terranes than zones of simple shear.

In total, inclusions in the melange have apparently formed through the combination of three fundamental processes including: 1) layer-parallel extension or true boudinage; 2) sedimentary processes; and 3) most dominant, layer-parallel shearing and cataclastic flow. In any single outcrop or inclusion, it is usually difficult, if not impossible, to establish which process was responsible for the formation of the stratigraphic disruption. In the Ghost Rocks Formation, however, it is the common occurrence of cataclastic web structures, layer-parallel shear zones, and con-

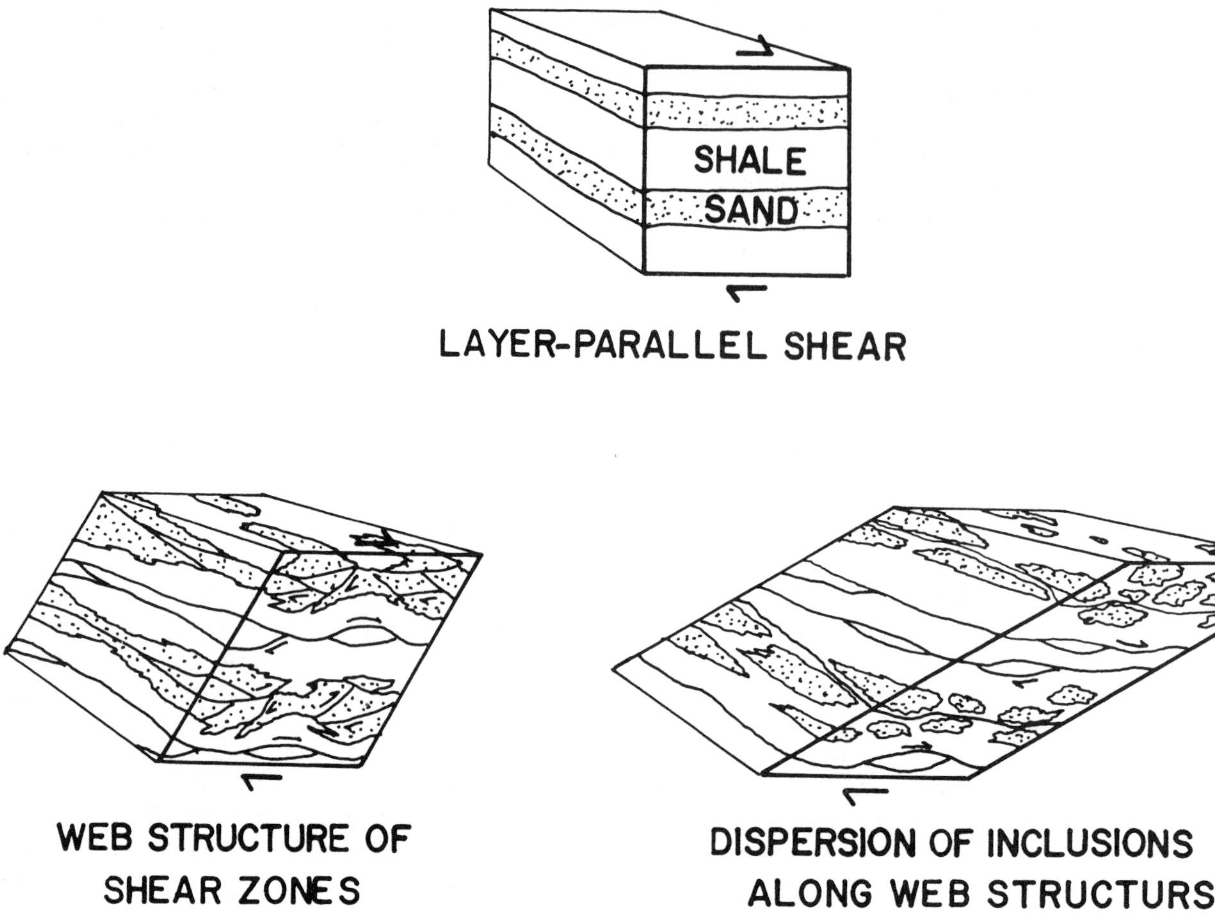

Figure 24. Diagram showing sequential development of sandstone-shale melange through layer-parallel shear and dispersion. The most significant aspects of this deformation are: 1) the dominance of approximately layer-parallel shear, and 2) the disruption of the sandstone layers by a dense web of shear zones. See text for discussion of possible strain histories related to this deformation.

sistent asymmetries that suggest the dominance of layer-parallel shearing as the disruptive mechanism. Consequently, the D_1 structures in the melange appear to represent a B-tectonite in which linear elements (e.g., fold axes, rotation axes, and pinch and swell structures) parallel a direction of extension, but are oriented perpendicular to the inferred direction of tectonic transport (Lister and Price, 1978; Turner and Weiss, 1963; Hobbs and others, 1976).

SECOND DEFORMATION (D_2)

D_2 is characterized by a closely spaced planar cleavage (S_2) that generally dips steeply to the northwest (Figure 4). This cleavage is the dominant structural feature at most outcrops of the melange (Figure 3), but its presence is difficult to detect because S_2 often parallels the lithologic layering, S_1. S_2 and S_1 are parallel in the upper (northwest) parts of both the Bayu and Boots Blues sections except in a narrow zone near the tuff horizons (Figure 3). Here the tuffs dip moderately to the northwest and are cross-cut by the steep northwest-dipping S_2 cleavage. Elsewhere in the Boots Blues section, S_1 and S_2 are parallel and the two cleavages cannot be distinguished in the field. In the lower part of the Bayu section, D_2 has folded S_1 in several places and S_1 and S_2 foliations can easily be distinguished (Figures 3A, 4, and 6). S_2 is parallel to the axial planes of these folds. The style and form of these folds are difficult to discern, however, because single sandstone layers cannot be traced through the fold closures and the fold hinge areas are often obscured by faulting. The parallelism of the S_1 foliation structurally above and below the folds and the apparent absence of repeated horizons (see section on volcanic rocks) suggest, however, that the folds are open and asymmetric. This fold style is also indicated by detailed maps of D_2 folds (Figure 4C). I observed no isoclinal D_2 folds. These observations suggest that S_2 was superposed on S_1 at a moderate to low angle.

In the hinge and short limb areas of the D_2 folds, S_1 and S_2 intersect at a high angle and the two foliations produce a pencil cleavage, L_2, that parallels the D_2 fold axes (Figures 3A, 6, and 7). In areas where S_1 and S_2 intersect at a low or moderate angle, a weak to strong lineation is apparent (Figure 7). This lineation, the pencil cleavage, and the fold axes plunge gently to the southwest and parallel the long axes (L_1) of inclusions within S_1. At a mesoscopic scale, therefore, S_1 and S_2 are coaxial but not coplanar deformations.

Microscopically, S_2 is characterized by thin, discrete zones of concentrated opaques and reoriented and/or recrystallized phyllosilicates. In the shales, these seams of opaques and phyllosilicates cross-cut the structures in the larger, folded, and swirled phyllosilicates that mark S_1. Where S_1 and S_2 are parallel, S_2 is distinguished by the increased concentration of opaques and phyllosilicates on the opposite sides of the sandstone inclusions. Where S_1 and S_2 occur at a moderate angle, S_2 is a zonal to rough crenulation cleavage (Gray, 1977). In these cleavage zones, the S_1 phyllosilicates are kinked and reoriented and are more closely spaced than outside these zones. Spacing of separate crenulation cleavage folia is irregular due to heterogeneities caused by the sandstone inclusions and earlier D_1 structures, but in homogeneous areas the folia, are usually separated by 1 to 2 millimeters. In some areas, D_2 folds have developed without an associated axial cleavage. These folds are kink-like and have wavelengths of several millimeters. The axial planes of the folds parallel S_2 in nearby areas. S_2 is rarely present within the sandstone inclusions, but where present, it is defined by concentrations of opaques that separate indented and sutured detrital quartz and feldspar grains. These zones often occur within fine-grained D_1 zones of cataclasis that roughly parallel S_2 outside the sandstone. The sandstones are otherwise little affected internally by D_2. I interpret the zones and seams of concentrated phyllosilicates and opaques that define S_2, respectively, as reoriented and/or neocrystallized grains and dissolution residues. S_2 is, therefore, believed to have formed as a result of pressure solution. Quartz, chlorite, and locally calcite fibers are present in pressure shadows and in extensional veins and apparently represent precipitated material.

D_1 swirls and folds of phyllosilicates have also commonly behaved as rigid inclusions during D_2. These zones of reoriented, relatively large phyllosilicates are unaffected internally by S_2, but they are bounded on two parallel sides by zones of concentrated opaque minerals and fine-grained phyllosilicates. These zones of opaque grains are identical to and parallel the S_2 zones that bound the sandstone inclusions. Apparently, the swirled and folded phyllosilicates (D_1 folds) are more competent and/or less dissolvable than the phyllosilicates oriented parallel to one another.

A preliminary study of orthogonally oriented thin sections reveals the presence of quartz fibers in the pressure shadows of some framboids of diagenetic pyrite. Although the fibers in the melange record multiple stages of extension, preliminary studies indicate that the latest and most significant period of growth occurred during the formation of S_2. These S_2 fibers record a coaxial period of extension oriented parallel to the D_2 fold axes and pencil cleavages. The fibers plunge moderately to the southwest. They also parallel a D_1 extension direction as indicated by the orientation of calcite fractures in the sandstone inclusions. In general, the fibers suggest that S_2 is approximately parallel to the XY plane of the finite strain ellipsoid, although more complete studies (in progress) are needed to confirm this.

CONCLUSIONS AND THE SIGNIFICANCE OF D_2

The parallelism of D_1 and D_2 extension directions suggests that the two deformations are kinematically related. One possibility is that the S_2 cleavage (possibly the XY plane of the finite strain ellipsoid) represents the plane of flattening associated with shearing deformations, the "S" plane of Berthe and others (1979) and Simpson and Schmidt (1983), and the "schistosity" of Ramsay and Graham (1981). The S_1 foliation would correspond to the "C" plane of these same workers. S_2 dips more steeply to the northwest than does S_1 and, therefore, is geometrically consistent with this interpretation as well as with the northwest-directed underthrusting inferred from S_1 structures. S_2 may have developed late in this shearing deformation, because the physical properties of the sediments in the accretionary prism were evolving and pressure solution may not have been possible during S_1.

Alternatively, S_2 cleavage may represent a regional-scale slaty cleavage locally associated with large thrust zones (e.g., Moore, 1978). This large thrust zone is not exposed in the Ghost Rocks Formation, but may be buried by the partially obducted and accreted Eocene Sitkalidak Formation (Figure 1; Moore and Allwardt, 1980). A large thrust zone may have developed in the later stages of D_1 (perhaps for reasons given above) as a result of the emplacement of the Eocene Sitkalidak Formation. With either alternative, however, the direction of extension in S_2 should be to the northwest, the direction of transport, and not to the southwest as indicated by the fibers (this apparent problem is discussed below).

These two alternatives have not been resolved. I prefer the first interpretation because the extension directions for S_1 and S_2 are parallel and because the emplacement of the Sitkalidak Formation correlates better with deformations recorded in the younger coherent terranes (Byrne, 1982a, 1982b and in preparation).

If a relationship between S_1 and S_2 is accepted, then the rocks in the melange apparently experienced a substantial amount of lateral flow parallel to the continental margin during both D_1 and D_2. Lateral flow during progressive shear, although difficult to explain kinematically because of room problems, is not uncommon. Compton (1980) has documented lateral flow in nappe structures in the Great Basin. He believes the nappes formed as a result of regional doming that could accommodate both down-dip transport and lateral flow of the nappe structures. Flattening strains also have been documented recently in a mylonitic shear zone in Arizona (Bykerk-Kauffman, 1983, as discussed above), and Weiss (1954) has described regional extensional strains that occur perpendicular to the inferred transport direction (i.e., B-tectonites) in rocks of the central Mojave desert in California. Lister and Price (1978) have discussed the strains and fabrics that might develop in these kinematic regimes.

Cowan (1982, in press) has proposed a model involving flattening strains and lateral flow for the Franciscan Complex. Based on the oblate shape of melange inclusions, Cowan argued that the melange had been extended in all directions parallel to bedding. He also suggested that the melange was deformed during vertical collapse of the continental margin, causing the margin to flow in all directions. Cowan, however, had no qualitative or quantitative estimates of the amount or direction of extension.

It is noteworthy that both the Franciscan Complex and the Ghost Rocks Formation may have experienced movement of material parallel to the presumed continental margin. In contrast to the Franciscan Complex, the Ghost Rocks Formation melange terranes contain a regionally important linear fabric. The melange has apparently not flowed uniformly in all directions. Similarly, there is no indication that the melange terranes moved and spread laterally down a regional bulge or dome. Apparently lateral, subhorizontal extension occurred in the melange as a consequence of progressive shear, perhaps representing a deformation as complex as rolling dough in one direction (see Figure 7 in Lister and Price, 1978, and discussion by Weiss, 1954).

ACKNOWLEDGMENTS

This paper is part of a PhD thesis completed at the University of California at Santa Cruz, and I greatly appreciate the stimulating environment the people there have provided. Casey Moore, Eli Silver, and Othmar Tobisch kindly reviewed earlier drafts of this paper. Casey Moore is truly an outstanding advisor and his example of enthusiasm and dedication has been appreciated both in and away from Santa Cruz. I learned a great deal from him. Gene Gonzales' skill of making beautiful, ultra-thin-sections, even on "ugly" rocks, made much of this project possible. Darrel Cowan has been a continuous source of encouragement and insightful queries. Ed Beutner is in part responsible for getting me started in melange terranes and I cannot thank him enough for his continued support. He also has kept me clear-thinking in these messed-up rocks. Mary Reid provided several summers of assistance in the field and never-ending encouragement to look at the details. Mark Brandon's help and discussions in the field are also appreciated.

Nick Rast, Bill Bosworth, and one other reviewer made many helpful comments on an early draft of this paper. Support was provided by National Science Foundation grants NSF 80-07464 (to Casey Moore) and EAR-8306578 (to Byrne). Mobil Oil Company, Union Oil, Amoco Production, the U.S. Geological Survey, and the U.S. Coast Guard also provided financial support. Liz Story (at Cornell University) and Judy Botelho (at Brown University) typed various versions of this paper. Part of this paper was also completed while I was a Shell Faculty Fellow at Brown University.

REFERENCES CITED

Armentrout, J., 1979a, Cenozoic stratigraphy, Kodiak Island archipelago, Alaska: Geological Society of America Abstracts with Programs, v. 11, p. 66.

—— 1979b, Paleocene plutonism in southern Alaska [abs.]: Pacific Science Congress, 14th, Khabarovsk, USSR, Proceedings.

Aydin, A., 1978, Small faults formed as deformation bands in sandstone: Pure and Applied Geophysics, v. 116, p. 913–930.

Aydin, A., and Johnson, A., 1978, Development of faults as zones of deformation bands and as slip surfaces in sandstone: Pure and Applied Geophysics, v. 116, p. 931–942.

Bachman, S. B., 1978, A Cretaceous and early Tertiary subduction complex, Mendocino coast, northern California, *in* Howell, D. G., and McDougall, K. A., eds., Mesozoic Paleography of the Western United States: Pacific Section of the Society of Economic Paleontologists and Mineralogists, p. 419–429.

Beach, A., 1977, Vein arrays, hydraulic fractures and pressure-solution structures in a deformed flysch sequence, S.W. England: Tectonophysics, v. 40, p. 201–225.

—— 1980, Numerical models of hydraulic fracturing and the interpretation of syntectonic veins: Journal of Structural Geology, v. 2, p. 425–438.

Berthe, D., Choukroune, P., and Jegonzo, P., 1979, Orthogneiss, mylonite and non-coaxial deformation of granites: The example of the South Armorican shear zone: Journal of Structural Geology, v. 1, p. 31–42.

Bykerk-Kauffman, A., 1983, Early to mid-Tertiary east-directed shear in the Buehman canyon area: eastern Santa Catalina Mountains, Arizona: Geological Society of America Abstracts with Programs, v. 15, p. 5.

Byrne, T., 1982a, Structural evolution of coherent terranes in the Ghost Rocks Formation, Kodiak Island, Alaska, *in* Leggett, J., ed., Trench and Forearc Sedimentation and Tectonics: Geological Society of London, p. 229–242.

—— 1982b, Contrasting styles of deformation in an ancient accretionary prism: Implication for subduction zone kinematics: Geological Society of America Abstracts with Programs, v. 14, p. 457.

Cobbold, P., Cosgrove, J., and Summers, J., 1971, The development of internal structures in deformed anisotropic rocks: Tectonophysics, 12, p. 23–53.

Compton, B., 1980, Fabrics and strains in quartzites of a metamorphic core Raft River Mountains, Utah, *in* Crittenden, M., Coney, P., and Davis, G., eds., Cordilleran Metamorphic Core Complexes: Geological Society of America Memoir, v. 153, p. 385–398.

Cowan, D. S., 1974, Deformation and metamorphism of the Franciscan subduction zone complex northwest of Pacheco Pass, California: Geological Society of America Bulletin, v. 85, p. 1623.

—— 1978, Origin of blueschist-bearing chaotic rocks in the Franciscan complex, San Simeon, California: Geological Society of America Bulletin, v. 89, p. 1415–1423.

—— 1982, Deformation of partly dewatered and consolidated Franciscan sediments near Piedras Blancas Point, California, *in* Leggett, J. K., ed., Trench-Forearc geology: Sedimentation and tectonics on modern and ancient active plate margins: Geological Society of London Special Publication 10, p. 439–457.

Cowan, D. S., in press, Structural styles in Mesozoic and Cenozoic melanges in the western Cordillera of North America: Geological Society of America.

Critteden, M., Coney, P., and Davis, G., 1980, Cordilleran metamorphic core complexes: Geological Society of America Memoir, v. 153, 490 p.

Etheridge, M., 1983, Differential stress magnitudes during regional deformation and metamorphism: Upper bound imposed by tensile fracturing: Geology, v. 11, p. 231.

Fergusson, C. L., 1984, The Gundahl Complex of the New England fold belt, eastern Australia: A tectonic melange formed in a Paleozoic subduction complex: Journal of Structural Geology, v. 6, p. 257–271.

Frost, E., and Martin, D., 1982, Mesozoic-Cenozoic tectonic evolution of the Colorado river region, California, Arizona, and Nevada, Cordilleran Publishers: San Diego, 608 p.

Gray, D., 1977, Some parameters which affect the morphology of crenulation cleavages: Journal of Geology, v. 85, p. 763–780.

Hill, M. D., 1978, Volcanic and plutonic rocks of the Kodiak-Shumagin Shelf, Alaska: Subduction deposits and near-trench magmatism [Ph.D. thesis]: Univ. California, Santa Cruz.

Hobbs, B., Means, W., and Williams, P., 1976, Outline of Structural Geology: New York: John Wiley and Sons, 571 p.

Johnson, A., and Aydin, A., 1982, Analysis of faulting in porous sandstones: Journal of Structural Geology, in press.

Lister, G., and Price, G., 1978, Fabric development in a quartz feldspar mylonite: Tectonophysics, v. 49, p. 37–78.

Logan, J. , Friedman, M., Higgs, N., Dengo, C., and Shimamoto, T., 1979, Experimental studies of simulated gouge and their application to studies of natural fault zones, *in* Speed, R., Sharp, R., and Evernden, J., eds., Analysis of Actual Fault Zones in Bedrock: United States Geological Survey, Open File Report 79–1239.

Lundberg, N., and Moore, J., 1981, Structural features of the Middle America Trench slope off southern Mexico, Leg 66 DSDP: Initial Reports of the Deep Sea Drilling Project, v. 66, Washington (U.S. Government Printing Office).

Moore, G. W., 1969, New formations of Kodiak and adjacent islands, Alaska: United States Geological Survey Bulletin, v. 1274–A, p. 4228.

Moore, J. C., 1978, Orientation of underthrusting during latest Cretaceous and earliest Tertiary time, Kodiak Islands, Alaska: Geology, v. 6, p. 206–213.

Moore, J. C., and Allwardt, A., 1980, Progressive deformation of a Tertiary trench slope, Kodiak Islands, Alaska: Journal of Geophysical Research, v. 85, p. 4741–4756.

Moore, J. C., and Connelly, W., 1979, Tectonic history of the continental margin of southwestern Alaska: Late Triassic to earliest Tertiary: *in* The Relationship of Plate Tectonics to Alaska Geology and Resources, p. H1-29, Alaska Geological Society, Anchorage, Alaska.

Moore, J. C., Byrne, T., Reid, M., Gibbons, H., Plumley, P., and Coe, R., 1983, Paleogene evolution of the Kodiak Islands, Alaska: Implications of ridge subduction in southern Climes: Tectonics, v. 2, p. 265–293.

Platt, J., 1983, Progressive refolding in ductile shear zones, Journal of structural Geology, v. 5, p. 619–622.

Plumley, P., Coe, R., Byrne, T., Reid, M., and Moore, J., 1982, Paleomagnetism of volcanic rocks of the Kodiak Islands indicates northward latitudinal displacement: Nature, v. 300, p. 50–52.

Plumley, P., Coe, R., and Byrne, T., 1983, Paleomagnetism of the Paleocene Ghost Rocks Formation, Prince William Terrane, Alaska: Tectonics, v. 2, p. 295–314.

Ramsay, J. G., 1967, Folding and fracturing of rocks: McGraw-Hill Book Co., San Francisco, p. 568.

—— 1980, Shear zone geometry: A review: Journal of Structural Geology, v. 2, p. 83–100.

Ramsay, J. G., and Graham, R. H., 1970, Strain variation in shear belts: Canadian Journal of Earth Sciences, v. 7, p. 786–813.

Reches, Z., 1978, Analysis of faulting in three-dimensional strain field: Tectonophysics, v. 47, p. 109–129.

Reed, B. L., and Lanphere, M. A., 1974, Chemical variations across the Alaska-Aleutian range batholith: Journal of Research, United States Geological Survey, v. 2, p. 343–352.

Reid, M. R., and Gill, J. B., 1980, Near trench volcanism, Kodiak Island, Alaska, Implications for ridge-trench encounter: Geological Society of America Abstracts with Programs, v. 12, p. 148.

Sanderson, D. J., 1974, Boudinage and apparent stretching lineation developed in folded rocks: Journal of Geology, v. 82, no. 5, p. 651–661.

Schamel, S., 1974, Eocene subduction in central Liguria. Italy: Yale University [Ph.D. thesis]: University Microfilms: 74–25, p. 792.

Silver, E. A., and Beutner, E. C., 1980, Melanges: Geology, v. 8, p. 32.

Simpson, C., and Schmid, S., 1983, An evaluation of criteria to deduce the sense of movement in sheared rocks: Geological Society of America Bulletin, v. 94, p. 1281–1288.

Smith, R. B., 1977, Formation of folds, boudinage and mullions in non-Newtonian materials: Geological Society of America Bulletin, v. 88, p. 312–320.

Stone, D., Panuska, B., and Packer, D., 1981, Paleolatitudes versus time for southern Alaska: Journal of Geophysical Research, v. 87, p. 3697–3707.

Turner, F. and L. Weiss, 1963, Structural Analysis of Metamorphic Tectonites: McGraw Hill, New York, 545 p.

Tchalenko, J., 1970, Similarities between shear zones of differential magnitudes: Geological Society of America Bulletin, v. 81, p. 1625–1640.

Vollmer, F., and Bosworth, B., in press, Formation of melange in a foreland basin overthrust setting, *in* Raymond, L. A., ed., Melanges: Their nature, origin and significance: G.S.A. Special Paper 198 (this volume).

Wallace, W., and Engebretson, D., 1984, Relationship between plate motions and Late Cretaceous to Paleogene magmatism in southwestern Alaska: Tectonics, v. 3, p. 295–315.

Weiss, L., 1954, A study of tectonite style: University of California Publication of Geological Sciences, v. 30, p. 1–102.

MANUSCRIPT ACCEPTED BY THE SOCIETY JULY 12, 1984

Printed in U.S.A.

Geological Society of America
Special Paper 198
1984

Formation of melange in a foreland basin overthrust setting: Example from the Taconic Orogen

F. W. Vollmer*
William Bosworth*
Department of Geological Sciences
State University of New York at Albany
Albany, New York 12222

ABSTRACT

The Taconic melanges of eastern New York developed through the progressive deformation of a synorogenic flysch sequence deposited within a N-S elongate foreland basin. This basin formed in front of the Taconic Allochthon as it was emplaced onto the North American continental shelf during the medial Ordovician Taconic Orogeny. The flysch was derived from, and was subsequently overridden by the allochthon, resulting in the formation of belts of tectonic melange. An east to west decrease in deformation intensity allows interpretation of the structural history of the melange and study of the flysch-melange transition. The formation of the melange involved: isoclinal folding, boudinage and disruption of graywacke-shale sequences due to ductility contrasts; subaqueous slumping and deposition of olistoliths which were subsequently tectonized and incorporated into the melange; and imbrication of the overthrust and underthrust sedimentary sections into the melange. The characteristic microstructure of the melange is a phacoidal conjugate-shear cleavage, which is intimately associated with high strains and bedding disruption. Rootless isoclines within the melange have apparently been rotated into an east-west shear direction, consistent with fault, fold, and cleavage orientations within the flysch. The melange zones are best modeled as zones of high shear strain developed during the emplacement of the Taconic Allochthon. Total displacement across these melange zones is estimated to be in excess of 60 kilometers.

INTRODUCTION

The association of melanges with subduction zone deformational processes and accretionary prisms is now widely accepted, although this is clearly not the only tectonic environment for melange formation (e.g., Saleeby, 1979; Page and Suppe, 1981; Jacobi, this volume). The basic phenomena involved in the formation of melanges in subduction zones include the formation of a sedimentary trough in the overthrust zone, infilling of this basin with sediments (flysch), and subsequent progressive deformation of the sediment pile. Additional complications result from the imbrication and incorporation of portions of the overthrust and underthrust sheets into the accretionary complex. Olistostromes may occur in association with the flysch deposits and in some cases appear to play an important role in the formation of melange (e.g., Robertson, 1977). Complex deformational histories, including strike-slip deformation and accretion of exotic terranes, may be expected along convergent plate boundaries (Hamilton, 1980; Karig, 1980). Several problems exist in studying melange formation in this environment. Initial deformation in a trench environment can be studied only through geophysical methods, coring, or submersible observation (e.g., White and Ross, 1970; Moore, 1979; Heezen and Rawson, 1977). Most melanges currently exposed on land have been through a protracted history that may have involved several episodes of deformation not directly related to the initial formation of the melange (e.g., Aalto,

*Present addresses: Vollmer, Department of Geology, State University of New York College at New Paltz, New Paltz, New York 12561. Bosworth, Department of Geology, Colgate University, Hamilton, New York 13346.

1981). Additionally, melange contacts are commonly faulted and do not preserve the transition from undeformed rock units into melange (e.g., Moore and Wheeler, 1978; Moore and Karig, 1980).

Melanges may also form, however, in other tectonic regimes involving thrusting. Ophiolitic melanges commonly form during the obduction of ophiolites onto continental margins, although this may not necessarily involve the incorporation of large amounts of flysch into the melange (Gannser, 1974; Williams, 1977; Cohen and others, 1981). Melange may also form through syndepositional deformation of flysch in front of non-ophiolitic thrust nappes, such as the North Penninic Melange of the Prealps (Homewood, 1977). Similarly, during the medial Ordovician Taconic Orogeny, a series of thrust nappes, the Taconic allochthons, were emplaced onto the North American continental shelf. In front of these nappes were deposited transgressive flysch sequences subsequently overridden by the nappes. The interaction of sedimentation and overthrust deformation produced melanges along the western margins of many of these nappes (Williams, 1979; St. Julian and Hubert, 1975; Bosworth and Vollmer, 1981). The Taconic Allochthon of eastern New York, Vermont, Massachusetts, and Connecticut is believed to have been emplaced during the attempted subduction of the North American continental margin into an east-dipping island arc subduction zone, with subsequent migration of the accretionary environment onto the continental shelf (Rowley and Kidd, 1981). The resulting external flysch basin, or foreland basin (Dickinson, 1977), has preserved the transition from undeformed flysch into highly disrupted melange.

Mapping has been done within several subareas in the Hudson Valley lowlands of eastern New York (Fig. 1). This mapping has mainly involved detailed structural analysis and the study of facies distributions. Subareas were chosen to include deformation intensities varying from undeformed flysch to highly disrupted melange. Key localities along the front of the Taconic Allochthon have also been mapped in detail. If we assume that spatial sequences showing increasing deformation intensity, in equivalent rock types, represent the increasing deformation of a rock body through time, we may infer the deformation history of these rocks. This paper discusses the evolution of folds and fabric in the Taconic melange, and the relative roles of tectonic dismemberment, imbricate thrusting, and olistostrome deposition in its formation.

REGIONAL GEOLOGIC SETTING AND STRUCTURAL POSITION OF THE MELANGE

The Taconic Orogen can be divided into three geologic zones or provinces (Rowley and Kidd, 1981), from west to east: 1) Cambro-Ordovician rifted continental margin, including transported sediments of the continental rise sequence (eastern New York and western Vermont, Massachusetts, and Connecticut); 2) suture zone of medial Ordovician age (west side of the Connecticut River synclinorium); and 3) Ordovician volcanic arc terrane (exposed along the eastern edge of Zone 2 and within the Bronson Hill anticlinorium). These correspond approximately to the Humber, Dunnage, and western edge of the Gander zones of Williams (1978). The Taconic Allochthon and associated medial Ordovician flysch lie within Zone 1 (Fig. 1). The allochthon is composed of slightly to moderately metamorphosed argillites and lesser silty or shaly sandstones and limestones, interpreted to represent a transported continental rise sedimentary sequence (Bird and Dewey, 1970). The graywackes, siltstones and shales of the Taconic flysch are found as the youngest unit of the allochthonous terrane, and to the west of and beneath the allochthon. These flysch units were deposited prior to and during emplacement of the Taconic Allochthon onto the North American Cambro-Ordovician shelf (Zen, 1961). Shelf carbonate rocks and sandstones are currently visible in windows through the flysch and on the perimeter of the Adirondack Grenville basement dome northwest of the Taconic Allochthon (Fig. 1).

The central suture zone of the orogen is composed of metamorphosed and highly deformed schists, phyllites and lesser gneisses, with included slivers of serpentinized mafic and ultramafic rocks (Doll and others, 1961; Merguerian, 1979; Merguerian and others, 1982). Glaucophane and omphacite-bearing amphibolites have been found locally within this terrane (Laird and Albee, 1975; Laird, 1981). The boundaries and many internal lithologic contacts of this zone are complex thrust faults or major thrust zones (Ratcliffe and Hatch, 1979; R. Stanley, mapping progress) some with demonstrable Taconic movement (Ratcliffe and Mose, 1978). This amphibolite schist assemblage is interpreted to represent a Taconian suture zone, with shards of dismembered ophiolite as remnants of the early Paleozoic Iapetus Ocean (Merguerian, 1979; Rowley and Kidd, 1981).

The eastern zone of the Taconic Orogen is a volcanic terrane consisting of metamorphosed mafic and felsic volcanic rocks, tuffs, and intermediate composition plutonic rocks (Thompson and others, 1968; Robinson and others, 1979). The basement of this complex is Precambrian gneiss dated at 620 to 600 m.y. (Besancon and others, 1977; Aleinikoff and others, 1979; Robinson and others, 1979) and therefore distinctly younger than the 1.1 billion-year-old Grenville basement of Zone 1 (Hills and Gast, 1964; Silver, 1968; Bickford and Turner, 1971). This eastern terrane is interpreted to represent a volcanic island arc that became sutured to the Paleozoic North American plate in medial Ordovician times (Rowley and others, 1979; Rowley and Kidd, 1981; Stanley and Ratcliffe, 1980).

Analogy with modern convergent plate boundary settings provides a tectonic interpretation for the Taconian geologic provinces. The polarity of the orogen (passive margin in the west, arc in the east) and location of the Taconian suture (central zone) suggest collision above an east-dipping subduction zone (Chapple, 1973), with westward-directed obduction of continental rise sediments and fragmented ophiolite onto the North American continental shelf. This scenario is supported by the westward transgressive nature of the Taconic flysch (Rickard, 1973; Fisher, 1977), which marks the migration of the trench environment

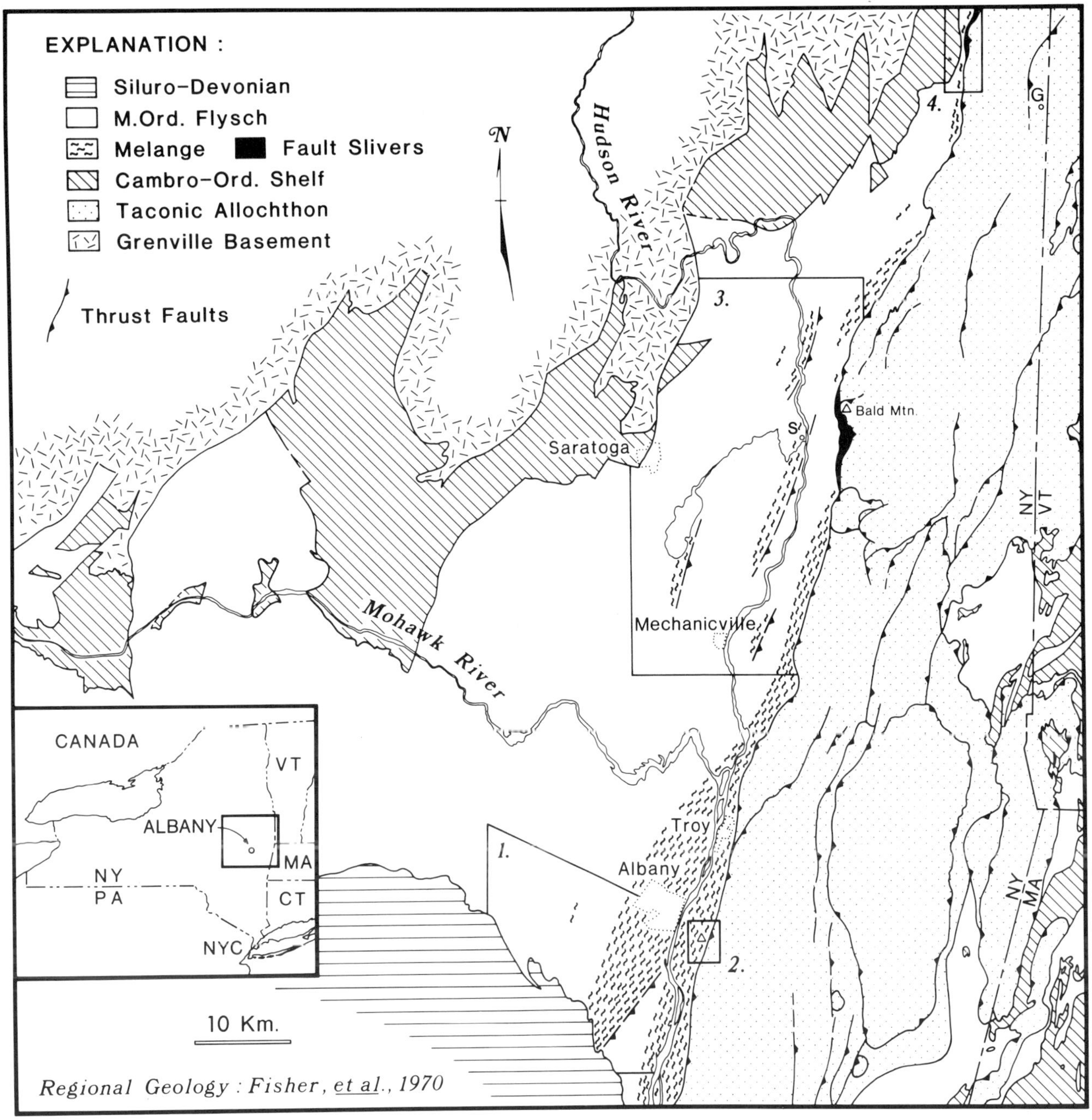

Figure 1. Regional geology of eastern New York. Numbered boxes indicate areas mapped for the present study (1 = Albany area, 2 = vicinity of Rysedorph Hill, 3 = Schuylerville area, 4 = Mettawee River section). Numerous thrust faults suggest internal imbrication of the Taconic Allochthon (stipple pattern). The fold-thrust mode of deformation continues into the flysch (unornamented), where major fault zones are delineated by melange (curl pattern). Modified from Fisher et al., 1970. (S = Schuylerville, G = Granville).

onto the downwarped passive continental margin, a depositional/structural setting that may best be referred to as a "foreland basin" (Dickinson, 1977). Detailed stratigraphic relationships of the Taconic flysch and their bearing on the Taconian arc-continent collision can be found in Rowley and Kidd (1981).

The Taconic flysch can be subdivided into three informal lithologic units: a sequence of interbedded shales, siltstones, and graywackes, with pebbly and rarely bouldery mudstone horizons; a dominantly shaly section, with local thin-bedded siltstones and fine-grained sandstones; and finally, melange (Bosworth and Vollmer, 1981; Vollmer, 1981a; Bosworth, 1980). The interbedded sequence represents the flysch "proper"; sediments deposited by turbidity currents in a longitudinal north-south elongate basin, derived from the arc terrane and advancing allochthons in the east (Rowley and Kidd, 1981, and references therein). Once sediments reached this basin, their transport direction changed to follow its north-south strike (Vollmer, 1981a; similarly observed for the Martinsburg Fm. by McBride, 1962). The dominantly shale unit is probably transitional to the interbedded unit, and corresponds to the more distal, fine-grained deposits of the turbidity flows. This unit is present to some extent within the deformed section of the flysch (Bosworth and Vollmer, 1981, Fig. 5) but becomes dominant farther west, where it is largely unaffected by Taconian deformation.

The final unit, melange, is restricted to flysch structurally beneath the allochthonous terrane or near its western edge. Known occurrences of Taconic melange and areas mapped by the authors are shown in Figure 1. The term *melange* is used in the manner suggested at the 1978 Penrose Conference on melanges (Silver and Beutner, 1980) and refers to a "mappable, internally fragmented, and mixed rock body containing a variety of blocks, commonly in a pervasively deformed matrix." A variety of melange types can be recognized in the Taconic flysch, and it is important that they each be distinguished in any discussion of the evolution or structural significance of melange in the Taconic Orogeny.

Most of the Taconic melange is composed of blocks and clasts of graywacke, siltstone, slate, and argillite in a phacoidally cleaved matrix (discussed below). This lithofacies is equivalent to the "broken formation" of Hsü (1968). This melange type occurs as much as 15 kilometers west of the present allochthon boundary (Fig. 1; Bosworth and Vollmer, 1981, Fig. 5; see also Berry, 1962). Other melange lithofacies of the Taconic flysch are distinguished by the presence of blocks of an exotic nature. Near the allochthon, the melange commonly contains pebble- to boulder-size clasts of quartzite, limestone, dolostone, limestone conglomerate, and chert. These lithologies can be correlated with units of the allochthonous Taconic sequence and with portions of the autochthonous carbonate shelf (Zen, 1967; Bird and Dewey, 1975). In less deformed outcrops, the exotic clasts are generally rounded and occur within an uncleaved mudstone matrix. These deposits are interpreted to have originated as olistostromes, as the most proximal sedimentary facies of the flysch depositional environment.

Some exotic blocks in the flysch are of an unknown affinity. This is particularly true for the melange just south of Schuylerville, New York, at Stark's Knob. Here a single knocker of pillow basalt (~100 m × 30 m × ?) sits within highly contorted shale, accompanied only by blocks of argillite and wacke, and rare clasts of chert. An additional complication arises in that graptolites obtained from shale near Stark's Knob are of early Ordovician age (Ruedemann, 1914; Berry, 1962); these shales are also interpreted to be caught up within the melange as exotic blocks, as flysch deposition in all other preserved parts of the orogen occurred in medial Ordovician times. Further discussion of the structural relationships observed at Stark's Knob can be found in Ruedemann (1914) and Bosworth (1982). The basalts were probably extruded onto the continental shelf in early Ordovician times, and subsequently imbricated into the flysch-melange structural sequence during the Taconian collisional event.

ALBANY QUADRANGLE

Exposures of the Ordovician rocks within the Albany 15 minute quadrangle (Fig. 1, area 1 and Fig. 2) exhibit a wide range of deformation intensities, generally increasing from west to east, within a sequence of shale, siltstone, and graywacke (Vollmer, 1981a). This range of deformation intensities provides an opportunity to study the response of these rock types to increasing strains, and the effect of lithologic variations on deformational style. The rocks of this area have been separated into seven structural domains, shown in Figure 2. Stereograms for these seven structural domains are given in Figure 3. The western flysch domain (domain 1, Figs. 2, 3) comprises mainly near-horizontal sequences of graywacke, siltstone, and shale. Shear zones are rare

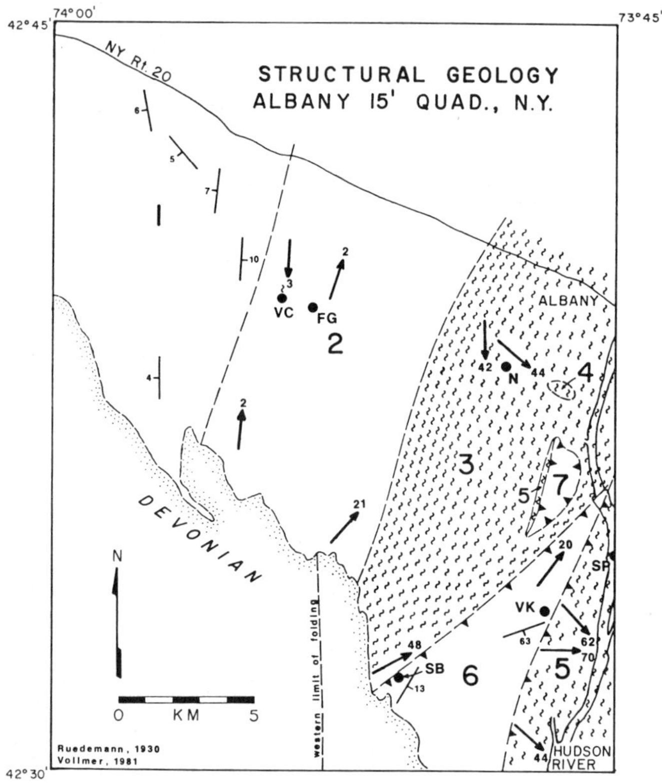

Figure 2. Structural geology of the Albany 15 minute quadrangle, eastern New York. Location is given in Figure 1. Structural symbols represent mean orientations for outcrop areas. Strike and dip symbols represent bedding; arrows signify fold hinge lines with plunges. Curl pattern = melange; no pattern = bedded flysch; stipple pattern = edge of Siluro-Devonian unconformity. Structural domains indicated: 1 = western flysch, 2 = eastern flysch, 3 = Normansville melange, 4 = Normans Kill gorge melange, 5 = eastern melange, 6 = Vloman Kill graywacke slice, 7 = Glenmont chert block. FG = Font Grove Creek, N = Normansville, SB = South Bethlehem, SP = Staats Point, VC = Vly Creek, VK = Vloman Kill.

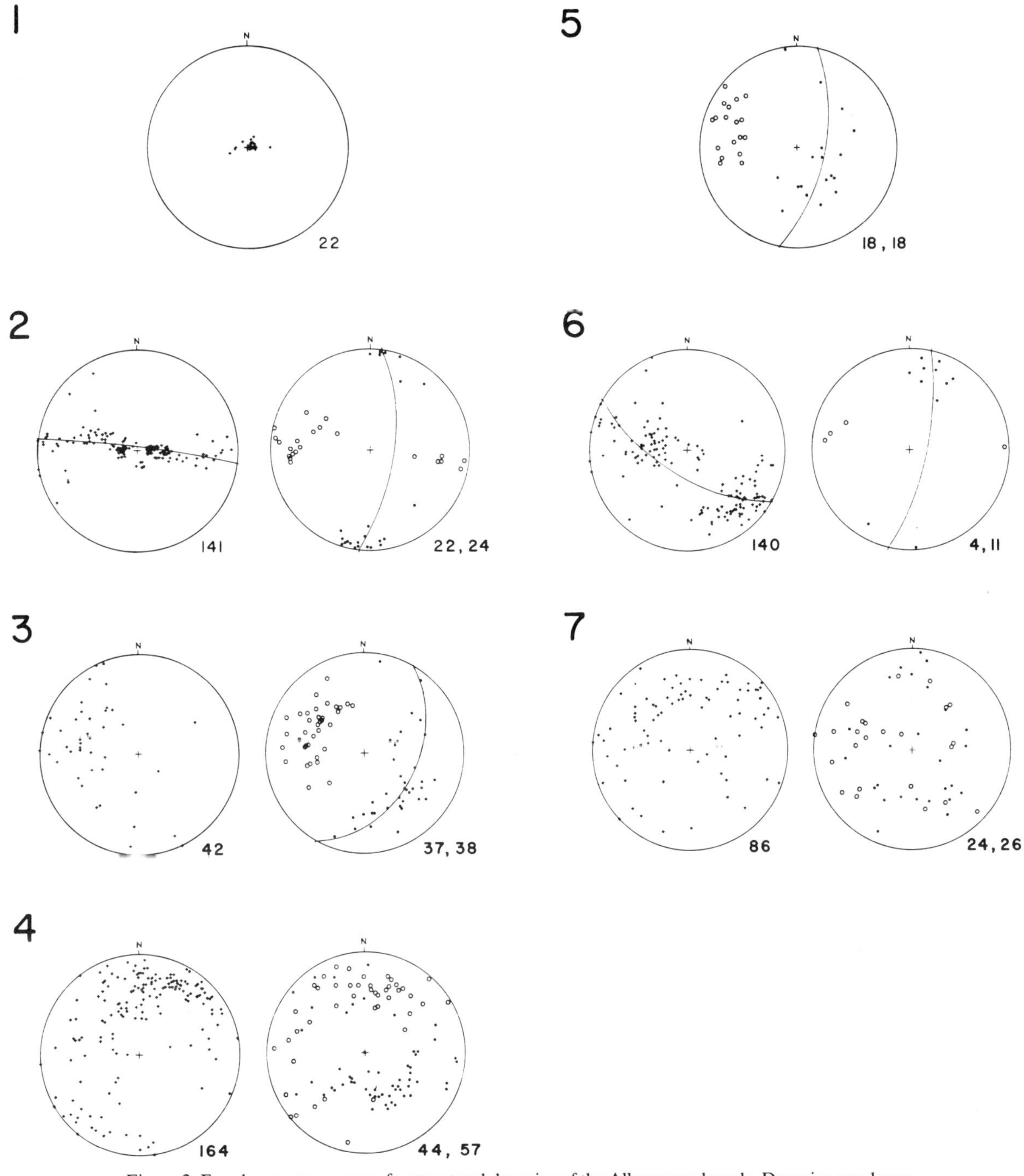

Figure 3. Equal-area stereograms for structural domains of the Albany quadrangle. Domains are shown in Figure 2. The left stereogram for each domain shows poles to bedding. The right stereogram shows poles to fold axial planes (open circles) and fold hinge lines (closed circles). Note change in fabric, particularly from domain 1 to 3.

in this sequence; however, fibrous slickenside steps (Durney and Ramsay, 1973) and a southeast-dipping internal cleavage within a one-centimeter-wide subhorizontal shear zone show northwest-directed thrusting. Gentle bedding dips here suggest long wavelength, low-amplitude folding. The eastern flysch domain (domain 2, Figs. 2, 3) also comprises interbedded graywacke, siltstone, and shale; however, the strata here form asymmetric folds predominantly with east-dipping axial planes and subhorizontal hinge lines (Fig. 4). The folds are mainly open, although several moderately plunging isoclinal folds occur near the eastern boundary of this domain. Half-wavelengths of the folds range from over 400 meters to one meter. Kink folds occur in some

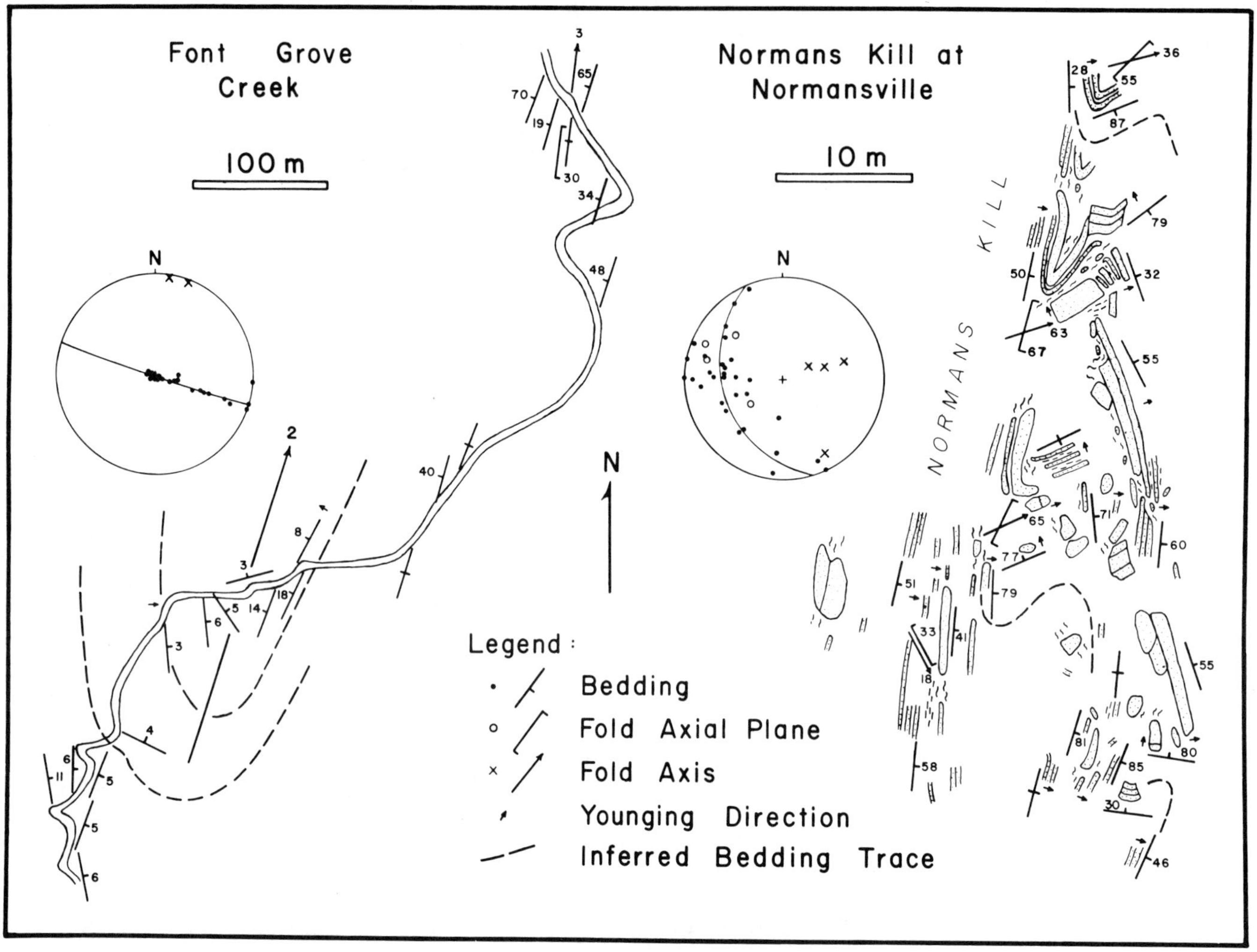

Figure 4. Structural maps of two outcrop areas within the Albany Quadrangle (Fig. 2) illustrating structural styles found in the eastern flysch (domain 2) and Normansville melange (domain 3). Outcrops at Font Grove define a large syncline with a horizontal axis, and an east-dipping axial plane. At Normansville, mapping has shown a series of dismembered asymmetric folds with steeply SE-plunging axes. Shaly sections in Normansville area show well-developed phacoidal cleavage.

outcrops, generally associated with north-northeast-striking thrust faults. These faults have moderate dips to the east and west and are narrow with minor offsets. The sense of relative motion can be determined from offsets and bedding flexures. One final, important structural feature of the eastern flysch is an approximately 20-meter-wide high strain zone exposed along Vly Creek (Fig. 2). This zone lies within otherwise only moderately folded strata, which steepen abruptly near the margins of the zone. The zone consists of dismembered arenite beds within phacoidally cleaved argillite. One small fold measured within this zone plunges 28 degrees southeast (domain 2, Fig. 3). This is the westernmost-known occurrence of the structural fabric characterizing the melange zones to the east.

The Normansville melange zone (domain 3, Figs. 2, 3) is characterized by rootless isoclinal folds, disrupted bedding, and phacoidally cleaved argillite. Rootless folds here show a variety of hinge line orientations, although they mainly plunge southeast and have moderately southeast-dipping axial planes (domain 3, Figs. 3, 4). The phacoidal cleavage also dips moderately southeast, subparallel to fold axial planes. Narrow faults commonly dip southeast with southeast or east-southeast-trending striations (Fig. 5). The eastern melange zone (domain 5, Figs. 2, 3) has a similar structural fabric, although bedding is seldom present. Here, blocks within the phacoidally cleaved argillite, normally less than 30 centimeters across, include dolomitic sandstone and some limestone, as well as abundant graywacke. Less deformed portions of this domain, particularly on the east side of the Hudson River at Staats Point and along Route 9J, are clearly olistostromic pebbly mudstones. These outcrops are characterized by rounded clasts within an uncleaved mudstone matrix (see Bird and Dewey, 1975, Fig. 6; Fisher, 1977, Figs. 65-67).

Included within our study are exposures of melange on the east side of the Hudson River in the vicinity of Rysedorph Hill (Fig. 1, area 2). This locality is particularly noted for the occur-

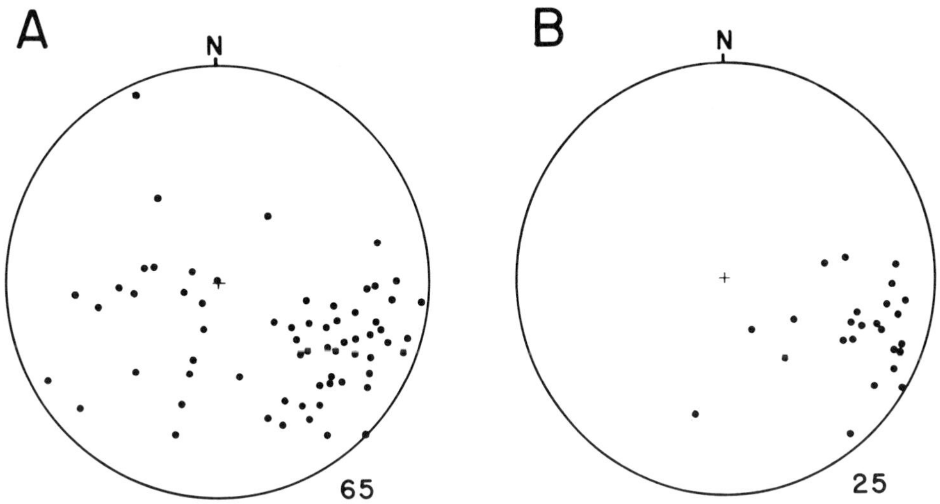

Figure 5. Equal-area stereograms of slickenside striations. Stereogram A shows data from the Albany Quadrangle (Fig. 1, area 1); stereogram B shows data from the Schuylerville area (Fig. 1, area 3). Overthrusting can be demonstrated on many of the fault surfaces by offset, bedding flexure, or stepped fibrous slickensides.

rence of large blocks of shelf-derived carbonate lying within phacoidally cleaved shales and siliceous argillites. Of particular importance is the presence of blocks, some greater than 10 meters in length, of a sand-matrix, limestone-pebble conglomerate, the "Rysedorph Hill Conglomerate" of Ruedeman (1901, 1914, 1930). The fauna of the matrix dates the conglomerate as Mohawkian (medial Ordovician), whereas the limestone pebbles span from Mohawkian to Champlainian (Chazyan), and purportedly through the Canadian (early Ordovician) (Ruedemann, 1901). Chazyan carbonates are not known from eastern New York, south of Ticonderoga, New York (Fisher and others, 1970; Fisher, 1977), and the Rysedorph Hill Conglomerate lithology has currently been identified only as slivers within the melange. We interpret the conglomerate as an outer shelf deposit that may have been transported considerable distance to its present position.

The Vloman Kill slice (domain 6, Figs. 2, 3) comprises interbedded graywackes, siltstones, and shales. Here the proportion of graywacke is substantially greater than in the surrounding melange. These rocks have been referred to the Austin Glen graywacke and are correlative with similar rocks extending south to the type locality. In this area, folds are again fairly open and hinge lines plunge gently to the north-northeast. Bedding is generally continuous, although often complexly broken and faulted. A moderately well-developed slaty cleavage forms divergent fans in the siltstones. Shortening associated with this cleavage development has apparently resulted in the buckling and thrusting of early bedding-parallel extensional fibrous veins, suggesting lithification prior to cleavage development.

Structurally, the most complex area comprises outcrops in and around the Normans Kill gorge (domain 4, Figs. 2, 3), a classic graptolite locality (Hall, 1847; Ruedemann, 1930; Riva, 1974) and the type locality of the Normanskill shale and the *Nemagraptus gracilis* graptolite zone (Ruedemann, 1930). These rocks are now ascribed to the Austin Glen Formation of the Normanskill Group (Fisher, 1977). Downward-facing folds, bedding-slaty cleavage relationships, and folded axial planes show that at least two phases of folding have occurred here. The second event apparently refolded early folds about a southeast-plunging axis. Detailed mapping at this locality has shown extensive dismemberment, with large structurally coherent blocks of folded strata juxtaposed along fault zones and dispersed within a disrupted shaly matrix. Rarely, an east-dipping slaty cleavage can be found in siltstone beds.

The final structural domain, the Glenmont block (domain 7, Figs. 2, 3), consists of folded and faulted green and black interbedded siliceous argillite and chert, structurally discontinuous from the graywacke-shale sequences. This large area of outcrop in the vicinity of Glenmont, New York, forms a block or klippe within the melange. The deformation within this block is characterized by at least two phases of folding. Apparently a late set of folds with an east-dipping axial planar cleavage was superimposed upon an early set of folds with variable orientations. Abundant faults have southeast plunging, fibrous, stepped slickensides, showing a southeast over northwest thrust sense (Fig. 5). This area is also a classic graptolite locality (Ruedemann, 1930; Riva, 1974), and the rocks may be assigned to the Mount Merino Formation. Another smaller block or sliver of chert occurs within graywacke-shale melange just west of Normansville on the Normans Kill. The Mount Merino Formation occurs within the upper portion of the Taconic sequence, directly below the Austin Glen (Pawlet of the northern Taconics) graywacke, and is also common as blocks or klippen within the Taconic melange. The type locality at Mount Merino is apparently such a block (Bird,

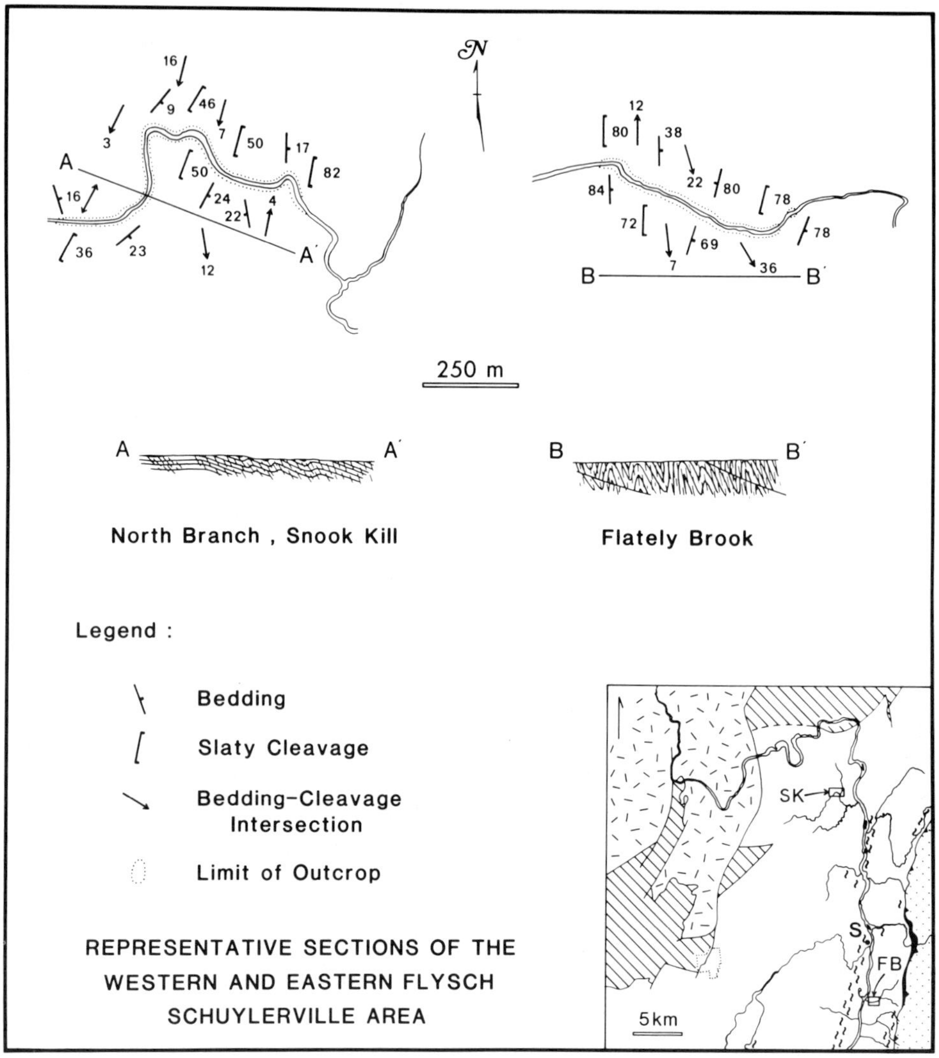

Figure 6. Structural measurements and interpreted cross section for representative areas of the western and eastern flysch, Schuylerville area (Fig. 1, area 3). Numbers with bedding and cleavage symbols give dip; those with bedding-cleavage intersection give plunge. Patterns used in location inset are same as in Figure 1. S = Schuylerville, SK = Snook Kill, FB = Flately Brook.

1969). Clasts of this lithology are common within the graywackes, and some cobble-sized fragments are found within the melange.

SCHUYLERVILLE AREA

Structures observed in the Taconic flysch of the Schuylerville area north of Albany, New York (Fig. 1, area 3), are similar in form and areal arrangement to those described above. A general increase in the intensity of deformation occurs from west to east, progressing from a gently folded section of bedded shales, siltstones, and graywackes; through a terrane composed of tightly folded strata cut by numerous zones of disrupted bedding within contorted and phacoidally cleaved shales; to a complex zone of melange and fault zones at the base of the allochthon (Bosworth, 1980). An important characteristic of the Schuylerville area is that the melange is not confined to the zone at the base of the allochthon, but is present as major structural and stratigraphic discontinuities within the flysch for as much as 15 kilometers west of the allochthon boundary (Bosworth and Vollmer, 1981).

Lesser degrees of deformation within the flysch are well illustrated by the river section on the north branch of the Snook Kill (Fig. 6). Over a distance (west to east) of less than one kilometer, bedding planes change from nearly flat-lying to involvement in gentle to open folds with moderately east-dipping axial planes and approximately horizontal hinge lines. Wavelengths are typically a few tens to one to two hundred meters, and amplitudes are generally no more than 10 to 20 meters. Weak slaty cleavage is present in the shale beds at the western edge of the section, dipping roughly 35° to the east. This slaty cleavage increases in strength to the east, being developed in siltstone and silty arenite as well, its dip increasing to 45-50° east and then becoming very steep at the east end of the section. No disrupted beds are found in this section, and the slaty cleavage is the only recognized tectonic foliation. Minor thrust faults are found parallel to the long east-dipping limbs of anticlines and cutting across section in the hinge zones of folds. On thick wacke beds, these commonly appear as grooved and stepped fibrous slickenside surfaces with east-over-west displacement.

East of the Snook Kill section, isolated outcrops commonly contain boudinaged wacke beds, isolated fold hooks, or simply lens-shaped pods of siltstone, wacke, and chert. Where outcrops are continuous or closely spaced, melange zones have been mapped (Fig. 1; see also Bosworth and Vollmer, 1981, Fig. 5), but many narrow or discontinuous zones have undoubtedly been missed. Figure 1 should be construed to represent only some of the most important melange zones and is a simplification of the actual geometry of the flysch-melange section. Structures of the melange are similar to those observed in domains three and five in the Albany Quadrangle. The fabric of the matrix is an anastomosing, east-dipping cleavage with numerous down-dip striations and commonly a polished or "glazed" appearance. Blocks and clasts within the melange are dominantly of "non-exotic" origin (found as bedded components of the flysch), except where previously noted.

The two most extensive melange zones lie in north-south elongate belts west of the Hudson River and at the base of the allochthon (Figs. 1, 7). In the vicinity of Bald Mtn. (Fig. 1), the melange is observed only locally. Here the boundary fault is marked by large slivers of shelf carbonate surrounded by lesser amounts of poorly exposed, phacoidally cleaved shale (Ruedemann, 1914; Rodgers and Fisher, 1969). These Bald Mtn. carbonates and the enclosing shale satisfy the definition of melange as used in this paper, and are referred to as fault slivers simply because of their sheet-like geometry (Bosworth, 1980) and obvious association with the frontal thrust of the Taconic Allochthon. The lithologies present in these slivers are also found as pebbles and cobbles within olistostromic horizons of the flysch near the allochthon. The large slivers are interpreted to be the source of the smaller cobbles in the depositional environment of the flysch (see further discussion in Rowley and Kidd, 1982). Although the margins of individual blocks are tectonized, primary sedimentary features are easily recognized in their interiors and on this basis and from the enclosed faunas, the carbonates have been dated as medial Mohawkian and late Canadian (early and medial Ordovician) in age (Ruedemann, 1914; Grabau, 1936; Chisick and Bosworth, 1984). Blocks of Rysedorph Hill Conglomerate are also present at Bald Mtn. (Ruedemann, 1914).

The section of flysch between the Schuylerville and Bald Mtn. melanges provides a good example of the flysch in a highly deformed state, yet not completely disrupted. The outcrops along Flately Brook have been chosen as an example of this terrane (Fig. 6), which is not well represented in the Albany area. Here hinge lines of folds have shallow to moderate plunges (7 to 36°), whereas axial planes and slaty cleavage dip steeply east (70° east to vertical). Fold asymmetry is less pronounced than in the western flysch, and fold tightness and amplitude have increased markedly. Numerous small thrust faults cut across bedding in this section, with east-over-west displacements dominating. Narrow zones of phacoidally cleaved shale containing small (generally < 1 m) clasts of arenite crop out along strike and farther east from the Flately Brook section. Clasts of well-cleaved slate, similar to the adjacent undisrupted slates, are also found within these narrow melange zones.

METTAWEE RIVER SECTION

Exposures along the Mettawee River in northern Washington County (Fig. 1, area 4) provide an opportunity to observe shelf carbonate lithologies imbricated within the flysch section. The largest carbonate slivers occur in a belt at the base of the allochthon (Fisher and others, 1970), similar in form to the Bald Mtn. terrane described above. Lithologically and faunally, the slivers studied thus far are equivalent to autochthonous Chazyan (early medial Ordovician) rocks exposed farther north along the shores of Lake Champlain (Selleck and Bosworth, 1984). Bedding within the slivers is tightly folded, an easily recognized anticline being exposed at the Mettawee River. Whether this folding pre-dates, is synchronous with, or post-dates final emplacement of the allochthon cannot be determined at the Mettawee locality.

Beneath the frontal slivers, there occurs a melange terrane containing smaller blocks of carbonate rock, quartz arenite, wacke, and abundant pebbly mudstones. Some of the pebbles are lithologies found in the larger blocks and slivers. The melange becomes more discontinuous and is replaced by bedded flysch to the west, away from the allochthon boundary. Structurally, the flysch is quite complicated here, with some refolding of early folds and slaty cleavage. At several narrow disrupted zones within the flysch, small slivers of hard limestone and calcareous arenite are caught up within a phacoidally cleaved matrix. Fold and small-scale fault orientation data are generally similar to those observed in the Schuylerville area and domains 2 (flysch) and 3 and 5 (melange) in the Albany Quadrangle.

MICROSTRUCTURES OF THE MELANGE MATRIX

A characteristic feature of the Taconic melanges is the pervasive, sub-planar fabric developed within the shaly matrix. This fabric is defined in hand specimen by the parallel alignment of lens or phacoid-shaped shale and siltstone clasts, which commonly have striated and polished surfaces. In thin section, these surfaces appear as narrow, anastomosing cleavage lamellae. X-ray diffraction studies indicate that the cleavage lamellae are composed mainly of illite and chlorite. This fabric is here referred to as "phacoidal" cleavage, reflecting the presence of lens or phaciod-shaped domains between cleavage folia. Observations made thus far have not suggested any major differences in the microstructure of the different melange lithofacies, although the cleavage formed within olistostromal horizons appears to be complicated by an early primary deformation. This preliminary

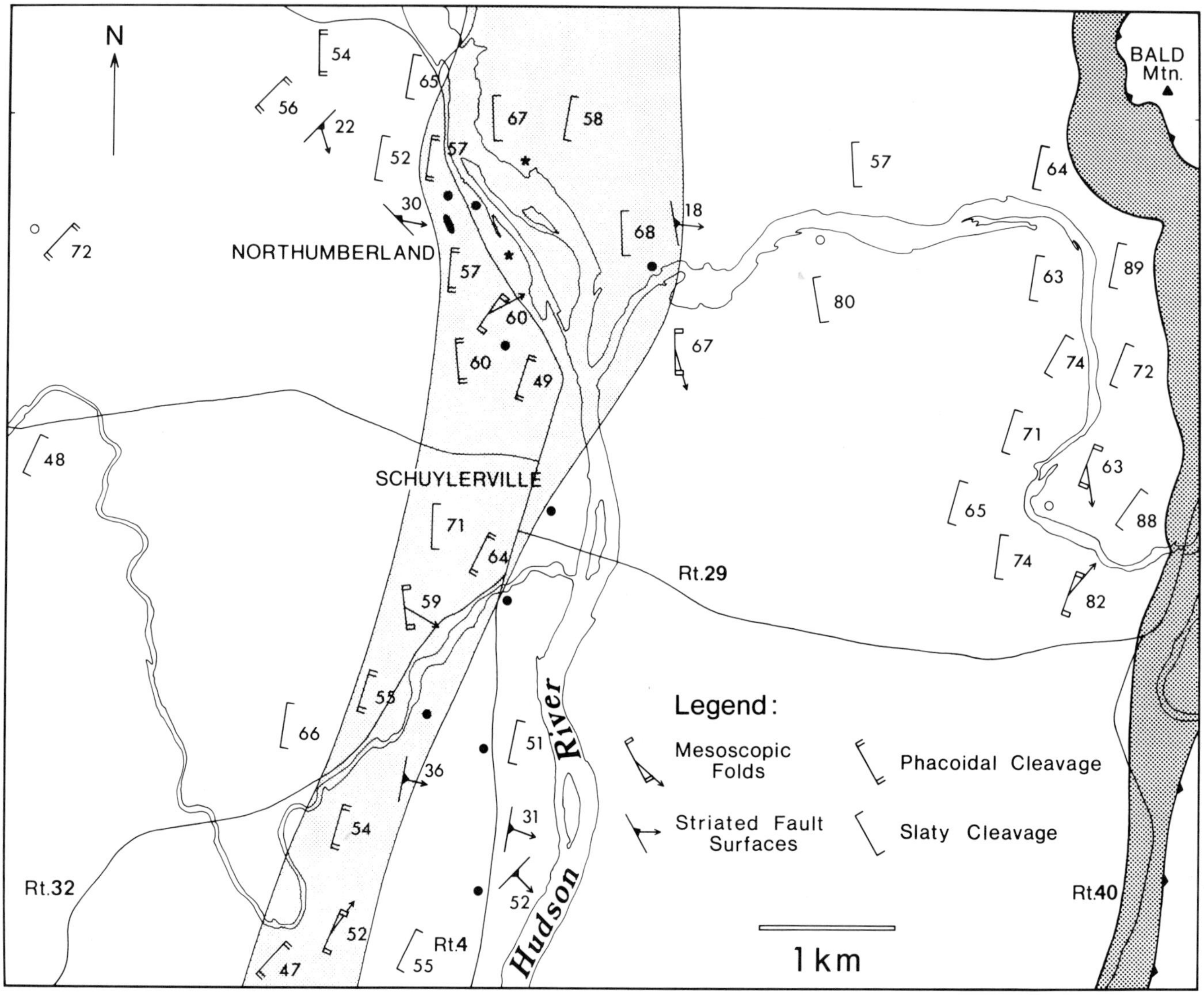

Figure 7. Structural data from the Schuylerville melange zone and adjacent flysch. Numbers with each symbol give dip of structural element, arrows indicate orientation of fold hinge lines on axial planes and striations on fault surfaces. Light stipple patterns = melange; heavy stipple pattern = fault rocks at base of Taconic Alochthon (Bald Mtn. Terrane carbonates); flysch is unornamented. Elliptical black body at Northumberland is Stark's Knob (pillow basalts). Graptolite fossil localities from Ruedemann (1914) are depicted as follows: asterisk = early Ordovician; solid circle = older medial Ordovician (probable *Nemagraptus gracilis* zone of Riva, 1974); open circle = younger medial Ordovician (probable *Diplograptus multidens* zone of Riva, 1974). From Bosworth, 1982. Used with permission of Vermont Geology.

account attempts only to relate the general results obtained from a number of melange localities chosen for their particularly good exposure or structural significance. We have looked mainly at the matrix of the graywacke-shale melange to avoid the effects of complex primary fabric.

Several dominant folia orientations are readily observed in outcrop, dipping moderately or steeply to the east. Significant populations of lamellae with other orientations, however, are also present. Lamellae can be measured in oriented hand samples to produce a three-dimensional statistical picture of the anastomosing cleavage network, or measurements can be taken from thin sections cut in various orientations to construct a number of two-dimensional pictures (Fig. 8). The phacoids enclosed by the cleavage lamellae are composed of shale, argillite, siltstone, wacke, or chert, depending on the melange lithofacies considered.

In hand specimen, shale phacoids are commonly crudely polyhedral and many appear rhombohedral in cross section, particularly where observed parallel to strike. The phacoid surfaces are polished and exhibit fine striations generally well developed on east-dipping surfaces, but also common in other orientations.

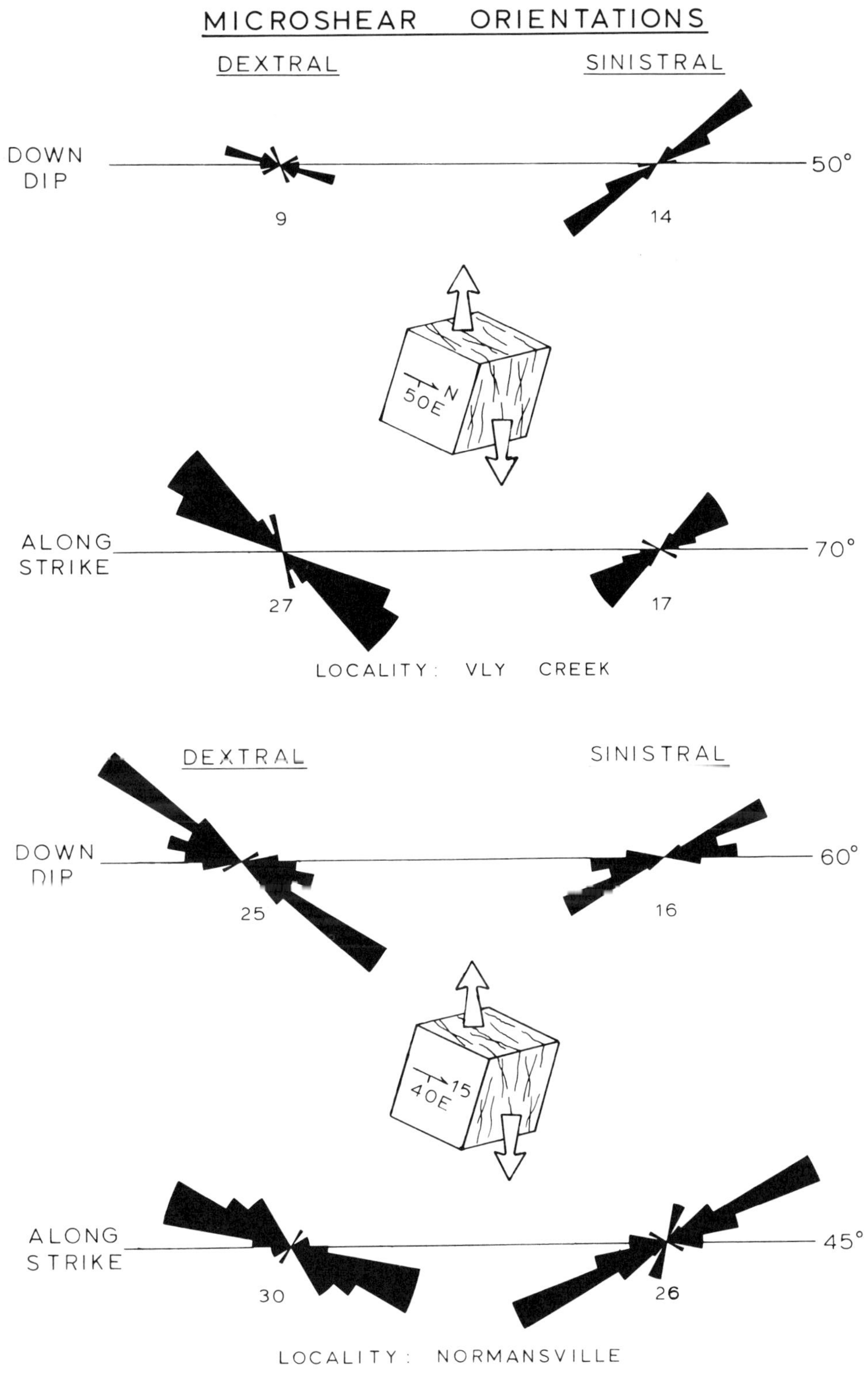

Figure 8. Microshear orientations from two localities within the Albany Quadrangle; see Figure 2 for locations. Microshear orientations were measured in thin section only where sense of shear could be determined from offset or flexure of primary phyllosilicate fabric. Horizontal lines represent approximate traces of mesoscopic foliation. Approximate angles between the two sets are given at right.

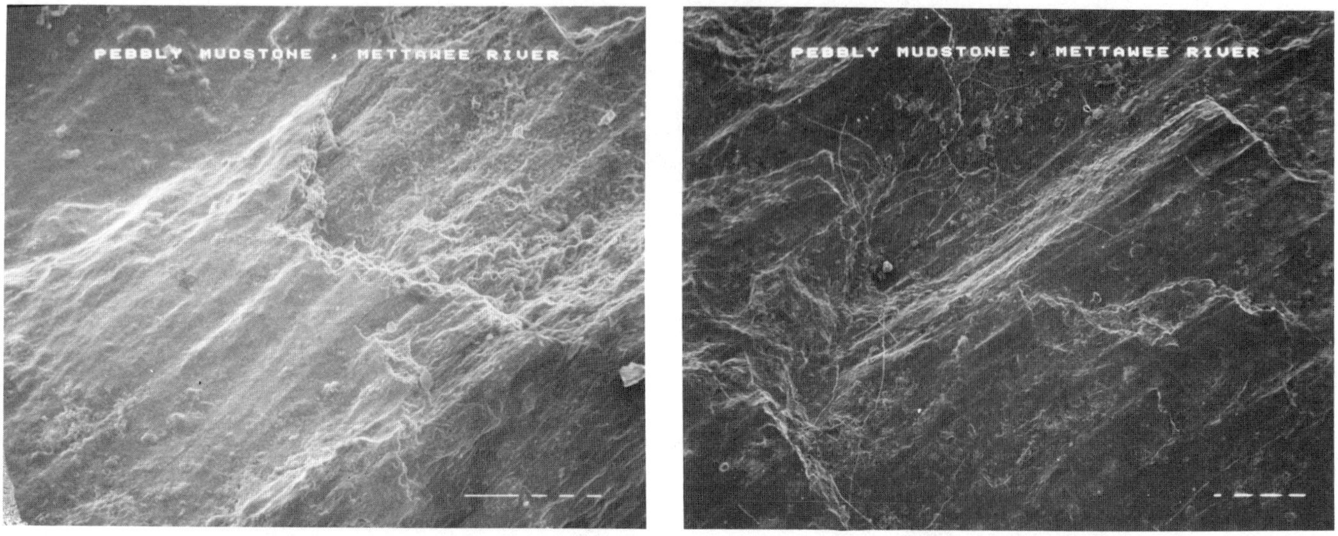

Figure 9. Scanning electron micrographs of phacoidal cleavage lamellae surfaces. Scale bar is bar farthest to the left in series and equals 100μ. Samples from pebbly mudstone horizon in melange on Mettawee River section (Fig. 1, area 4).

Scanning electron microscopy of the lamellae surfaces reveals a micromorphology corresponding to that observed in hand specimen, with microsteps, tracks of asparities that project from adjacent phacoids, and abundant striations (Fig. 9). In thin section, the phacoid boundaries appear as thin lamellae across which offsets are observed. They vary in character from crenulations to sharp, narrow shears (Bosworth and Vollmer, 1981, Fig. 11; Fig. 10). Measurable offsets across the shears are up to several millimeters, although more commonly no lithologic correlation can be made at the scale of the thin section. Sense of offset can be determined in most cases by the displacement of markers and by the bending of the pre-cleavage fissility into the zone. Measurement of the orientations of these microshears in sections cut perpendicular to the cleavage, both parallel and perpendicular to strike, shows two discrete populations in each section. The sense of shear on these surfaces indicates that these populations form conjugate normal sets of microshears (Fig. 8). The obtuse angles of intersecting sets are bisected by the normal to the mesoscopic cleavage. Although it is difficult to determine the relative orientations of the microshear planes in three dimensions, it is apparent from hand sample observations that the phacoids are bounded on all sides by shear surfaces, and that a minimum of four shear sets is required to define their crudely polyhedral shapes. Mutual cross-cutting relationships suggest that the conjugate microshears form coevally. Also, many of the microshear surfaces are distinctly non-planar, and in some cases curvature of a single surface may result in the opposite sense of shear along the same surface.

Orientations of these microshears are shown in Figure 8 for two samples from the Albany Quadrangle. The first is from the narrow, high strain zone at the Vly Creek locality; the second is from the Normansville locality within the main belt of tectonic melange. The angle between conjugate microshears varies between 45 and 70 degrees, but shows no consistent change from one locality to the other. A further important observation is from samples taken near the edge of the Vly Creek high strain zone (Fig. 10). The microshears in these specimens again have similar orientations and are distinguished only by their wider spacing and small offsets.

The origin of phacoidal cleavage through the progressive development and continued offset of conjugate microshears is suggested by the presence of widely spaced, small offset shear crenulations in the margins of the Vly Creek high strain zone (Fig. 10). Farther into the zone, a similar but more intense penetrative fabric defines the phacoidal cleavage. The relative orientations of the microshears remain relatively constant, although the total deformation, as shown by the disruption of bedding, increases significantly. The fabric of this zone appears identical to that typical of the main melange zones. This cleavage is also characterized by at least four distinct microshear orientations. It can be shown that eight slip systems, corresponding to four fault surfaces, are sufficient to accommodate a general, constant volume, three-dimensional deformation (Reches, 1978). It is suggested, then, that the formation of these microshear sets allows the melange matrix to accommodate a general, constant volume, three-dimensional state of strain. Deformation may then occur analogously to plastic flow by glide on crystal slip systems. Disruption of layering is caused by repeated offsets on the conjugate microshears. The relatively constant angle between microshear sets indicates that the largest component of strain is most likely taken up by slip along the phacoid boundaries, with a lesser component of internal deformation. This state of strain most likely corresponds to a flattened elongate ellipsoid ($\lambda_1 >> \lambda_2 > 1 > \lambda_3$), as shown by the alignment of fold axes and by the extensional character of the microshear sets in the plane of

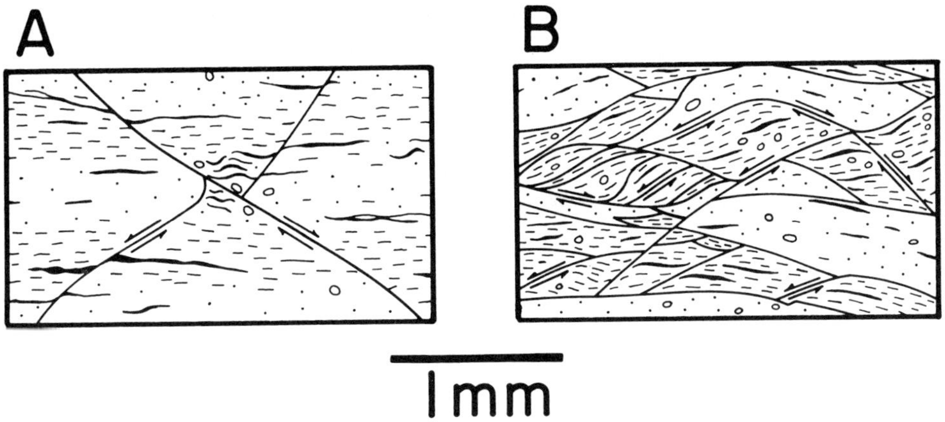

Figure 10. Line drawings from photomicrographs of thin sections from the Vly Creek high strain zone. A is from the margin of the zone; B is from the center of the zone. The fabric in B appears identical to that found within the main melange zones. Shear sense on many surfaces can be determined by flexure or offset of the primary fabric. Planar fabric is mainly clay, argillaceous fragments or fine-grained micas. Stipple represents siltier domains with quartz grains and lithic fragments in addition to argillaceous material.

the average foliation (Vollmer, 1981b). Phacoidal cleavage may develop preferentially in areas where strain rates are sufficiently high so that brittle failure is favored over diffusional processes.

As few constraints on the bulk strain history of these rocks are currently available, the possible significance of a non-coaxial (versus coaxial) deformation history in the formation of phacoidal cleavage currently cannot be properly assessed. However, it is clear from regional relationships that a deformation model approaching progressive simple shear is more reasonable than one invoking largely pure shear (a history of coaxial two- or three-dimensional strains). In particular, it is difficult to explain the emplacement of the Taconic alloththon onto the flysch sequence without the occurrence of large sub-horizontal shear displacements.

STRUCTURAL INTERPRETATION

Interpretation of the structures within the study area relies on across-strike variations in deformation style, and estimation of relative strains based on fold tightness, degree of bedding disruption, and microstructural interpretation of the melange matrix. Structures observed in the west, in the western and eastern flsych, suggest a component of horizontal shortening increasing to the east. Asymmetry of the folds and overthrust sense suggest a horizontal east-over-west (in the north) or southeast-over-northwest (in the south) shear component. Towards the east and into the melange, folds plunge more steeply east and southeast (Figs. 2, 3, 4, 6, 7). This change is associated with an increase in fold tightness and degree of boudinage, suggesting reorientation of the folds with increasing strain. This progression is shown in Figures 2 and 3 as a general change in fold orientation from west to east associated with a change from open to isoclinal folds. Folded extensional fibrous veins observed at several localities within the melange demonstrate some degree of lithification prior to folding (Bosworth and Vollmer, 1981, Fig. 8). It is suggested that this change in fold orientations towards the Taconic Allochthon indicates the rotation of fold hinge lines into a shear direction, and that this was related to the west-northwest-directed emplacement of the Taconic Allochthon during medial Ordovician times. This is consistent with thrust fault, fold, and cleavage orientations and the orientation of foliations in narrow east-dipping shear zones. The most intense deformation of the flysch is recorded in the extreme east, where bedding is completely disrupted to form the melange. Melange occurs first as narrow shear zones, as at Vly Creek, and then as broad belts just to the west of and along the front of the Taconic Allochthon.

Graptolite faunas found within this area also provide some insight into the deformational history. Most of the eastern rocks, including those of the Normans Kill gorge area, the Glenmont block (Mount Merino chert), the Austin Glen graywacke, and large sections of the Schuylerville melange are of *Nemagraptus gracilis* zone age (Riva, 1974; Riva, 1982, personal communication; also Berry, 1962; Fisher and others, 1970), corresponding to the late Llandeilian to early Caradocian (early medial Ordovician to earliest late medial Ordovician). However, several localities within the Normansville melange zone of the Albany Quadrangle give graptolites indicative of the younger *Corynoides americanus* zone (Riva, 1982, personal communication). One locality is within the city of Albany north of the Normans Kill gorge; the other is just west of Normansville on the Normans Kill. Farther west, at Vly Creek and in the vicinity, collections give *Climacograptus spiniferus* zone age graptolites, the youngest known within the deformed rocks (latest Caradocian; latest medial Ordovician). A similar younging in the age of the flysch is also observed west of the Schuylerville melange, although both the older and younger flysch crop out to the east (Fig. 7, and discussion in Bosworth, 1982). This suggests that, in general, older rocks progressively surface to the east, and the imbrication has resulted in the juxtaposition of rock units of various ages (see also Rickard and Fisher, 1973). Similar relationships have been observed along the Mohawk River gorge and near Quebec City, where Austin Glen and Mount Merino correlative units have been thrust over, and imbricated into, the younger flysch (Riva, 1982, personal communication).

Paleogeographic reconstructions (Rodgers, 1970; Fisher, 1977; Rowley and Kidd, 1981) suggest that carbonate deposition continued on the Paleozoic continental shelf during *N. gracilis* zone age, while deposition of flysch from an approaching eastern source had just begun on the continental rise sequence far to the east. This suggests that the cherts and graywackes of *N. gracilis* zone age, at least, have been thrust a minimum of 60 kilometers, the distance east-southeast to the present eastern extent of the Cambro-Ordovician carbonate sequence. It is suggested then, that the depositional-deformational environment associated with melange formation migrated across the slope-shelf boundary and continued, albeit weakly, until *C. spiniferus* zone age, or late Caradocian times. Although Ordovician units younger than this age are apparently undeformed, these units occur only farther west, beyond the limit of major deformation. An upper age limit for the formation of the melange in this area, however, is demonstrated by the unconformity exposed along the Spryt Creek, near South Bethlehem. Here thick beds of Austin Glen graywacke grade westward into highly deformed melange. The melange contains blocks of graywacke, some siliceous argillite, and steeply plunging rootless folds. Unconformably overlying the phacoidally cleaved melange fabric are Helderberg Group dolostones and limestones of latest Silurian to early Devonian ages (Ruedemann, 1930; Rickard, 1975). The carbonates are gently folded with some bedding plane thrusts, but are clearly unaffected by the fabric of the melange.

DISCUSSION

The structures and lithofacies of the Taconic flysch are interpreted in a model invoking progressive deformation of synorogenic deposits in a foreland basin overthrust environment. The central premise of this model is that the observed west-to-east spatial increase in intensity of deformation represents, as well, the temporal increase in deformation that occurred at any given locality within the flysch terrane. The more advanced stages of deformation resulted in the disruption of the bedded flysch sequence and the formation of melange. Final steepening of structures through time is suggested by the steepening of structures toward the allochthon. This steepening may have occurred through rigid body rotation along curved fault planes, or through an increased component of horizontal shortening (Bosworth and Vollmer, 1981).

Three processes are believed to have contributed to the formation of the Taconic melange: boudinage and general disruption of alternating graywacke-shale sequence; sub-aqueous slumping and deposition of olistostromic horizons that were subsequently tectonized and incorporated into the melange; and imbrication of the overthrust and underthrust sedimentary sections. The progressive disruption of the bedded flysch sequence can be readily inferred in the field. Attenuation and fracturing of less ductile wacke and siltstone beds leads to boudinage and the initial breakup of the sedimentary sequence. This can be observed at numerous localities in the flysch, and particularly at the margins of the mappable melange zones and on perimeters of some larger blocks already enclosed within melange. The sedimentary order is further destroyed and beds become untraceable as those boudins experience greater separation and rotation, the surrounding shales or slates taking on the characteristics of phacoidally cleaved melange matrix.

The sedimentary origin of some blocks within the melange is also apparent. The occurrence of well-rounded blocks of limestone and other lithologies within some melange outcrops suggests mechanical abrasion in the sedimentary environment. Pebbly mudstone horizons within the relatively undeformed sections of the flysch demonstrate that very coarse sedimentation took place at least locally within the flysch depositional environment. Tectonized olistoliths, for example in the eastern melange of the Albany Quadrangle, are generally distinguished in the field by their diversity of clast lithologies, and by the rarity or absence of disrupted bedding. A reasonable interpretation for the structural significance of the olistoliths is that they were generated by slope instabilities at active, emergent faults in the flysch basin (Bosworth and Vollmer, 1981; see also Zen, 1967, and Bird, 1969). The olistostromes would then be overridden by the upper thrust slice (or underthrust with the lower sheet) and incorporated into the fault zone. Finally, imbrication has led to the juxtaposition of rock units of various ages resulting in a general west to east stacking of progressively older units. Imbrication of the underlying shelf sequence has also occurred, leading to the incorporation of slivers of carbonate rock. The interplay of sedimentary and tectonic processes envisioned for the Taconic foreland basin is summarized in Figure 11.

The melange zones of the Hudson River Valley, then, occur as high strain zones across and within which rock units of various ages and lithologies have been imbricated and interspersed. In particular, regional relationships suggest that these melange belts represent shear zones associated with the emplacement of the Taconic Allochthon during medial Ordovician times (Bosworth and Vollmer, 1981; Bosworth, 1982). Displacements in excess of 60 kilometers across portions of the melange are indicated by paleogeographic reconstructions and the juxtaposition of various age units. These melange zones may have initiated as narrow high strain zones similar to that observed at Vly Creek. The fabric within the melange is dominated by boudinage, plunging rootless isoclines, and a phacoidal microshear cleavage. The pervasive microshear fabric demonstrates large shortening strains perpendicular to the fabric, and to the axial planes of the isoclines. This shortening was accommodated by movement on conjugate slip systems, analogous to bulk plastic flow. Deformation apparently progressed beyond the stage of isoclinal folding to that of extreme attenuation of isoclinal folds. The plunging fold axes are best

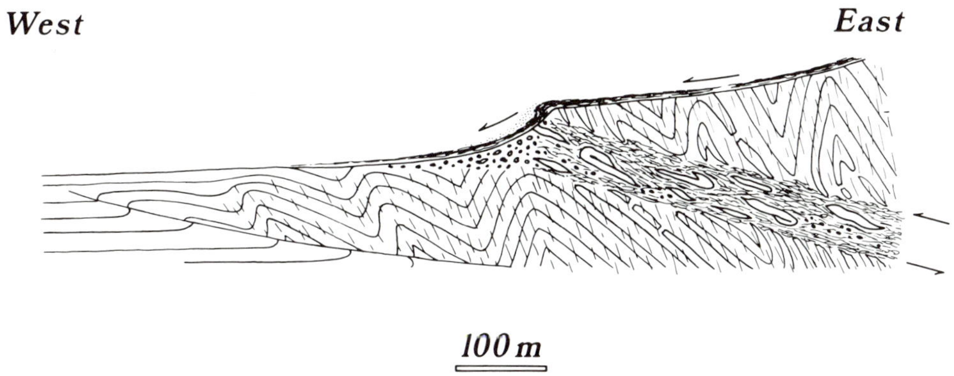

Figure 11. Model for the origin of melange in a foreland overthrust setting. Turbidity flows and slumping (single arrows) deliver sediments to the foreland basin, where they may be carried laterally along the basin axis. Coarse material (pebbly and bouldery mudstones—stipple pattern) is deposited near active fault scarps, where it is incorporated into evolving melange zones. Shearing across the melange zones (paired arrows) results in progressive boudinage and eventual disruption of the bedded flysch sequence, producing the bulk of the melange blocks and clasts and the anastomosing melange fabric. Slaty and phacoidal cleavage is interpreted to form coevally, in differing deformational environments.

explained through the rotation of initially horizontal hinge lines. In general, plunges of fold axes increase with increasing strain, as indicated by fold tightness and degree of boudinage. Within the melange zones, these axes form a mesoscopic fabric plunging moderately to the east or east-southeast, parallel to slickenside striations on fault surfaces. The direction of this linear fabric is interpreted to be the direction of maximum elongation in the melange. Although late horizontal shortening and further imbrication is believed to have subsequently rotated this fabric to steeper orientations, its trend is reasonably interpreted to represent the direction of transport of the Taconic Allochthon.

The amount of strain required to rotate the fold axes cannot be directly calculated without a more complete knowledge of the deformation path. Theoretical and experimental modeling of fold axes rotation within a body undergoing progressive bulk simple shear show that significant alignment of axes near the shear direction does not occur below shear strains of 10 or 20 (Skjernaa, 1980; Cobbold and Quinquis, 1980). Any additional component of extension in the intermediate strain direction should decrease the amount of rotation for a given shear strain. Therefore, if we assume a deformation approximating simple shear, plus some amounts of flattening, as a reasonable model, shear strains on the order of $\gamma = 10$ may be regarded as a conservative estimate for the melange zones. Although the thicknesses of the individual melange zones are poorly constrained, their cumulative outcrop width (w) (Fig. 1) is approximately 5 to 10 kilometers. The melange fabric commonly dips 45° or more to the east, and with high shear strains this fabric should be sub-parallel to the boundaries of the zones. The cumulative melange zone thickness should then be $w \cdot \cos 45°$, or 3½ to 7 kilometers. With a shear strain (γ) of 10, displacements of 35 to 70 kilometers could be accommodated across the melange high strain zones. In its original, more nearly horizontal attitude, this magnitude of lateral transport is in good agreement with that estimated from previous paleogeographic arguments.

CONCLUSIONS

The Taconic melanges were formed in a foreland basin environment through the deformation of a synorogenic flysch sequence. This flysch sequence was deposited in a north-south elongate basin in front of the Taconic Allochthon as it was emplaced onto the North American continental margin during the medial Ordovician Taconic orogeny. The flysch was derived from, and was subsequently overridden by the allochthon, resulting in the formation of belts of tectonic melange. The flysch sequence comprises three main lithologic units: a distal, dominantly shale unit; an interbedded graywacke and shale unit; and proximal pebbly mudstones. All of these units have locally been tectonized into melange. Tectonization of pebbly mudstone horizons has resulted in the formation of melange with "exotic" blocks. These blocks may be correlated with sedimentary units found within the allochthon and the underlying carbonate shelf, and were probably derived from exposed fault scarps formed during overthrusting.

An east to west increase in deformation intensity as well as local variations in deformation intensity have been used to infer the structural history of the melanges and to study the flysch-melange transition. Formation of melange from the three main lithologic units involved initial folding and thrusting, followed by tightening of the folds to form isoclines. This main phase of folding occurred after some degree of lithification of the sediments. Continuing deformation resulted in boudinage due to lithologic ductility contrasts, and the formation of a phacoidal conjugate-shear cleavage. The formation of this cleavage is believed to allow accommodation of a general deformation. Original layering within the matrix is disrupted through repeated slicing on conjugate microshear planes. A final process that contributes to the observed disruption in the melange is large-scale imbricate thrusting. This has resulted in the incorporation of larger blocks, slivers, and klippen of chert, carbonate rock, and several exotic rock types of unknown origin into the melange terrane.

High strains are indicated within the melange by isoclinal folding, boudinage, and pervasive extensional microfaulting. Transitions from flysch into melange show a progressive change in fold orientation, from horizontal north-northeast to steeply southeast plunging. This reorientation is suggested to be due to rotation of fold axes into the shear direction with increasing strains. This interpretation is consistent with the vengence of cleavage, folds, and faults throughout the area. The melange zones, then, are best explained as zones of high shear strain associated with the emplacement of the Taconic Allochthon. Total displacement across these melange zones is estimated to be in excess of 60 kilometers.

ACKNOWLEDGMENTS

The authors would like especially to thank W. D. Means, W.S.F. Kidd, and D. B. Rowley for their kind assistance in this study. Discussions with P. Hudleston, J. Riva, S. Chisick, and B. Selleck helped to clarify various aspects of the study. P. Hudleston critically reviewed an early draft of this work. The authors would also like to thank M. Cloos, B. E. Lorenz, and an anonymous reviewer for comments. Bosworth was funded through NSF graduate fellowship, a Penrose Grant from the Geological Society of America, and a William and Flora Hewlett Foundation Grant of Research Corporation. Vollmer received support through NSF Grant EAR 8007812 to Means. Field support was provided by the State University of New York at Albany. The staff of Colgate University and the University of Minnesota kindly assisted in the preparation of the present work.

REFERENCES CITED

Aalto, K. R., 1981, Multistage melange formation in the Franciscan Complex, northernmost California: Geology, v. 9, p. 602–607.

Aleinikoff, J. N., Zartman, R. E., and Lyons, J. B., 1979, U-Th-Pb geochronology of the Messabesic Gneiss and the granite near Milford, south-central New Hampshire: New evidence for Avalonian basement and Taconic and Alleghanian disturbances in eastern New England: Contributions to Mineralogy and Petrology, v. 71, p. 1–11.

Berry, W.B.N., 1962, Stratigraphy, zonation and age of the Schaghticoke, Deepkill and Normanskill shales, eastern New York: Geological Society of America Bulletin, v. 73, p. 695–718.

Besancon, J. R., Gaudette, H. R., and Naylor, R. S., 1977, Age of the Messabesic Gneiss, southern New Hampshire: Geological Society of America Abstracts with Programs, v. 9, p. 242.

Bickford, M. E., and Turner, B. B., 1971, Age and probable anatectic origin of the Brant Lake Gneiss, southeastern Adirondacks, New York: Geological Society of America Bulletin, v. 82, p. 2333–2342.

Bird, J. M., 1969, Middle Ordovician gravity sliding, Taconic region, *in* Kay, G. M., ed., North Atlantic Geology and Continental Drift: American Association of Petroleum Geologists, Memoir 12, p. 670–686.

——and Dewey, J. F., 1970, Lithosphere plate-continental margin tectonics and the evolution of the Appalachian Orogen: Geological Society of America Bulletin, v. 81, p. 1031–1060.

——and Dewey, J. F., 1975, Selected localities in the Taconics and their implications for the plate tectonic origin of the Taconic region, *in* Ratcliffe, N. M., ed., New England Intercollegiate Geological Conference, 67th Annual Meeting, Guidebook, p. 87–121.

Bosworth, W., 1980, Structural geology of the Fort Miller, Schuylerville and portions of the Schaghticoke 7½' quadrangles, eastern New York, and its implications in Taconic geology [Ph.D. thesis, Part I]: Albany, State University of New York, 237 p.

——1982, Evolution and structural significance of master shear zones within the parautochthonous flysch of eastern New York: Vermont Geology (symposium on Taconic geology), v. 2, p. 6–13.

——and Vollmer, F. W., 1981, Structures of the medial Ordovician flysch of eastern New York: Deformation of synorogenic deposits in an overthrust environment: Journal of Geology, v. 89, p. 551–568.

Chapple, W. M., 1973, Taconic Orogeny: Abortive subduction of the North American continental plate?: Geological Society of America Abstracts with Programs, v. 5, p. 573.

Chisick, S. A., and Bosworth, W., 1984, The Bald Mountain carbonates: A lithologic and faunal correlation with the peritidal Lower Ordovician (Cassinian) Fort Cassin-Providence Island Formations: Geological Society of America Abstracts with Programs, v. 16, p. 9.

Cobbold, P. R., and Quinquis, H., 1980, Development of sheath folds in shear regimes: Journal of Structural Geology, v. 2, p. 119–126.

Cohen, C. R., Schweichert, R. A., and Odom, A. L., 1981, Age of emplacement of the Schistes Lustres nappe, alpine Corsica: Tectonophysics, v. 73, p. 267–283.

Dickinson, W. R., 1977, Tectono-stratigraphic evolution of subduction-controlled sedimentary assemblages, *in* Talwani, M., and Pitman, W. C. III, eds., Island Arcs, Deep Sea Basins and Back-Arc Basins: Maurice-Ewing Series 1, American Geophysical Union, Washington, D.C., p. 33–40.

Doll, C. G., Cady, W. M., Thompson, J. B., Jr., and Billings, M. P., 1961, Centennial geologic map of Vermont: Vermont Geological Survey.

Durney, D. W., and Ramsay, J. G., 1973, Incremental strains measured by syntectonic crystal growths, *in* DeJong, K. A., and Scholten, R., eds., Gravity and Tectonics: New York, Wiley, p. 67–96.

Fisher, D. W., 1977, Correlation of the Hadrynian, Cambrian and Ordovician rocks in New York State: New York State Museum and Science Service Map and Chart Series 25, 75 p.

——Isachsen, Y. W., and Rickard, L. V., 1970, Geologic map of New York State: New York State Museum and Science Service Map and Chart Series 15.

Gannser, A., 1974, The ophiolitic melange, a world-wide problem on Tethyan examples: Eclogae Geologicae Helvetiae, v. 67, p. 479–507.

Grabau, A. W., 1936, Palaeozoic formations in the light of the pulsation theory, vol. II (p. 370–389): Peking, University of Peking Press, 793 p.

Hall, J., 1847, Descriptions of the organic remains of the lower division of the New York System: Paleontology of New York, v. 1, 338 p.

Hamilton, W., 1980, Complexities of modern and ancient subduction systems, *in* Continental Tectonics, National Council Geophysics Study Committee: National Academy of Sciences, p. 33–44.

Heezen, B. C., and Rawson, M., 1977, Visual observations of the sea floor subduction line in the Middle America trench: Science, v. 196, p. 423–426.

Hills, F. A., and Gast, P. W., 1964, Age of pyroxene-hornblende granitic gneiss of the eastern Adirondacks by the rubidium-strontium whole rock method: Geological Society of America Bulletin, v. 75, p. 759–766.

Homewood, P. W., 1977, Ultrahelvetic and North-Penninic flysch of the Prealps: a general account: Eclogae Geologicae Helvetaie, v. 70, p. 267–270.

Hsü, K. J., 1968, Principles of melanges and their bearing on the Franciscan Knoxville paradox: Geological Society of America Bulletin, v. 79, p. 1063–1074.

Jacobi, R. D., 1984, Modern submarine sediment slides and their implications for melange and the Dunnage Formation in north-central Newfoundland, *in* Raymond, L. A., ed., Melanges: Their nature, origin, and significance: Geological Society of America Special Paper 198 (this volume).

Karig, D. E., 1980, Material transport within accretionary prisms and the "Knocker" problem: Journal of Geology, v. 88, p. 27–39.

Laird, J., and Albee, A. L., 1975, Polymetamorphism and the first occurrence of glaucophane and omphacite in northern Vermont: Geological Society of America Abstracts with Programs, v. 7, p. 1159.

——1981, High-pressure metamorphism in mafic schist from northern Vermont: American Journal of Science, v. 281, p. 97–126.

McBride, E. F., 1962, Flysch and associated beds of the Martinsburg Formation (Ordovician), central Appalachians: Journal of Sedimentary Petrology, v. 32, p. 39–91.

Merguerian, C., 1979, Dismembered ophiolite along Cameron's Line-West Torrington, Connecticut: Geological Society of America Abstracts with Programs, v. 11, p. 45.

——Baskerville, C. A., and Okulewicz, S., 1982, Cameron's Line in the vicinity of New York City: Geological Society of America Abstracts with Programs, v. 14, p. 65.

Moore, G. F., and Karig, D. E., 1980, Structural geology of Nias Island, Indonesia: Implications for subduction zone tectonics: American Journal of Science, v. 280, p. 193–223.

Moore, J. C., 1979, Middle American trench: Geotimes, September, p. 20–22.

——and Wheeler, R. L., 1978, Structural fabric of a melange, Kodiak Islands, Alaska: American Journal of Science, v. 278, p. 739–765.

Page, B. M., and Suppe, J., 1981, The Pliocene Lichi melange of Taiwan: Its plate-tectonic and olistostromal origin: American Journal of Science, v. 281, p. 193–227.

Ratcliffe, N. M., and Hatch, N. L., Jr., 1979, A traverse across the Taconide Zone in the area of the Berkshire Massif, western Massachusetts: *in* Skehan, J. W., and Osber, P. H., eds., The Caledonides in the U.S.A., Geological Excursions in the Northeast Appalachians: Weston, Mass., Weston Observatory, p. 175–224.

——and Mose, D. G., 1978, Probable Taconic age of the Middlefield thrust zone, eastern margin of the Berkshire massif, Massachusetts, on the basis of Rb/Sr geochronology of intrusive granitic rocks: Geological Society of America Abstracts with Programs, v. 10, p. 81.

Reches, Z., 1978, Analysis of faulting in three-dimensional strain field: Tectonophysics, v. 47, p. 109–129.

Rickard, L. V., 1973, Stratigraphy and structure of subsurface Cambrian and Ordovician carbonates of New York: New York State Museum and Science Service, Map and Chart Series 18, 57 p.

——1975, Correlation of the Silurian and Devonian rocks in New York State:

New York State Museum and Science Service, Map and Chart Series 24.
——and Fisher, D. W., 1973, Middle Ordovician Normanskill Formation, eastern New York: age, stratigraphic and structural position: American Journal of Science, v. 273, p. 580–590.
Riva, J., 1974, A revision of some Ordovician graptolites of eastern North America: Paleontology, v. 17, p. 1–40.
Robertson, A.H.F., 1977, The Moni melange, Cyprus: An olistostrome formed at a destructive plate margin: Journal of the Geological Society of London, v. 133, p. 447–466.
Robinson, P., Thompson, J. B., Jr., and Rosenfeld, J. L., 1979, Nappes, gneiss domes and regional metamorphism in western New Hampshire and central Massachusetts: *in* Skehan, J. W., and Osberg, P. H., eds., The Caledonides in the U.S.A., Geological Excursions in the Northeast Appalachians: Weston, Mass., Weston Observatory, p. 93–174.
Rodgers, J., 1970, The tectonics of the Appalachians: New York, Wiley, 271 p.
Rodgers, J., and Fisher, D. W., 1969, Paleozoic rocks in Washington County, New York, west of the Taconic Klippe, *in* Bird, J. M., ed., NEIGC Guidebook, p. 6-1-6-12.
Rowley, D. B., Kidd, W.S.F., and Delano, L. L., 1979, Detailed stratigraphic and structural features of the Giddings Brook slice of the Taconic Allochthon in the Granville area, *in* Friedman, G. M., ed., New York State Geological Association and NEIGC Guidebook, p. 186–242.
——and Kidd, W.S.F., 1981, Stratigraphic relationships and detrital composition of the medial Ordovician flysch of western New England: implications for the tectonic evolution of the Taconic Orogeny: Journal of Geology, v. 89, p. 199–218.
——and Kidd, W.S.F., 1982, Stratigraphic relationships and detrital composition of the medial Ordovician flysch of western New England: Implications for the tectonic evolution of the Taconic Orogeny: A reply to John Rodgers: Journal of Geology, v. 90, p. 219–226.
Ruedemann, R., 1901, Trenton conglomerate of Rysedorph Hill, Rensselaer County, New York: New York State Museum Bulletin 49, p. 3–14.
——1914, *in* Cushing, H. P., and Ruedemann, R., Geology of Saratoga Springs and vicinity: New York State Museum Bulletin 169, 177 p.
——1930, Geology of the Capital District: New York State Museum Bulletin 285, 218 p.
St. Julien, P., and Hubert, C., 1975, Evolution of the Taconic Orogen in the Quebec Appalachians: American Journal of Science, v. 275A, p. 337–362.
Saleeby, J., 1979, Kaweah serpentinite melange, southwest Sierra Nevada foothills, California: Geological Society of America Bulletin, Part I, v. 90, p. 29–46.
Selleck, B., and Bosworth, W., 1984, Allochthonous Chazyan (early medial Ordovician) Carbonates in eastern New York: Tectonic and paleoenvironmental interpretation: American Journal of Science, in press.
Silver, E. A., and Beutner, E. C., 1980, Penrose Conference report: Melanges: Geology, v. 8, p. 32–34.
Silver, L. T., 1968, A geochronologic investigation of the anorthosite complex, Adirondack Mountains, New York, *in* Isachsen, Y. W., ed., Origin of anorthosites and related rocks: New York State Museum and Science Service Memoir 18, p. 233–252.
Skjernaa, L., 1980, Rotation and deformation of randomly oriented planar and linear structures in progressive simple shear: Journal of Structural Geology, v. 2, p. 101–109.
Stanley, R., and Ratcliffe, N. M., 1980, Accretionary collapse of the western margin of Iapetus in central New England during the Taconic orogeny: Geological Society of America Abstracts with Programs, v. 12, p. 85.
Thompson, J. B., Jr., Robinson, P., Clifford, T. N., and Trask, N. J., Jr., 1968, Nappes and gneiss domes in west-central New England: *in* Zen, E-An, White, W. S., Hadley, J. B., and Thompson, J. B., Jr., eds., Studies of Appalachian Geology: Northern and Maritime: New York, Wiley, p. 203–218.
Vollmer, F. W., 1981a, Structural studies of the Ordovician flysch and melange in Albany County, New York [MS thesis]: Albany, State University of New York, 151 p.
——1981b, Significance of small scale structures for the deformation history of the Taconic melange, eastern New York: Geological Society of America Abstracts with Programs, v. 13, p. 574.
White, R. S., and Ross, D. A., 1979, Tectonics of the western Gulf of Oman: Journal of Geophysical Research, v. 84, p. 3479–3489.
Williams, H., 1977, Ophiolitic melange and its significance in the Fleur de Lys Supergroup, northern Appalachians: Canadian Journal of Earth Science, v. 14, p. 987–1003.
——1978, compiler. Tectonic lithofacies map of the Appalachian Orogen: Memorial University of Newfoundland, 2 sheets.
——1979, Appalachian orogen in Canada: Canadian Journal of Earth Science, v. 16, p. 792–807.
Zen, E-An, 1961, Stratigraphy and structure at the north end of the Taconic Range in west-central Vermont: Geological Society of America Bulletin, v. 72, p. 293–338.
——1967, Time and space relationships of the Taconic Allochthon and Autochthon: Geological Society of America Special Paper 97, 107 p.

MANUSCRIPT ACCEPTED BY THE SOCIETY JULY 12, 1984

Printed in U.S.A.

Geological Society of America
Special Paper 198
1984

Flow melanges and the structural evolution of accretionary wedges

Mark Cloos
Department of Geological Sciences
The University of Texas at Austin
Austin, Texas 78713

ABSTRACT

When subduction rates are several cm/yr or more, various geologic and heat flow considerations indicate the portion of the accretionary wedge beneath the overriding plate will consist of two structural "layers." The upper layer is composed of schistose rocks directly in contact with, and heated by the base of the overriding plate. The lower layer is composed of fluid-rich, undercompacted, less heated mud and sand. With continued convergence, large shear strains are concentrated in the lower layer and tectonic flow melange develops. The portion of the accretionary wedge that accumulates in front of the overriding plate can be, as in the Franciscan of California, largely composed of subduction-generated melange blanketed by bedded slope sediments. Mappable units within an accreted melange wedge are defined by variations in the type and relative numbers of clasts. The population of clasts in flowing melange can vary where high-pressure metamorphic rocks are only locally plucked from the base of the overriding plate, where upflow ceases, and where features such as seamounts are dismembered as they locally impinge the accretionary complex. Melange intrusions, in the form of dikes and sills, can also form mappable melange units within an accretionary wedge.

INTRODUCTION

For nearly the last 20 years, bodies of rock containing a mixture of blocks set in a matrix often have been called melanges (see Raymond, this volume, for a review of terminology). Melanges develop from a variety of mixing processes in many different environments. One end member in the range of formative processes is purely sedimentary melanges, which form at or near the sediment-water interface. Mixing results from downslope, often turbulent flowage that is driven directly by gravity. The assemblage of included clasts reflects the diversity of rocks exposed in the source area (Abbate et al., 1970). The other extreme is purely tectonic melanges, which can form at great depth as well as near the surface. In this case, mixtures of rocks result from interleaving of slices of rock in fault zones (Cowan and Miller, 1981) or large amounts of nonturbulent, tectonically driven flowage (Cloos, 1982a). Purely tectonic melanges are a manifestation of subcrustal movements, and the diversity of clasts reflects the diversity of rock types present at depth or added to the system during deformation. Certainly, many melanges result from a combination of tectonic and sedimentary processes. The best example of such interplay is the case where tectonic uplift produces the surface slope required for slumping or sedimentary gravity flows. Commonly, the recognition of the primary importance of sedimentary mixing processes is difficult, because subsequent tectonic deformation is pervasive and of large magnitude (Hsu, 1974). This is likely in the subduction zone environment.

It is generally agreed that both sedimentary and tectonic melanges develop at convergent plate margins. However, the relative importance of the end member processes has proved difficult to demonstrate. Nevertheless, a major criterion used for the recognition of ancient subduction zones is a linear belt of melange that is at least a few km wide and tens of km long. Unfortunately, the true distribution of melange in actively accreting margins is still poorly understood, because growing accretionary wedges are largely underwater. Seismic reflection profiles across wedges typically show an uppermost blanket, up to a km thick, of well-bedded slope sediments. This has been confirmed by deep sea drilling. Locally, thicker packets of sediments accumulate in slope basins (Moore and Karig, 1976). The "basement" beneath the bedded sediments generally lacks decipherable acoustic returns (Seely et al., 1974; Karig and Sharman, 1975). Seismic reflection profiles suggest, and the limited numbers of drill and dart cores indicate, that surficial olistostromal processes are a relatively minor aspect of sedimentation atop accretionary wedges (Hamilton, 1979; Karig, 1980). In the same trench systems where olistostromal processes appear to be minor, the basement rock exposed at uplifted forearc highs has been characterized as melange. Examples of melange from active margins include the Oyo Complex of Nias Island southwest of Sumatra (Moore and Karig, 1980), the Uyak Complex on Kodiak Island, south of Alaska (Connelly, 1978), and the Basal Complex of Barbados east of the Lesser Antilles (Speed and Larue, 1982). Hamilton (1979), as part of an extensive survey of the geology of the Indonesian region, discussed tracts of subduction-generated melange exposed in Timor and smaller islands to the east. By far the best exposed, most complete and thoroughly studied, inactive accretionary wedge is the Franciscan Complex of California, which has an extensive melange belt (Hamilton, 1969; Ernst, 1970). Although most of the discussion of melanges in this paper is directed towards the Franciscan Complex, similar processes are likely to be active at other converging margins at one time or another.

The Franciscan melanges are characterized by a diverse assemblage of rock types, which have variable degrees of metamorphic recrystallization and are immersed in a pelitic matrix. The major difference between the Franciscan melange belt and melanges described from active subduction complexes is that high P/low T blueschist facies metamorphic blocks are widely distributed in the Franciscan melanges (Coleman and Lanphere, 1971). It is important to note that these distinctive blocks, which have received so much attention because of their coarse grain size and unusual mineral assemblages, are volumetrically insignificant (Bailey et al., 1964).

An important aspect of the Franciscan Complex is that detailed mapping of the tens of km wide central melange belt has shown that it can be subdivided. Because the matrix is poorly exposed, subdivisions are largely based upon estimates of the relative abundance of the different types of blocks. Mappable subunits have been called "melange units" (Hsu, 1969) and have lateral continuity varying from hundreds of meters to tens of km along strike. Studies that subdivided or discuss subdivisions in the melange belt include those of Hsu (1969), Maxwell (1974), Gucwa (1975), and Crawford (1976).

The existence of mappable melange units has been used as an argument for sedimentary origin of melange. This is an important conclusion, because a sedimentary origin places major constraints on the nature of the environment of deposition, the evolution of source areas, and the magnitude of uplift of previously subducted materials. In contrast, a tectonic origin of the melange units places major constraints on the mechanics of growth and movement in the accretionary wedge. In either case, potentially the most informative melange units are those containing the highest pressure exotic blocks, such as blueschist or eclogite.

One purpose of this paper is to present several tectonic explanations for the origin of melange units within the Franciscan melange belt. These explanations are in accord with the flow melange model of Cloos (1982a) and are to be considered one possible explanation for their origin. Each occurrence of melange must be evaluated independently. Olistostromes formed during periods of subduction must be present in the Franciscan Complex, but they appear to be a relatively minor contributor to the overall chaos. Inasmuch as clasts of blueschist, eclogite, and amphibolite form a distinctive class of blocks, which is readily identified and used to differentiate melange units in some areas (e.g., Hsu, 1969; Gucwa, 1975), the cause of their metamorphism and relationship to the finely crystalline mud matrix of the melanges requires clarification. Therefore, the first part of this paper will discuss a general model for the structural and metamorphic evolution of subduction complexes that relates early "high-grade" metamorphism to later melange formation. This paper is not intended as a comprehensive review of the melange problem in the Franciscan Complex. The processes discussed here are just a few of the many likely to be active in growing accretionary wedges.

EVOLUTION OF ACCRETIONARY WEDGES BENEATH THE OVERRIDING PLATE

The style of deformation at deep levels of an accretionary wedge must evolve with time. There are several reasons for this. First, such parameters as the rates and direction of convergence, dip of the descending plate, and type and volume of subducted sediment will change during the millions of years that a typical subduction zone is active. Second, and probably most important because it must occur in all subduction zones, is that the thermal structure of the complex must change. When rates of convergence are several cm/yr, the overriding plate must cool as heat is conducted down towards the cold descending plate. Also, heat continues to flow up and out of the base of the forearc basin. Thus, given constant or increasing subduction rates, the overriding plate is hottest and thermal gradients most extreme during the initial stages of subduction (cf. Oxburgh and Turcotte, 1974; Wang and Shi, 1984). Consequently, the hottest portion of a growing accretionary wedge is that which accreted early and is in contact with the base of the overriding plate (Platt, 1975; Brown et al., 1982).

When subduction begins, sediments on top of the descending plate are offscraped at the front and carried beneath the overriding plate. Some of the subducted sediment is accreted to, and recrystallized at, the base of the hanging wall (Figure 1). During a single period of convergence, schistose rocks will develop only in the part of the sediment pile that is directly beneath the hot hanging wall and is heated to sufficiently high temperatures (>200-250°C) to cause thermally driven coarsening of grain size. This zone is a metamorphic aureole or "layer" at the base of the hanging wall. The accreted aureole provides "thermal insula-

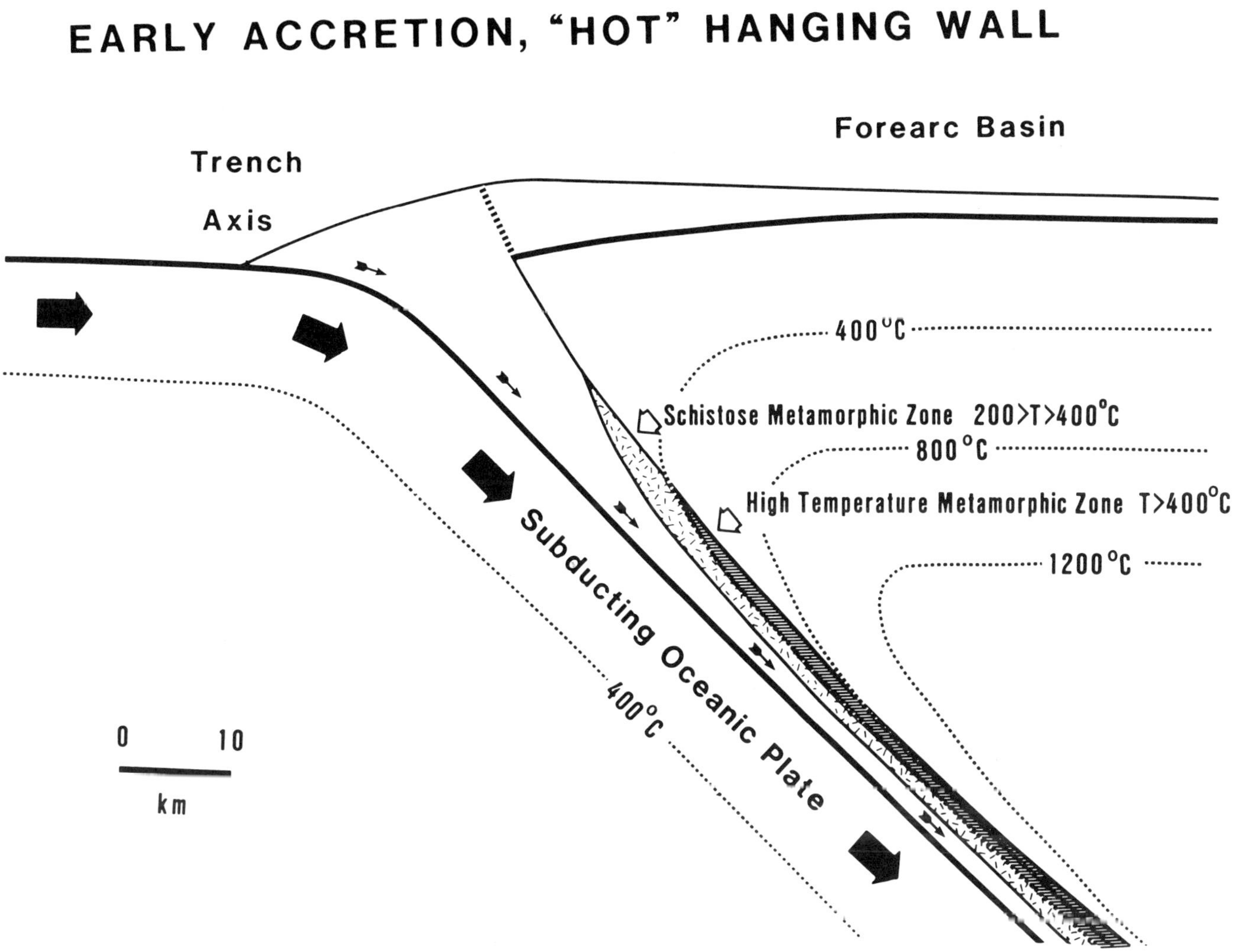

Figure 1. Schematic cross section illustrating formation of metamorphic aureole at base of the overriding plate during the initial stages of subduction.

tion" for the later subduction of sediment. Eventually, sediment is subducted to great depth, but at such low temperatures (Bird, 1978; Wang and Shi, 1984) that thermally driven textural changes are minor. Because of lower viscosity, the continued shear strain from convergence is concentrated in the lower, cold "layer" of fluid-rich, feebly recrystallized sediment. When this happens, tectonic flow melange develops. Under appropriate conditions, some of the melange upwells (Figure 2, Cloos, 1982a), and then dewaters, compacts, and lithifies near the surface. In this situation, the accretionary wedge grows trenchward as the dewatered melange becomes more viscous, and effectively decouples (cf. Kehle, 1970) from the underlying "flow channel" of lower viscosity, more recently subducted, fluid-rich sediment. Eventually, bedded slope sediment blankets the accreted melange wedge of subducted materials. Large-scale, gravity-driven slumping, continued movement deep in the accreted melange wedge, and, locally, olistostromes cause variable degrees of disruption of the bedded, nonsubducted slope sediments. Locally, mappable melange bodies may be created by surficial processes within the nonsubducted sediment cover.

At the deeper levels beneath the hanging wall, flow melange is also slowly accreted to the base of the upper metamorphic aureole. Upwelling melange can slowly drag portions of the metamorphic aureole towards shallower levels (see Cowan and Silling, 1978). Most important, continued convergence causes the upper metamorphosed layer to slowly cool to the same temperature as the upwelling melange.

At times of low sediment supply at the trench axis, faster convergence rates, or a change in plate dip, subduction erosion of the accretionary wedge can occur with the lowermost accreted materials removed first. In the extreme case, both metamorphic aureole and portions from the base of the hanging wall can be tectonically eroded away (Figure 3). This is currently happening along much of the Peru-Chile trench system (Coulbourn, 1981).

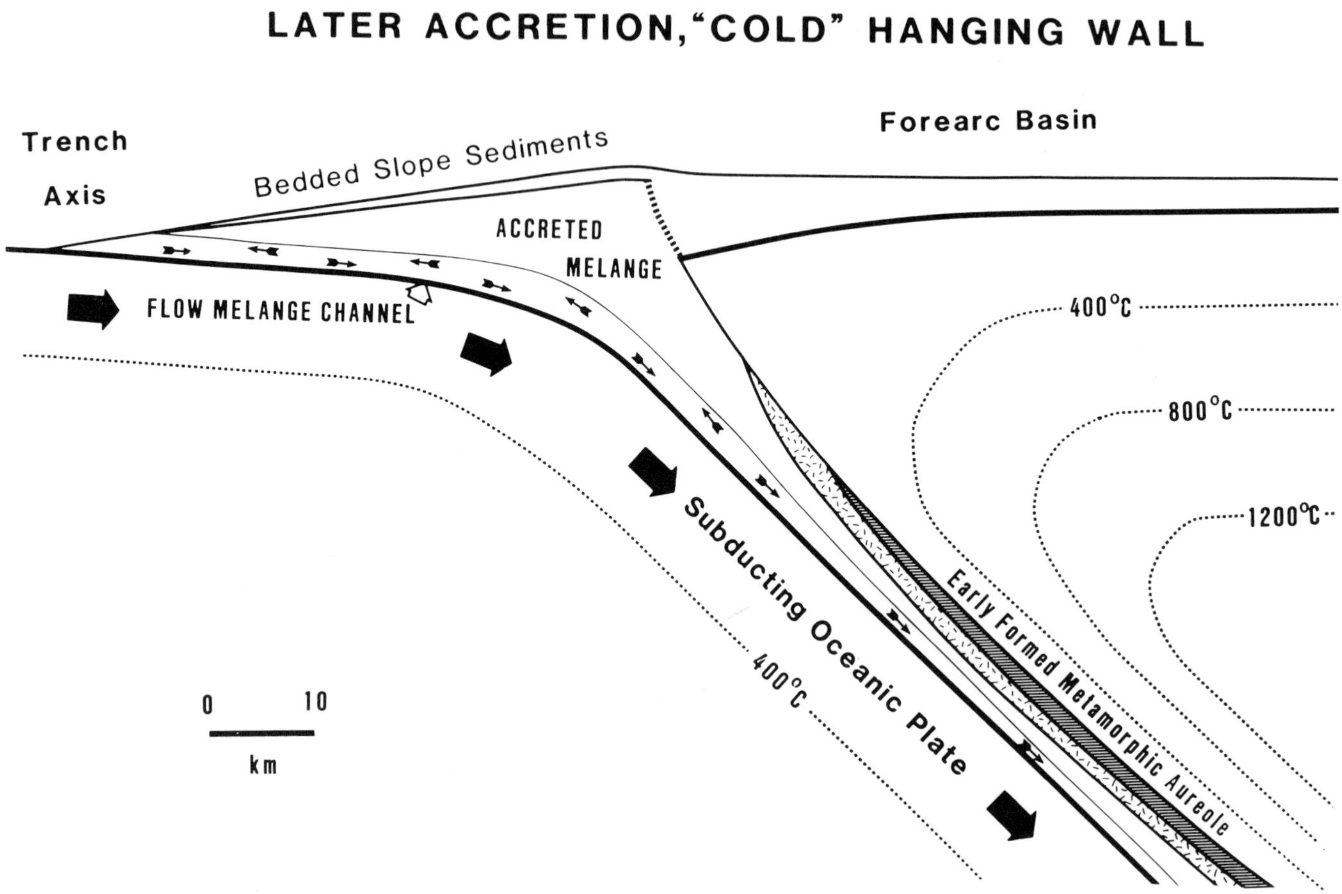

Figure 2. Schematic cross section of a mature convergent plate margin. The hanging wall has cooled, and an accretionary wedge of melange has developed. Compare with Figure 11 in Moore et al. (1982).

In this case, accretionary melange formed during later periods of subduction would be directly juxtaposed against the base of the overriding plate.

In any case, when subduction stops or slows, the deep levels of the accretionary wedge heat up as the "normal" geothermal gradient is reattained. The cold melange layer recrystallizes. Mafic blueschists metamorphose to amphibolites. Long after subduction stops, unaltered blueschists will be preserved only in the shallow level portions of the accretionary wedge. In short, the major conclusion based upon heat flow considerations is that when subduction erosion is not a major effect, the portion of the accretionary wedge beneath the overriding plate will have two fundamental "layers," an upper schistose, visibly metamorphosed layer and a lower, apparently "unmetamorphosed" melange layer. The nature of these two layers will be discussed in the next two sections, with general consideration of their relationships in the Franciscan Complex of California, where the shallowest portion of the upper metamorphosed layer is the South Fork Mountain Schist and related rocks of the eastern belt and the lower layer is the more easterly part of the central melange belt. The high-grade blocks found in the western portion of the central belt were derived from deep levels of the upper metamorphosed layer.

Upper "metamorphosed" layer: The upper accreted layer consists of sandstone, shale, chert and other types of sediment deposited atop the plate before and during the initial stages of subduction. Fragments of volcanic rocks from the top of the descending plate or ultramafic rocks from the base of the overriding plate may become interleaved with the subducted material during initial convergence. Metamorphic recrystallization is driven by the conduction of heat outwards from the base of the hanging wall. If the overriding plate is young ophiolite (as it was at the beginning of Franciscan subduction), the initial vertical thermal gradient within the hanging wall can be over 50°C/km. More commonly the initial vertical thermal gradients are probably around 30°C/km. A coarsely schistose or gneissic foliation develops at temperatures of 200 to 250°C, whereas "high-grade" mineral assemblages, such as those found in eclogite and amphibolite, require temperatures around 500°C (Brown and Bradshaw, 1979). Assuming that the subducted material is heated only by thermal conduction, and that the sediments and hanging wall have similar thermal conductivities, are instantaneously juxtaposed, and the initial vertical thermal gradient is 40° C/km, then consideration of the nature of the heat flow in a semi-infinite half-space leads to the conclusion that subduction to depths of 10 km is required to develop a metamorphic schistosity, with depths

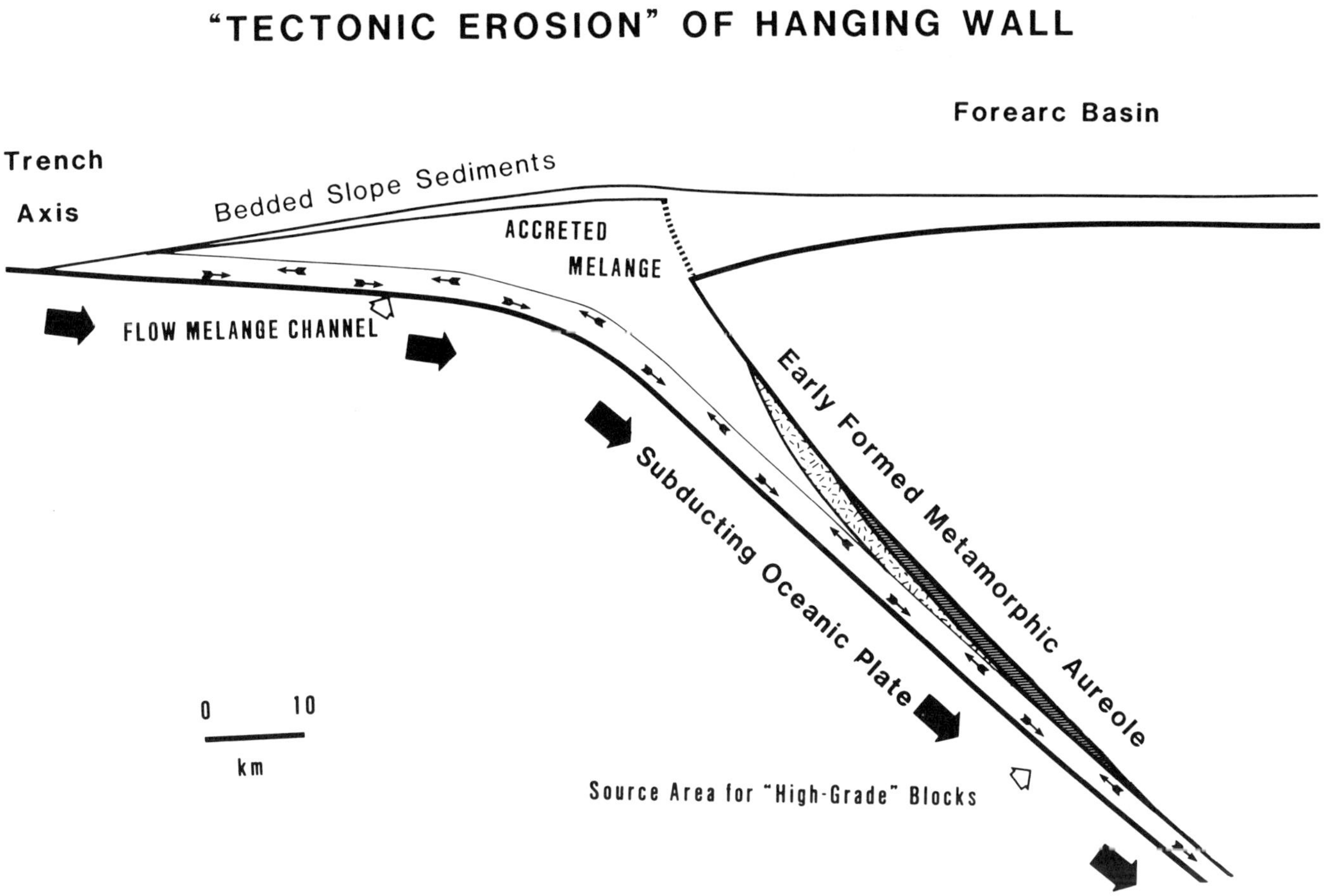

Figure 3. Schematic cross section of a convergent plate margin in which a portion of the metamorphic aureole has been removed by subduction erosion. The "high-grade" rocks are exposed at the top of the flow channel. At times of upwelling, blocks of blueschist or eclogite can be transported upwards.

of 25 km required to develop "high-grade" mineral assemblages. This conclusion agrees with pressure estimates from petrologic studies in the Franciscan (e.g. Brown and Bradshaw, 1979). If the subducted material is progressively heated during the downward journey, somewhat shallower depths are required to attain the required temperatures. The important feature is that, when convergence is at rates of several cm/yr or more, visibly metamorphosed rocks will develop only in rocks near or in contact with the base of the overriding plate and thus beneath the forearc basin.

The degree of recrystallization within the metamorphosed layer will be variable because of the extreme gradient in temperature between the hanging wall and the subducting plate. Subdivision of the upper layer can be based upon mineralogical (isograds, e.g., Ghent, 1965; Suppe, 1973) and textural (textural zones, e.g., Blake et al., 1967) criteria and is a direct manifestation of the flow of heat out the base of the hanging wall. Simultaneous or later movement in the accretionary wedge can cause faulting, with the result that major mineralogical and textural zones are no longer gradational (e.g., Worrall, 1981).

During the initial stages of convergence, the metasediments in the upper layer deform ductilely as they undergo syntectonic recrystallization. Even mafic rocks interleaved with the sediments become relatively deformable because of reaction-enhanced ductility (White and Knipe, 1978) during hydration and recrystallization to greenschist or blueschist assemblages. All rock types are ductilely folded and refolded (Wood, 1971; Bishop, 1977; Worrall, 1981). Because of syntectonic recrystallization and the development of a coarse grain size, the average viscosity of the overlying metamorphic layer will be considerably higher than that of the underlying zone of fluid saturated, unlithified sediments that are deposited in the trench axis and continually subducted. As a result of the viscosity contrast (which should increase as the hanging wall cools), the shear due to continued subduction becomes concentrated in the lower layer and a flow melange develops.

Lower melange layer: The lower melange layer will be composed of a mixture of material similar to that of the upper one, with the possible exception that slices from the base of the overriding plate may be absent. Mineralogical and textural variation within the melange layer will be minor because the temperatures are uniformly low throughout it (Cloos, 1983a). At low temperatures (<200-250°C), thermally driven coarsening of grain size is minor. The sediments deform ductilely, because the lack of induration, small grain size, and high fluid pressures (Cloos, 1982b) enable the sliding and rotation of grains past one another. In contrast, fragments of mafic intrusive rock or volcanic rock incorporated in the melange would behave as relatively rigid bodies as they undergo varying degrees of hydration and recrystallization at such low temperatures. Coarse-grained blocks of blueschist, eclogite, and amphibolite detached at deep levels from the base of the hanging wall and incorporated in the melange behave as even more rigid bodies, because their hydration and development of greenstone mineral assemblages is minor. As a result, there is a wide range of ductilities in the mixture of materials that makes up the melange layer. The distribution of strain is heterogeneous, with most strain being concentrated in the matrix of mudstone and unlithified sandstone.

The average viscosity of the lower melange layer will be significantly less than that of the overlying metamorphosed layers, because it is largely composed of unlithified, undercompacted sediments that were rapidly buried by subduction. Where mud is abundant, fluid pressures are at or near lithostatic values, because of low permeability and tectonically induced consolidation. Although partial dewatering occurs, high fluid pressures are maintained, because clay mineral dehydration and organic matter breakdown is delayed, due to the low geothermal gradients, until great depths are attained. Once the shear between the converging plates is concentrated in the low-viscosity channel of melange, continued convergence results in thorough mixing of the melange from the combined effect of nonuniform rates of flowage within the zone of flowing melange and continuous sinking and breakup of blocks (Cloos, 1982a). The true magnitude of strain of the matrix is not obvious, because the fabric attains a steady state during flow. Also, post-flowage dewatering and compaction modify the fabric.

MAPPABLE MELANGE UNITS IN ACCRETIONARY WEDGES

In the previous section, a general scenario for the structural and metamorphic evolution of accretionary wedges was presented. In this section, several mechanisms are proposed that can account for the tectonic generation of mappable melange units within an accretionary melange wedge. As mentioned, mappable melange units are usually defined by the relative abundance of different types of clasts.

Localized source of blocks

Sources of blocks in the melanges include materials deposited in the trench axis, materials transported in from the oceanic regime, or metamorphic rocks plucked from the aureole at the base of the overriding plate. Metamorphic blocks such as blueschist, eclogite, or amphibolite that sink into the melange or are plucked from the base of the hanging wall can be transported towards the surface during times of upflow (Figure 3, Cloos, 1982a). These distinctive block types would be concentrated in the parts of the melange "downstream" from the source area. If there are only a few small point sources, high-grade blocks would be concentrated in a few areas along strike of the melange belt. At times of oblique convergence, the distinctive blocks would be distributed along strike of the belt in the direction of relative motion. Such concentrations or localized distributions of distinctive types of blocks could be mapped as melange units. Examples of this type of melange unit may be the linear belts of melange mapped by Gucwa (1975). Similar melange units appear to extend from San Francisco northwards past the California-Oregon border. Overall, this process is very similar to that in glaciers, which pluck distinctive rock types from a single source area and then deposit them downstream in the moraines.

Another likely source of distinctive blocks is seamounts transported into the subduction zone. Seamounts would be disaggregated as they ride into the channel of flowing melange and fragments of seamount, altered to greenstone, would be abundant in the area of impingement with the accretionary wedge. Sufficiently tall seamounts may be detached and embedded in the wedge with (MacPherson, 1983) or without deep subduction. Generally, the basal portion will be disrupted. Associated rocks such as cherts interbedded with mafic flows or limestone atolls atop the seamount could also form a distinctive suite of blocks in the melanges. Greenstone, chert, or limestone derived from seamounts can be concentrated in certain areas of the melange belt during head-on convergence or distributed along strike during periods of oblique convergence.

Melange Intrusion

Another mechanism for generating mappable melange units by tectonic flow processes in an accretionary wedge is based upon the intrusion of low-density, fluid-rich, flow melange into denser, overlying dewatered and accreted melange. Diapirism of unlithified sediment in subduction complexes has also been discussed by Scholl and Marlow (1974) and Maxwell (1974). Melange intrusions may be dikes or sills (Figure 4). Some intrusions may rise to the surface and extrude tectonic melange onto the top of the accreted wedge (Cloos, 1983b). Melange intrusions are commonly observed on a small scale with fractured blocks intruded by mud-matrix melange (Cloos, 1982a). On a larger scale, melange intrusions are found cutting into the Cambria Slab, a trench slope basin near San Simeon, California (see Hall, 1974). Worrall (1981) has mapped the Oven Lid melange, which contains high-pressure blocks and has contacts that appear to crosscut the contacts of other formations in the eastern belt. This setting of melange can also be explained as an intrusion of flow melange from deeper levels. Narrow zones, which consist of melange and high-pressure blocks, are often simply called shear zones, but they may be melange intrusions from deeper levels. Examples include the shear zones mapped by McLaughlin et al. (1982) in the Coastal Belt and Koch (1966) in western Oregon. Melanges interlayered with bedded sediments may be intrusive sills. Possible examples are concordant melange sheets mentioned by Jones et

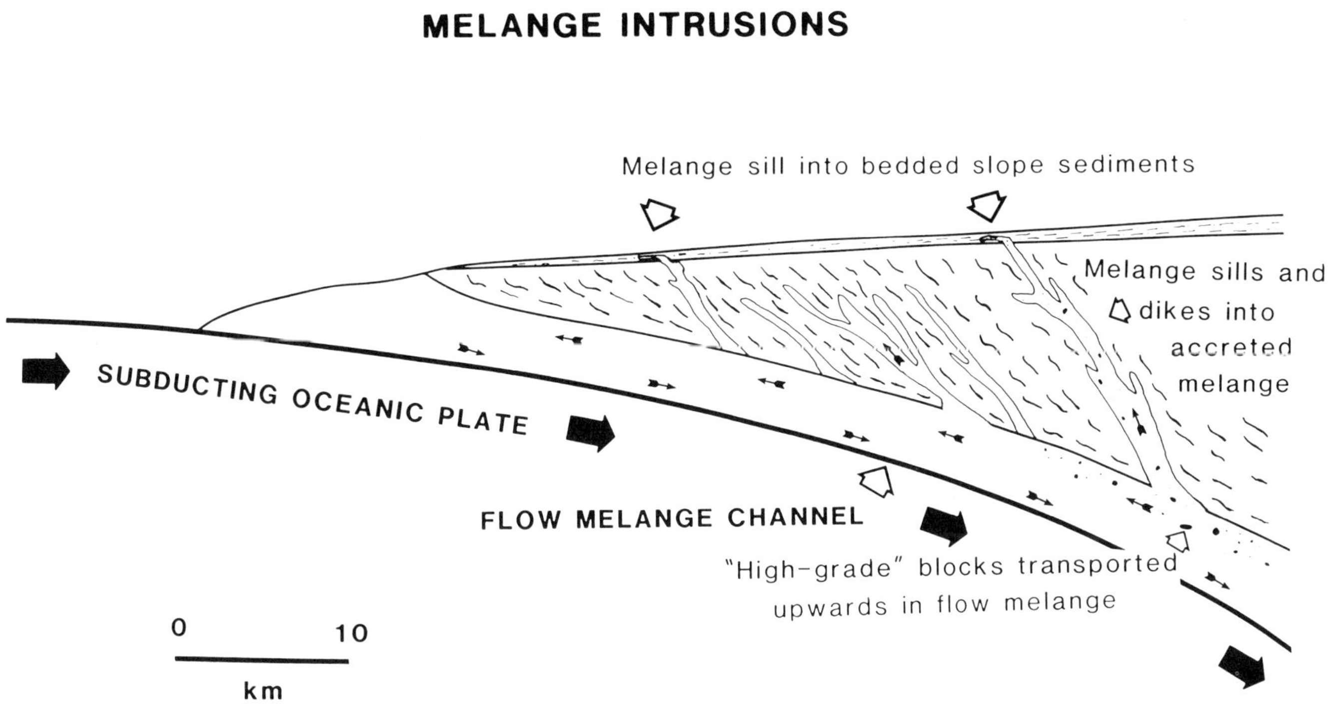

Figure 4. Diagram illustrating the intrusion of flow melange into shallow level portions of the accretionary wedge. Intrusions will contain exotic blocks if they are present in the flow melange channel. Melange intrusions that rise to the level of the bedded slope cover may spread as sills or extrude on the surface.

al. (1978) in the Geysers area north of San Francisco and mapped by Crawford (1976) in the Diablo Range.

Even in areas of good exposure, the presence of melange intrusions commonly will be difficult to demonstrate. Whereas melange dikes are obvious, melange sills in bedded sequences may be confused with surficial olistostromes (Cloos, 1983b). As with igneous intrusions in tectonically active areas, the intrusion processes may be repeated many times. Intrusions of melange into previously accreted melange are readily identifiable only where the clast populations are distinctly different. As discussed, the clast population in the flow melange channel that could feed intrusions can vary during a single period of convergence. Various scenarios can be envisioned where mappable melange units, formed by melange intrusion, are generated. For example, during the early stages of accretion, all mafic blocks in the melange could be greenstones containing minerals indicative of various degrees of high-pressure recrystallization. During later periods of subduction erosion, intrusive melange would be distinctive, because they lack greenstones bearing high-pressure minerals. As deep portions of the accretionary complex can be removed at this time, the top of the flow channel can become in contact with the distinctive suite of high-pressure metamorphic rocks formed during the initial stages of subduction when the hanging wall was hot (Figure 3). At subsequent stages of subduction accretion, upwelling melange can transport the high-grade blocks to shallow levels and then carry them as "xenoliths" into the shallow levels of the accretionary wedge. The variety of ways in which the clast population can vary during melange flow, coupled with multiple periods of melange intrusion, may result in complex interfingering of mappable melange units that are distinguishable only because of differing proportions of blocks.

SUMMARY

During convergence, the portion of the accretionary wedge beneath the overriding plate and forearc basin will have a basic two-layer structure. The upper layer consists of rocks that were accreted, heated by the hanging wall, and metamorphosed to coarse-grained schists. Once emplaced, the upper layer provides thermal insulation for the later subduction of fluid-rich, undercompacted, low-viscosity sediment at low temperature. With continued convergence, shear strains are concentrated in the lower layer and tectonic flow melanges develop. During times of upflow, high-pressure metamorphic rocks, formed during the initial stages of subduction, can be transported towards the surface. The clast population in the upwelling melange can vary along strike of the complex, because of localized plucking of high-pressure blocks from the metamorphic aureole at the base of the hanging wall and because of localized impingement and subsequent dismemberment of features such as seamounts. Consequently, the accreted melange wedge can be subdivided into mappable subunits on the basis of relative abundance of different types of clasts. Melange intrusions, as dikes or sills, can also form mappable melange units.

ACKNOWLEDGMENTS

This paper has been improved through constructive reviews by Greg Moore, Bob McLaughlin, and Karen Herrington. Discussions with Steve Lipshie, Bill Bruner, Bruce Nelson, Bill Carlson, and particularly Ron Shreve provided important insight.

REFERENCES CITED

Abbate, E., Bortolotti, V., and Passerini, P., 1970, Olistostromes and olistoliths: Sedimentary Geology, v. 4, p. 521–557.

Bailey, E. H., Irwin, W. P., and Jones, D. L., 1964, Franciscan and related rocks, and their significance in the geology of western California: California Division of Mines and Geology Bulletin 183, 177 p.

Bird, P., 1978, Stress and temperature in subduction shear zones: Tonga and Mariana: Geophysical Journal of the Royal Astronomical Society, v. 55, p. 411–434.

Bishop, D. G., 1977, South Fork Mountain Schist at Black Butte and Cottonwood Creek, northern California: Geology, v. 5, p. 595–599.

Blake, M. C., Jr., Irwin, W. P., and Coleman, R. G., 1967, Upside-down metamorphic zonation, blueschist facies, along a regional thrust in California and Oregon: United States Geological Survey Professional Paper 575-C, p. 1–9.

Brown, E. H., and Bradshaw, J. Y., 1979, Phase relations of pyroxene and amphibole in greenstone, blueschist and eclogite of the Franciscan Complex, California: Contributions to Mineralogy and Petrology, v. 71, p. 67–83.

Brown, E. H., Wilson, D. L., Armstrong, R. L., and Hanakel, J. E., 1982, Petrologic, structural and age relations of serpentinite, amphibolite and blueschist in the Shuksan suite of the Iron Mountain-Gee Point area, North Cascades, Washington: Geological Society of America Bulletin, v. 93, p. 1087–1098.

Cloos, M., 1982a, Flow melanges: Numerical modeling and geologic constraints on their origin in the Franciscan subduction complex: Geological Society of America Bulletin, v. 93, p. 330–345.

—— 1982b, Origin and importance of high fluid pressures during melange formation in the Franciscan complex of California: Geological Society of America Abstracts with Programs, v. 14, no. 4, p. 155.

—— 1983a, Comparative study of melange matrix and metashales from the Franciscan subduction complex with the basal Great Valley sequence, California: Journal of Geology, v. 91, p. 291–306.

—— 1983b, Reply to a discussion by Murphy and Aalto of the paper titled "Flow melanges: Numerical modeling and geologic constraints on their origin in the Franciscan subduction complex, California: Geological Society of America Bulletin, v. 94, p. 1243–1244.

Coleman, R. G., and Lanphere, M. A., 1971, Distribution and age of high-grade blueschists, associated eclogites, and amphibolites from Oregon and California: Geological Society of America Bulletin, v. 82, p. 2397–2412.

Connelly, W., 1978, Uyak complex, Kodiak Islands, Alaska: A Cretaceous subduction complex: Geological Society of America Bulletin, v. 89, p. 755–769.

Coulbourn, W. T., 1981, Tectonics of the Nazca plate and the continental margin of western South America, 18°S to 23°S: Geological Society of America Memoir 154, p. 587–618.

Cowan, D. S., 1982, Deformation of partly dewatered and consolidated Franciscan sediments near Piedras Blancas Point, California: Geological Society of London Special Publication, No. 10, p. 439–457.

Cowan, D. S., and Silling, R. M., 1978, A dynamic scaled model at trenches and its implications for the tectonic evolution of subduction complexes: Journal of Geophysical Research, v. 83, p. 5289–5396.

Cowan, D. S., and Miller, R. B., 1981, Deformational styles in two Mesozoic fault zones, western Washington, U.S.A.: Geological Society of London Special Publication, No. 9, p. 483–490.

Crawford, K. E., 1976, Reconnaissance geologic map of the Eyler Mountain quadrangle, Santa Clara and Alameda Counties, California: United States Geological Survey Miscellaneous Field Studies Map MF764.

Ernst, W. G., 1970, Tectonic contact between the Franciscan melange and Great Valley sequence, crustal expression of a late Mesozoic Benioff zone: Journal of Geophysical Research, v. 75, p. 886–902.

Ghent, E. D., 1965, Glaucophane-schist facies metamorphism in the Black Butte area, northern Coast Ranges, California: American Journal of Science, v. 263, p. 385–400.

Gucwa, P. R., 1975, Middle to Late Cretaceous sedimentary melange, Franciscan complex, northern California: Geology, v. 3, p. 105–108.

Hall, C. A., 1974, Geologic map of the Cambria region, San Luis Obispo County, California: United States Geological Survey Miscellaneous Field Studies Map MF599.

Hamilton, W., 1969, Mesozoic California and the underflow of the Pacific mantle: Geological Society of America Bulletin, v. 80, p. 2409–2430.

—— 1979, Tectonics of the Indonesian region: United States Geological Survey Professional Paper 1078, 345 p.

Hsu, K. J., 1969, Preliminary report and geologic guide to Franciscan melanges of the Morro Bay-San Simeon area, California: California Division of Mines and Geology Special Publication 35, 46 p.

—— 1974, Melanges and their distinction from olistostromes: Society of Economic Paleontologists and Mineralogists Special Publication 19, p. 321–333.

Jones, D. L., Blake, M. C., Jr., Bailey, E. H., and McLaughlin, R. J., 1978, Distribution and character of upper Mesozoic subduction complexes along the west coast of North America: Tectonophysics, v. 47, p. 207–222.

Karig, D. E., 1980, Material transport within accretionary prisms and the "knocker" problem: Journal of Geology, v. 88, p. 27–39.

Karig, D. E., and Sharman, G. F., 1975, Subduction and accretion in trenches: Geological Society of America Bulletin, v. 86, p. 377–389.

Kehle, R. O., 1970, Analysis of gravity sliding and orogenic translation: Geological Society of America Bulletin, v. 81, p. 1641–1664.

Koch, J. G., 1966, Late Mesozoic stratigraphy and tectonic history, Port Orford-Gold Beach area, southwestern Oregon Coast: American Association of Petroleum Geologists Bulletin, v. 50, p. 25–71.

MacPherson, G. J., 1983, The Snow Mountain volcanic complex: An on-land seamount in the Franciscan terrain, California: Journal of Geology, v. 91, p. 73–92.

Maxwell, J. C., 1974, Anatomy of an orogen: Geological Society of America Bulletin, v. 85, p. 1195–1204.

McLaughlin, R. J., Kling, S. A., Poore, R. Z., McDougall, K., and Beutner, E. C., 1982, Post-middle Miocene accretion of Franciscan rocks, northwestern California: Geological Society of America Bulletin, v. 93, p. 595–605.

Moore, G. F., and Karig, D. E., 1976, Development of sedimentary basins on the lower trench slope: Geology, v. 4, p. 693–697.

—— 1980, Structural geology of Nias Island, Indonesia: Implications for subduction zone tectonics: American Journal of Science, v. 280, p. 193–223.

Moore, J. C., Watkins, J. S., Shipley, T. H., McMillen, K. J., Bachman, S. B., and Lundberg, N., 1982, Geology and tectonic evolution of a juvenile accretionary terrane along a truncated convergent margin. Synthesis of results from Leg 66 of the Deep Sea Drilling Project, southern Mexico: Geological Society of America Bulletin, v. 93, p. 847–861.

Oxburgh, E. R., and Turcotte, D. L., 1974, Thermal gradients and regional metamorphism in overthrust terrains with special reference to the eastern Alps: Schweizerische Mineralogische und Petrographische Mitteilungen, v. 54, p. 641–662.

Platt, J. P., 1975, Metamorphic and deformational processes in the Franciscan complex, California: Some insights from the Catalina Schist terrane: Geological Society of America Bulletin, v. 86, p. 1337–1347.

Scholl, D. W., and Marlow, M. S., 1974, Sedimentary sequence in modern pacific trenches and the deformed circum-pacific eugeosyncline: Society of Economic Paleontologists and Mineralogists Special Publication, No. 19, p. 193–211.

Seely, D. R., Vail, P. R., and Walton, G. G., 1974, Trench slope model, *in* Burk, C. A., and Drake, C. L., eds., The Geology of the Continental Margins: New York, Springer-Verlag, p. 249–260.

Speed, R. C., and Larue, D. K., 1982, Barbados: Architecture and implications for accretion: Journal of Geophysical Research, v. 87, p. 3633–3643.

Suppe, J., 1973, Geology of the Leech Lake Mountain-Ball Mountain region, California: University of California Publications in the Geological Sciences, v. 107, 82 p.

Wang, C. Y., and Shi, Y., 1982, On the thermal structure of subduction complexes: A preliminary study: Journal of Geophysical Research, v. 89, p. 7709–7718.

White, S. H., and Knipe, R. J., 1978, Transformation- and reaction-enhanced ductility in rocks: Journal of the Geological Society of London, v. 135, p. 513–516.

Wood, B. L., 1971, Structure and relationships of late Mesozoic schists of NW California and SW Oregon: New Zealand Journal of Geology and Geophysics, v. 14, p. 219–239.

Wood, R. M., 1979, A re-evaluation of the blueschist facies, Part 1, The pressure-temperature hysteresis cycle: Geological Magazine, v. 116, p. 21–33.

Worrall, D. M., 1981, Imbricate low-angle faulting in uppermost Franciscan rocks, south Yolla Bolly area, northern California: Geological Society of America Bulletin, v. 92, p. 703–729.

MANUSCRIPT ACCEPTED BY THE SOCIETY JULY 12, 1984

Printed in U.S.A.

Geological Society of America
Special Paper 198
1984

Modern submarine sediment slides and their implications for melange and the Dunnage Formation in north-central Newfoundland

Robert D. Jacobi
Department of Geological Sciences
State University of New York at Buffalo
4240 Ridge Lea Road
Amherst, New York 14226

ABSTRACT

Modern submarine sediment slides produce two features: a slide *scar* that delineates a zone of removal and a *deposit* of slide material. The upturned edges of the slide scar form prominent scarps with relief of generally less than 100 m. The zone of deposition includes "hummocky," "blocky," and "debris flow" high-resolution (3.5 - 12 kHz) seismic facies. These echo types probably represent olistoliths, piles and/or folds of deformed sediment, and debris flow deposits, respectively. Cores and bottom photographs exhibit deformed, chaotic material typical of debris flow deposits. Sediment slides are common, not only on active margins, but also on passive margins. Slide complexes are not restricted to base-of-slope sites; rather, a single slide on a passive margin can stretch over 700 km from the continental shelf break to the abyssal plain.

Giant submarine sediment slides have implications for studies of melange. Sediment slides provide an extremely important mechanism for generating the internal chaos characteristic of melange. Because sediment slides are not restricted to convergence zones, the presence of a chaotic unit in the rock record does not imply, of itself, a paleosubduction zone. Other characteristics of the melange must be observed and studied before proposing a paleotectonic site of formation. These characteristics include clast lithology and structural fabric history.

The Cambro-Ordovician Dunnage Formation is a melange that crops out in north-central Newfoundland in the Dunnage tectono-stratigraphic zone. The Dunnage melange exhibits soft-sediment deformation features similar to those observed in cores raised from submarine sediment slides. These features are consistent with the interpretation that the chaotic nature of the Dunnage Formation formed initially by sediment sliding. The deformational features include pebbly mudstones with no cleavage and isoclinal folds that are overprinted at high angles by non-axial-planar cleavage. The original site of deposition of the Dunnage Formation is equivocal, but angular clasts in the Dunnage were derived from both an island arc and oceanic crust. This observation suggests that the heads of the slides were very close to subaerial exposures of a combined island arc and ophiolite terrane (perhaps similar to present-day Luzon). By Silurian times, the Dunnage Formation was involved in overthrusting which, considering regional relationships, was probably forearc thrusting.

INTRODUCTION

Melange is a chaotic mixture of rock fragments generally in a fine-grained matrix (see Raymond, this volume, for a review of terminology). Although many melanges exhibit cleavages (especially scaly cleavage), the chaotic nature of many melanges may not have formed initially by tectonic means. The purpose of this report is to examine a mode of sedimentary origin possible for many melanges. The first section of the report discusses large-scale submarine sediment slides which result in chaotic sedimentary deposits. The first section also details some of the implications of these slides for studies of the rock record, especially with respect to melange. The next section reviews criteria that may establish the petrogenesis and tectonic setting of various types of melange. In the third section, the Dunnage melange of north-central Newfoundland is discussed with regard to the considerations presented in the previous sections.

During the past twenty-five years, high-resolution seismic reflection profiles (3.5 and 12 kHz echograms) have recorded the upper 50 m of the seafloor in all parts of the ocean basins. As compilation efforts proceeded, it became apparent that the echo-character of the seafloor could yield much information concerning the near-bottom processes operating in the ocean basins (see review by Damuth, 1980b). One of these major processes is the downslope movement of sediment by means exclusive of turbidity flows, i.e., slump blocks, folds, and sediment slides. Slump blocks are slope failures in which sediment is downdropped and rotated, but contortion of internal layering is minimal. The slump block is bounded by a slump fault. In contrast, sediment slides are slope failures in which the sediment was deformed and fluidized during its transport downslope, resulting in highly deformed layering and ultimately a slurry, or debris flow (Embley and Jacobi, 1977). Between these two end members (slump and slide) is a range of features, including primarily slump folds. These are slope failures in which the sediment is slightly deformed into large-scale folds, as slow downslope transport (creep) occurs. Because modern submarine sediment slides result in features perhaps common to many melanges, several aspects of such slides are important, including: (1) the recognition of sediment slides by means of echograms; (2) the physical characteristics of submarine sediment slides; and (3) the distribution of modern slides. Studies of these aspects have implications for cleavage generation in melange and melange petrogenesis.

The formation of cleavage has been studied and debated for more than one hundred years, since Sorby (1853) first suggested that mechanical rotation of phyllosilicates in lithified sediments produced a slaty cleavage. In the past twenty years, three different mechanisms (dewatering, pressure solution, and recrystallization) have been advanced for the formation of slaty cleavage, based in large part on studies of the deformed and metamorphosed Martinsburg Formation (e.g., Maxwell, 1962; Alterman, 1972; Geiser, 1975; Groshong, 1976; Beutner and others, 1977; Knipe and White, 1977; Beutner, 1978; Gray, 1979; and Woodland, 1982). Recent studies utilizing thin sections, SEM, HVEM, and strain calculations generally suggest that pressure solution and recrystallization in the solution passageways may produce slaty cleavage in lithified sedimentary rocks experiencing metamorphism and relatively high strain rates (e.g., Knipe and White, 1977; Beutner, 1978; Woodland, 1982). Conversely, in regionally unmetamorphosed and undeformed rocks exposed on land, "incipient" cleavage is observed locally in the hinge zones of slump folds (e.g., Williams and others, 1969; Tyler, 1972; Corbett, 1973). In these examples, it appears that soft-sediment sliding and folding concomitant with dewatering and rotation of phyllosilicates must account for the formation of slaty cleavage. A similar debate involves the formation of scaly cleavage. Both tectonic processes (e.g., Cowan, 1974) and dewatering of unlithified sediments during sediment sliding (e.g., Lash, 1980) have been proposed for the origin of scaly cleavage. Future study of sediments in cores raised from modern submarine slides should provide a clear understanding of types of deformational fabrics that occur in unmetamorphosed, non-tectonic deep-sea sediment slides. These observed fabrics would function as a "baseline" that could be utilized to evaluate the origin of cleavage in melange.

The term "melange" has a very long and confusing history (see review by Raymond, this volume). Because of the long association of the term "melange" with the Franciscan Complex (e.g., Hsu, 1968, 1969, 1971, 1973, 1974; Cowan, 1974, 1978; Cowan and Page, 1975), to many geologists the term "melange" connotes a chaotic unit formed by shearing in a subduction-zone complex. This connotation was reinforced by application of the "melange"-subduction zone couplet to other chaotic units, even though the chaotic units were not clearly related to subduction (see review by Raymond, this volume). As shown here, extremely large tracts of chaotic sediments very similar to those in melange terranes can form by submarine sediment sliding. These slides occur in all tectonic environments, not just subduction zones. Cleavage imposed on the slide deposit at a later time would complete the transformation of such slide deposits into typical melange, without the chaotic unit ever having been involved in a subduction zone. Recognition of this problem forces two considerations. (1) The term "melange" should be used only as a non-genetic, field term for designation of chaotic units where the origin of the chaos is unclear. In chaotic units where it is evident that tectonic processes formed the chaotic nature, "tectonic melange" should be used, whereas "sedimentary melange" or "olistostrome" should be utilized if the chaotic unit originated as a sediment slide. (2) Because sheared chaotic units can occur in tectonic environments other than subduction zones, criteria must be established that allow discrimination between the various tectonic settings of melange formation.

The Dunnage Formation (Kay and Eldredge, 1968) is a melange that outcrops in north-central Newfoundland in the Dunnage tectono-stratigraphic zone of Williams (1978). This formation has intrigued geologists for the past 20 years. All investigators believed the melange was critical in unraveling the plate-tectonic history of central Newfoundland, but the exact nature of this plate-tectonic history has remained obscure and in dispute.

Generally, the Dunnage melange was assigned an origin in the trench of a westward dipping, subducting oceanic plate (e.g., Bird and Dewey, 1970; Horne, 1970; Kay, 1976). However, some workers believed the melange may have formed on the back-arc side of a volcanic arc (e.g., Jacobi and Schweickert, 1976; Hibbard, 1976). The forearc advocates based their conclusion on the fact that the Dunnage was a melange (*ipso facto,* the site of a paleosubduction zone). However, the lack of scaly cleavage in much of the Dunnage melange suggested to others that the chaotic nature of the Dunnage melange formed through soft-sediment sliding, not tectonic disruption via brittle shear. The final section of this report reviews published data and presents new data bearing on the genesis of the Dunnage melange.

MODERN SUBMARINE SEDIMENT SLIDES

Slide Scar

Submarine sediment slide complexes consist of a slide scar and a slide deposit. Slide scars are surfaces along which sediment has been removed by slide processes. The master slide scar commonly has a rectangular cross-section (Fig. 1). The base of the rectangle is formed by the floor of the slide scar where the scar parallels subbottom reflectors. The sides of the rectangle are formed by "erosional" scarps where the slide scar cuts steeply upsection. The erosional scarps are generally 30 to 100 m high, but range up to at least 250 m. Slopes on the erosional scarps are quite variable, from about 10° to more than 16°. In plan view, these master "erosional" scarps are commonly U-shaped and open downslope. The "erosional" scarps generally bound the zone of removal. In areas where the zone of removal is partly bounded by slump faults, the boundary scarps have a more complex origin, as they are partly fault scarps and partly upturned edges of slide scars. The downslope termination of the master slide scar has never been observed, because it is always buried by slide material. However, it must cut upsection at its downslope boundary because the related slide material rests on undisturbed sea floor downslope from the zone of removal.

Parallel "erosional" scarps (commonly no more than two) occur along the margin of some zones of removal and indicate either multiple slide movements within the same zone or more complex patterns of removal, e.g., incomplete removal near the edge of the zone. Considerations of morphologic and depositional details suggest that the former alternative (multiple slides) is the more likely explanation for some slides along the northwest African margin (Jacobi, 1976). However, side-scan sonar studies in other areas (Flood and others, 1979; Popenoe and others, 1982) suggest that incomplete removal of sediment is also locally important.

A downslope seismic reflection profile (3.5 kHz or 100 Hz) in the zone of removal reveals that the slide scar usually follows a prominent reflector. On the northwest African margin, a slide scar follows a prominent reflector across several undulating sediment waves, never cutting downsection on the relative bathymetric highs or cutting upsection on the relative lows (Jacobi and Hayes, 1982; Jacobi, 1982). This suggests that little energy is transferred across the slide surface from the slip of sediments above the scar to the sediments below the scar. Thus, little deformation of the autochthonous sediments should occur as a result of the debris sliding across the autochthonous sediments. A similar relationship is observed in olistostromes in Newfoundland, where discrete boulder beds of slide material have passed over turbidites without deforming the underlying units (e.g., breccia units within the Cooks Brook Formation, Port au Port Peninsula, Newfoundland).

The upslope portions of several slide scars do not have slide material deposited on them (e.g., northwest African margin; Jacobi, 1976). Instead, younger turbidites or hemipelagic sediments are deposited directly on the much older sediment beneath the scar, resulting in a hiatus of ~250,000 to ~2,500,000 years. This type of disconformity is difficult to distinguish in seismic sections and the rock record from the more common disconformities caused by tectonism, eustatic sea-level changes, or submarine canyons. Discrimination between the slide scar "disconformity" and broad disconformities caused by eustatic sea-level changes might be possible if: (1) the slide scar "disconformity" extends long distances down the slope and relatively short distances along the strike of the slope; (2) the "disconformity" cuts sharply upsection at its terminus; and (3) slide material is observed at the same horizon in other localities. However, none of these criteria differentiate slide scars from submarine canyons and channels. In addition, turbidites commonly infill zones of removal, and slide material is observed on the floors of canyons and channels. Thus, at present there appears to be no single criterion that can differentiate slide scars from submarine channels in the rock record. This difficulty is not surprising, since some of the submarine canyons appear to be initially carved into the continental slope by successive slides that result in headward erosion (Farre and others, 1983). One method that might differentiate slide scars from submarine canyons is based on local relief. Because slide scarps generally have relief of <250 m (and commonly ~50 m), if the stratigraphic relief on the slide scar "disconformity" in the rock record is greatly in excess of 250 m, the disconformity is probably not the result of a single slide scar. This problem in identifying the origin of the disconformity has been recognized both in modern ocean basins (e.g., the Albian-Coniacian "hiatus" in DSDP site 369, Lancelot and Seibold, 1977) and in the rock record, e.g., certain exposures of the Taconic "unconformity" in both Newfoundland (e.g., Jacobi and Schweickert, 1976) and western Massachusetts (Rowley, 1982).

Slide Deposits

The seismic facies of slide deposits include "hummocky" slide material, "blocky" slide material, and debris flow material (Figs. 1-4). The hummocky and blocky slide materials are characterized on 3.5 and 12 kHz echograms as irregularly spaced hyperbolae with varying vertex elevations. The local relief de-

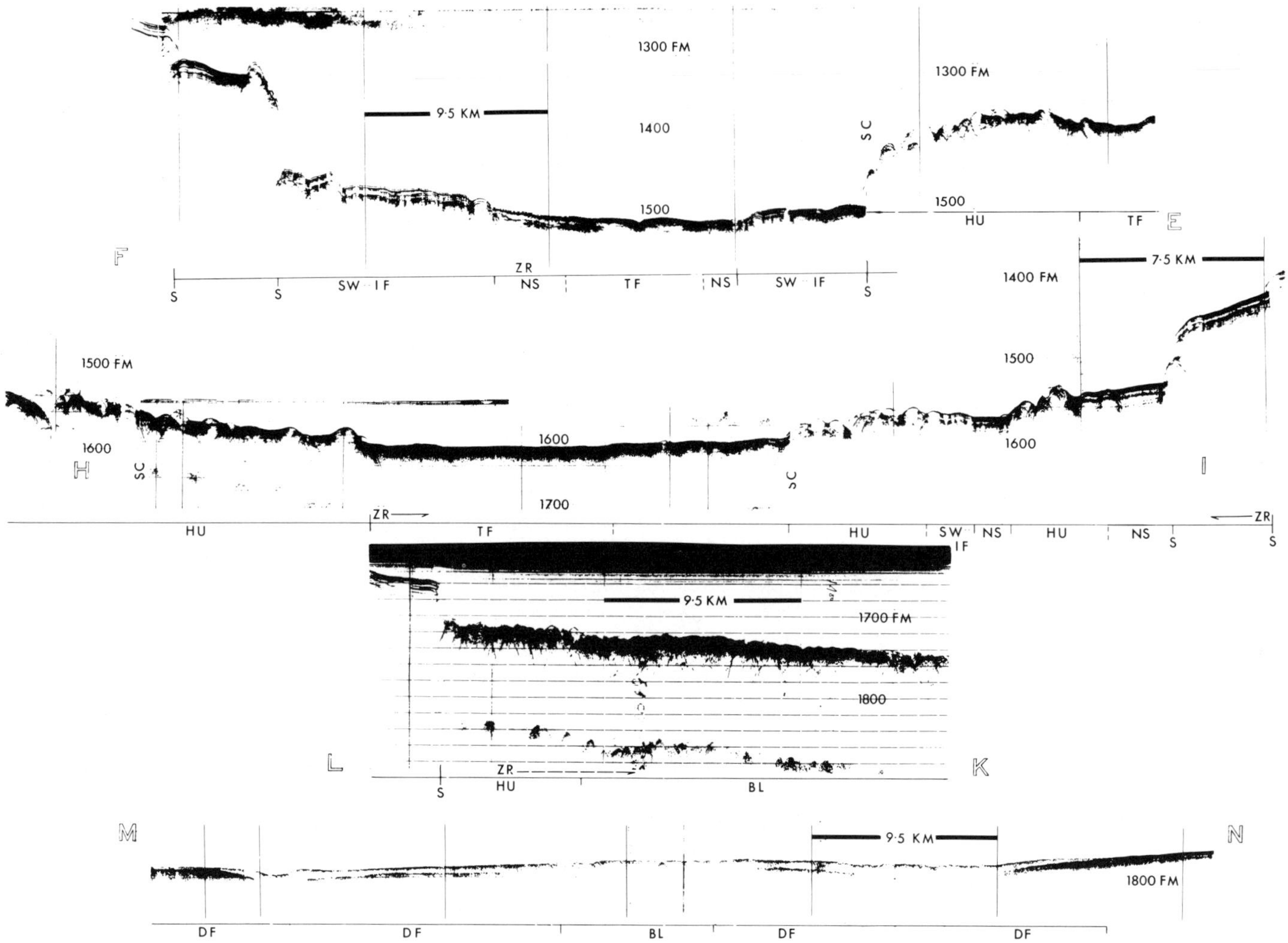

Figure 1. Four 3.5 kHz echograms across slide complexes on the northwest African continental margin northwest of Dakar, Senegal. Symbols refer to types of seismic facies (BL = blocky terrane, DF = debris flow, HU = hummocky terrane, NS = normal hemipelagic sedimentation, S = scarp, SC = scale change, SW-IF = sediment wave with irregular form, TF = turbidity flow pathway, ZR = slide scar, CC = ship's course change). Figure from Jacobi (1976).

fined by vertices of adjacent hyperbolae differentiates hummocky slide material from blocky slide material: hummocky material has ⩾40 m relief while blocky material has ⩽40 m relief. Debris flow deposits return weak, slightly prolonged, continuous echoes that delineate a transparent wedge of sediment.

Hyperbolae with high relief (⩾40 m) (hummocky slide deposits) occur generally near the "erosional" scarps. Horizontal multiple reflectors occur within the cross-sectional area outlined by some of these hyperbolae (Fig. 2). The source of high-relief hyperbolae is unclear, but the hyperbolae may represent either olistoliths or small zones of incomplete removal (Jacobi, 1976). The sequence of reflectors within the hyperbolae commonly can be correlated with sequences in the unremoved sections upslope from the erosional scarps, but a few incomplete and unrecognizable reflector sequences in the hyperbolae suggested to Jacobi (1976) that such hyperbolae from the northwestern African margin represent olistoliths. A similar conclusion was reached by Damuth (1980a) for like features in the South China Sea. However, within the zone of removal defined by the "master erosional" scarp, side-scan sonar studies have shown smaller "erosional" scarps extending downslope from, and perpendicular to, the master scarp (Flood and others, 1979; Popenoe and others, 1982). Back-to-back scarps of this nature could give rise to the hyperbolae with subbottoms as observed on surface-ship 3.5 kHz echograms. Thus, there is evidence that suggests the hyperbolae also could be the result of incomplete removal of sediment within the zone of removal.

Hyperbolae (⩾40 m) without associated subbottom reflectors are common in several slides (Figs. 1 and 2). These hyperbolae are probably reflections resulting from slide blocks in which some degree of homogenization of the sedimentary layering has taken place. This homogenization would tend to erase any inher-

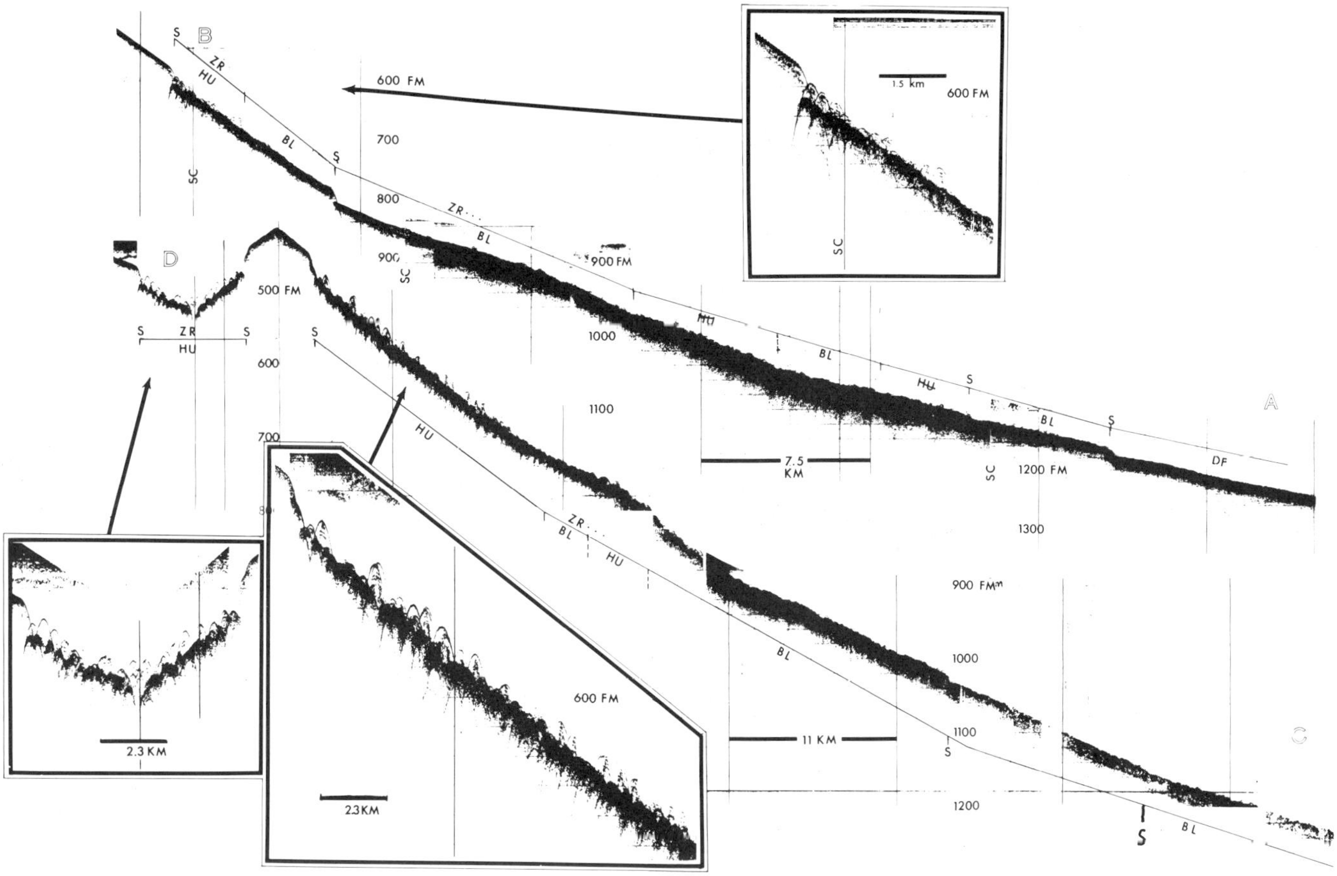

Figure 2. Downslope profiles of sediment slide complexes observed on 3.5 kHz echograms recorded on the northwest African continental margin northwest of Dakar, Senegal. Depth is uncorrected fathoms. Symbols are the same as those used in Fig. 1. Figure from Jacobi (1976).

ent V_p contrast. Small-scale roughness on the surface of the blocks could also account for the absence of subbottom reflectors (Embley, 1975). This roughness is probably indicative of deformation. In a slide near Dakar, Senegal, data obtained from circular ship tracks suggest that there is no preferred orientation of the large blocks (Jacobi, 1976).

The hummocky terrane characterized by hyperbolae lacking associated subbottom reflectors grades downslope into blocky terrane defined by hyperbolae $\leqslant$40 m relief (Fig. 2). The hyperbolae in the blocky terrane may represent blocks, piles, or both of sediment in which more mixing and liquefaction have taken place than in the hummocky terrane, but less mixing than in the debris flow, because the sediment, unlike debris flow material, is sufficiently stiff to form large blocks or piles.

The zone of debris flow deposits is a wedge of acoustically transparent slide material commonly $\leqslant$50 m thick. On the 3.5 kHz echograms, the debris flow deposits have a weak, slightly prolonged bottom echo with no internal reflectors and a few small hyperbolae tangent to the sea floor (Figs. 1, 3, and 4). Bottom photographs suggest that the weak, prolonged echo probably results from small-scale bumps on the surface of the debris flow that may be folds or other structures related to the flow of the material (e.g., Embley, 1976; Jacobi, 1976). The lack of internal reflectors is related to liquefaction and deformation of the sediment. Recent debris flow deposits have an upper surface that is slightly elevated above the surrounding undisturbed sea floor. The deposit does not have a ponded appearance as does a turbidity flow deposit (Figs. 1, 3, and 4). The slope of the zone boundary varies from quite steep (e.g., Embley, 1976) to very shallow on distal localities (e.g., Jacobi, 1976; Embley, 1980). The zone itself has little relief, except where it has covered older bottom features. These observations suggest that these deposits are too fluid to remain in blocks or discrete piles of sediment, but have not experienced sufficient liquefaction to become turbidity flows, i.e., the deposits are debris flow material. This conclusion is consistent with theoretical and experimental considerations of debris flows (e.g., Dott, 1963; Hampton, 1972; Middleton and Hampton, 1973) and with piston core studies of the submarine deposits (see following section). Both the piston core studies and the theoretical and experimental considerations suggest debris flows are gravity-flow sediment-water mixtures, or slurries, which result in deposits similar to unconsolidated pebbly mudstones.

The contact between a debris flow deposit and blocky slide material commonly is gradational (Fig. 2). However, some slide

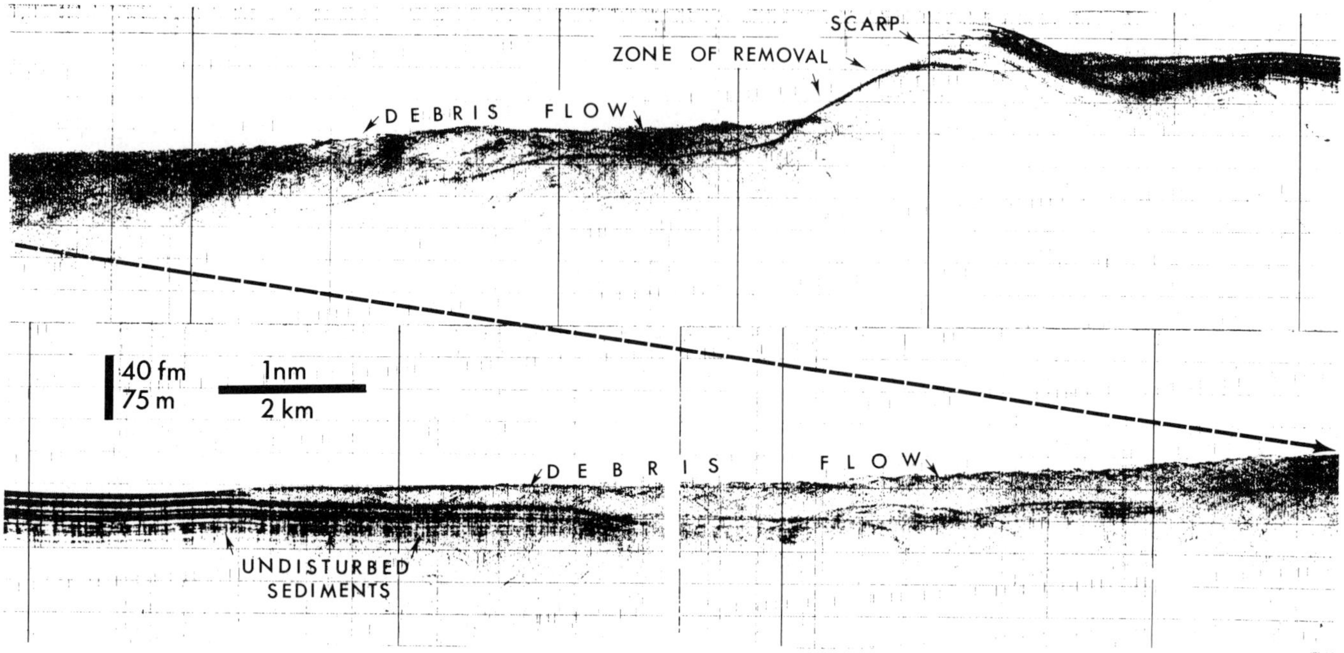

Figure 3. Sediment slide complex on an abyssal hill in the northwest Pacific (recorded on 3.5 kHz echogram). Figure from Jacobi and others (1980) and Damuth and others (1983).

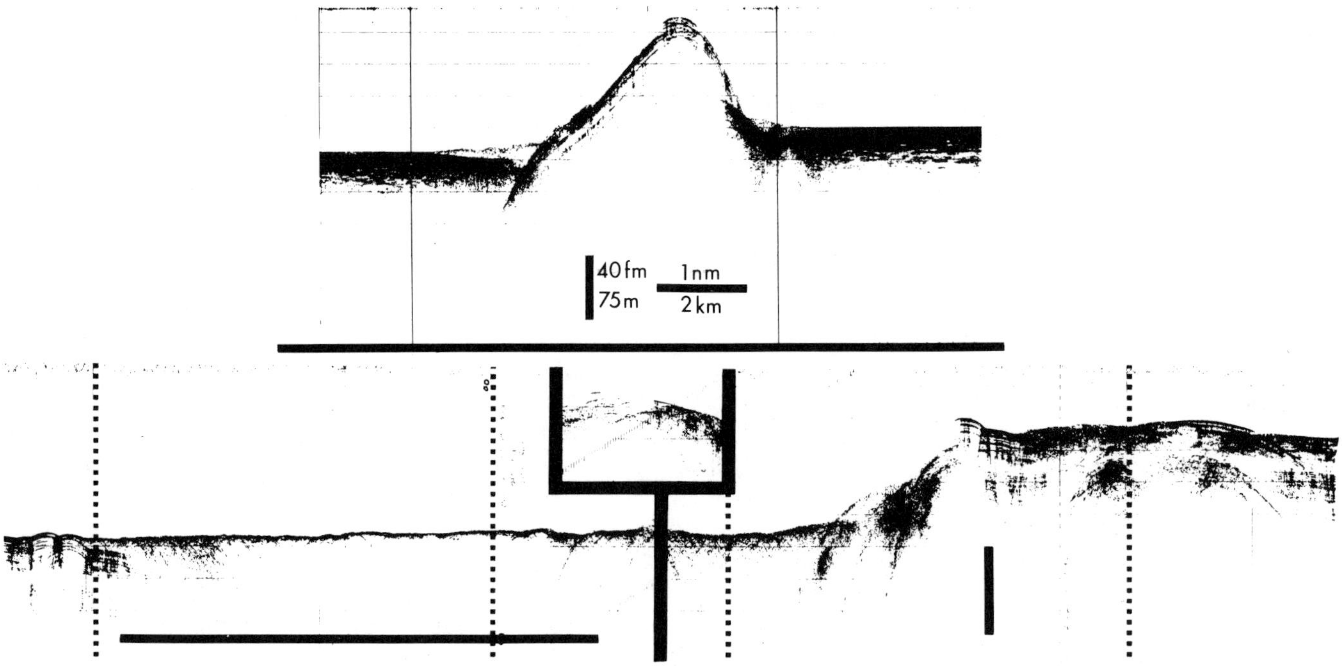

Figure 4. Sediment slide complexes on an abyssal hill in the South China Sea (upper profile) and along the Doldrums Fracture Zone in the southeastern North Atlantic (lower profile). In both profiles note the zone of removal (slide scar) bounded by the scarp on the left side of the positive bathymetric relief and the transparent wedge of debris flow deposits at the base of the slope. The insert in the lower profile displays the transparent wedge at 90° to the remainder of the profile. In the lower profile the horizontal bar represents 10 km, vertical bar = 100 m. Upper profile is from Damuth (1980a); lower profile is from Jacobi and Hayes, (1982).

complexes exhibit abrupt contacts between the blocky and debris flow material (e.g., Jacobi and Mrozowski, 1979; Flood and others, 1979; Embley, 1980). Regional map patterns in the western Atlantic (Vasallo and others, 1983) suggest this sharp contact may represent a hiatus, i.e., the blocky and debris flow materials are not part of the same localized flow. An alternative, but less likely, explanation is that a distinct threshold in liquefaction values is necessary for the transformation of blocky material to debris flow. However, in general, a typical complete downslope sequence of slide material is hummocky → blocky → debris flow material, all with gradational contacts (Fig. 2). This downslope sequence, with increasing subdued surficial relief, suggests increasing liquefaction in a downslope direction. Many slides, however, lack hummocky material, blocky material, or both.

Morgenstern (1967), Hampton (1972), and Middleton and Hampton (1973) suggested that turbidity flows should be generated at the leading edge of debris flows. On echograms, such turbidity flows are extremely difficult to differentiate from more recent flows that merely utilize an existing sediment slide depression. Mass-balance considerations by Jacobi (1976) showed that turbidity flows generated from debris flows on the African continental margin are of minor volumetric importance. However, large turbidity flow channels are observed at the base of isolated seamounts that do not reach wavebase. These turbidity flows could be generated by sediment slides, slumps, or both (Jacobi and Hayes, 1982; Jacobi, 1982). Stratigraphic studies of sediment cores from Rockall Bank (Flood and others, 1979), from the east margin of North America (Embley, 1980), and from the Mediterranean Sea (Kastens, 1981) suggest that turbidity flows were generated at the toes of debris flows in these areas.

Evidence from Bottom Photographs and Piston Core Data

Observation of the sediment slide deposits includes study of piston and box cores, study of bottom photographs, and studies from submersibles. Piston cores in the various zones of slide deposition commonly show various types of disturbed sedimentary layering (Fig. 5), whereas such structures are not evident in cores outside the slide areas. The disturbed layering includes dipping and contorted strata, isoclinal folds with horizontal axial surfaces, angular contacts, faults, rubble, and pebbly mud with pebbles oriented parallel to flow (e.g., Stanley, 1974; Embley, 1976, 1980; Embley and Jacobi, 1977; Jacobi, 1976; Jacobi and Mrozowski, 1979; Flood and others, 1979; and Jacobi and Hayes, 1982). Generally, cores raised from areas mapped as debris flow deposits contain sediments which exhibit high ductility, such as "flow folds," and pebbly mud. In contrast, cores taken in "blocky" slide material generally have deformed, but less ductile, sediment, i.e., brittle faults and angular folds are more common (Jacobi and Mrozowski, 1979; Jacobi and Hayes, 1982). Although based on a limited number of samples, these generalizations are consistent with the supposition that blocky material experienced less liquefaction than did debris flow material.

Mesoscopic pervasive cleavage is not observed in any of the cored debris flow material from the northwest African continental margin or the U.S. east coast continental margin. The lack of a comprehensive microscopic examination of modern submarine debris flow material precludes characterization of the possible cleavage fabrics present in the debris flow. Moore and Geigle (1974) presented data concerning non-tectonic soft-sediment deformational fabrics in a fold from the Mississippi Fan. In the core of this slump fold, thin-section study revealed that detrital phyllosilicates have an apparent preferred orientation parallel to the axial surface. Ongoing SEM and thin section research at SUNY-Buffalo may suggest that in modern, submarine debris flow material, only minimal axial-planar reorientation of phyllosilicates occurs in zones of highly strained, fairly coherent units (e.g., cores of tight isoclinal folds).

Bottom photographs of sediment slide deposits show small-scale, surficial roughness that probably results in the weak echo from these deposits (Embley, 1976, 1980; Embley and Jacobi, 1977; Flood and others, 1979; and Kastens, 1981). This roughness is atypical of undisturbed sea floor and is related to the flow of the sediment slide material. The roughness usually consists of linear mounds and troughs (surficial folds?) or small, linear, sharp "noses" (several centimeters thick) of debris flow materials (rubble?) which appear to have flowed out over the surrounding debris flow material or undisturbed sediments.

Distribution of Sediment Slides

Wherever passive continental margins have been studied, surficial sediment slides have been discovered, suggesting that such deposits typify passive margins. Embley and Jacobi (1977) and Embley (1980) reviewed the occurrences of most known major slides in the North and South Atlantic: on the southwest African margin, northwest African margin (Figs. 1 and 2), Madeira Rise, Rockall Bank, eastern North American margin, northern Gulf of Mexico, Amazon Cone, and southeastern South American margin. Additional localities on passive margins include the Norwegian-Greenland Sea (Damuth, 1978), Bay of Biscay (Kenyon and others, 1978), southeastern Mediterranean (e.g., Almagor and Garfunkel, 1979), and the South China Sea (Damuth, 1980a). Small sediment slides in the ocean basins are observed on seamounts, abyssal hills (Figs. 3 and 4), island pedestals, fracture zones (Fig. 4), and aseismic ridges, e.g., on Fiji Islands (Houtz, 1962) New Caledonia Basin, western Pacific (Bentz, 1974), Samoan Archipelago (Lonsdale, 1975), South China Sea (Damuth, 1980a), Indian Ocean (Abbott and others, 1981), and the northwest Pacific (Jacobi and others, 1980; Damuth and others, 1983).

Active margins have not been examined in any detail. However, it is apparent that they, too, have a great number of slide deposits. Sediment slides and slumps have been observed at several locations in the Pacific Ocean and Mediterranean Sea, e.g., the Scripps Canyon (Chamberlain, 1964; Dill, 1964); Valdez, Alaska (Coulter and Migliaccio, 1966); the eastern Aleutian Trench (e.g., Piper and others, 1973; von Huene, 1974); the Middle America Trench (e.g., Ross and Shor, 1965); North Is-

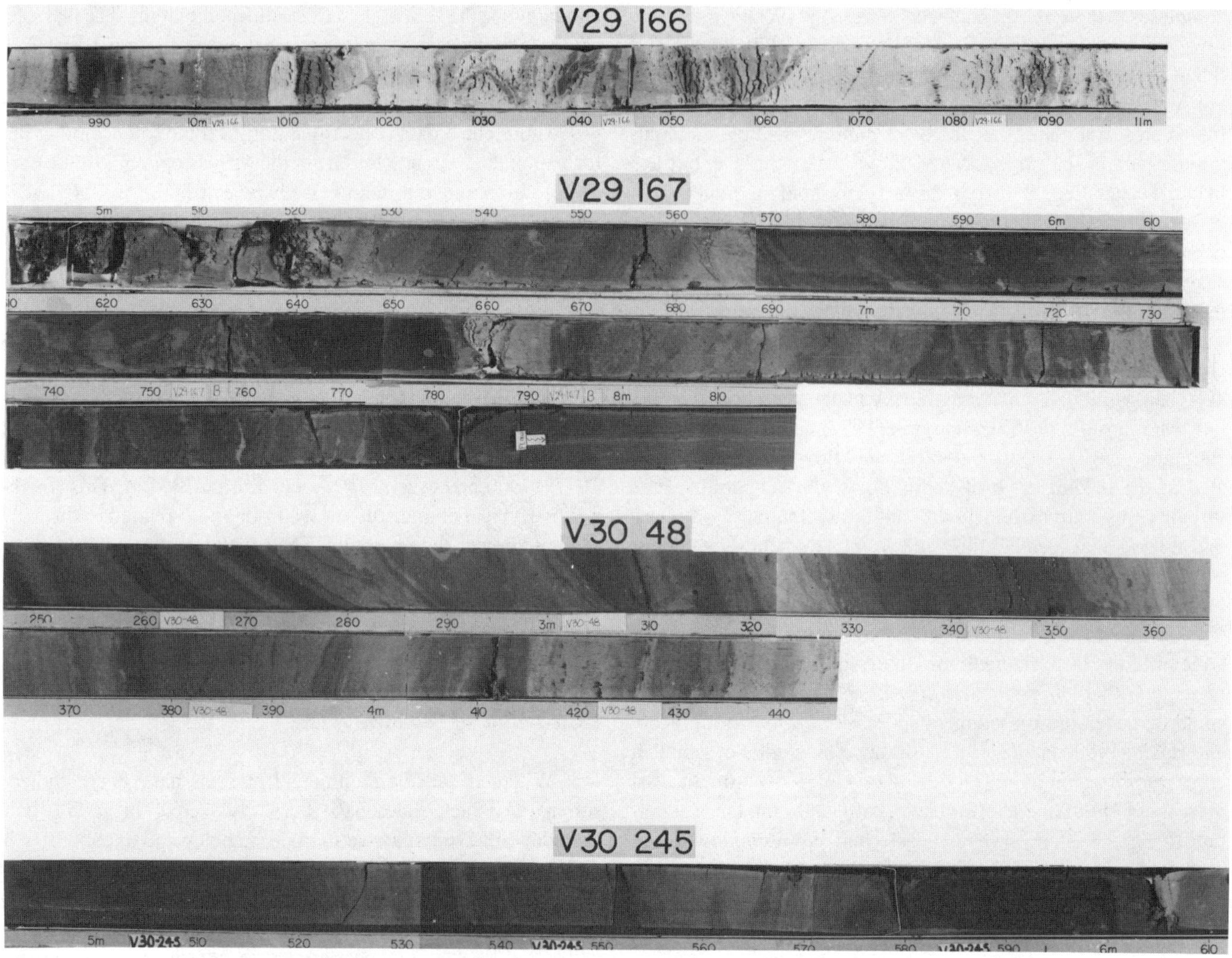

Figure 5. Sections of split piston cores retrieved from slide deposits located on the northwest African margin northwest of Dakar, Senegal. Highly disturbed layering is displayed by all cores. The disturbed layering consists of dipping strata in core 48 (up to 55° apparent dip), core 167 (vertical dip at 576 cm), and core 245 (25° at 525 cm). Well-defined isoclinal folds with horizontal axial surfaces are present in core 245 at 558 cm and 563 cm, core 48 at 290 cm, and core 167 at 752 and 778 cm. Core 166 has an angular isoclinal fold with a horizontal axial surface at 1030 cm. Angular contacts in core 166 at 1013-1020 cm and 1059-1065 cm are perhaps faults. Sharp angular contacts in core 167 at 637 cm and 656 cm are probably the contacts of an exotic block. Disrupted bedding forming rubble, pebbly mud, and wildflysch-like sediment are evident in several of the cores studied (e.g., in core 167 at 760 cm). Figure from Jacobi (1976).

land, New Zealand (Lewis, 1971); Baja, California (Normark, 1974); Sunda Trench (Moore and others, 1976); Peru-Chile Trench (Coulbourn and Moberly, 1977); Calabrian Ridge, Eastern Mediterranean Sea (Kastens, 1981); Aegean Sea Trough (Ferantinos and others, 1981); Bonin Trough, western Pacific (Jacobi and Mrozowski, 1979); and Palawan Trough, South China Sea (Damuth, 1980a).

Sediment slides occur initially on both steep and shallow slopes (>8° to <1°) and in both deep and shallow water (200-5400 m) (Embley and Jacobi, 1977; Embley, 1980). Individual slides occur in all submarine physiographic provinces and even onlap abyssal plains (e.g., Embley, 1980). Only one microphysiographic province appears to have anomalously few slope failures, the sediment wave provinces formed by bottom water flow (Jacobi and Hayes, in press). However, slides upslope from sediment wave provinces commonly result in slide material deposited over the sediment waves (e.g., Embley, 1980). In fact, several slides are so large that a single seismic facies of one sediment slide can be found in several physiographic provinces. This is demonstrated by a sediment slide near Senegal. Here the blocky slide material extends from the continental slope out onto the lower continental rise, a minimum distance of 500 km. Furthermore, the complete slide extends from the shelf break to the lower continental rise, a minimum distance of 700 km (Jacobi and Hayes, 1982). It must be emphasized that sediment slide deposits are not merely base-of-slope phenomena; rather, sediment slides are ubiquitous in the modern deep-sea environment.

Most of the surficial slides on passive margins are thought to have been emplaced in the late Wisconsin Glacial Stage during periods of lowered sea level and high sediment supply (e.g., Woodcock, 1979; Embley, 1980). Interestingly, IPOD cores on the northwest African margin (e.g., Ryan and others, 1978) contain thick Miocene slide deposits. The Miocene was also a time of lower sea level (e.g., Diester-Hass, 1977). However, seismic recognition of deeply buried slides is extremely difficult, with the result that we have, as yet, only a very limited knowledge of the extent of buried slide deposits. This present lack of knowledge precludes conclusions concerning the distribution of slides through geologic time and, thus, the relationship between sliding and periods of glaciation and lowered sea level. A few slide events on passive margins (e.g., Grand Banks slump, Heezen and Ewing, 1952) and many on active margins (e.g., Calabrian Ridge, Kastens, 1981) probably have earthquakes as a triggering mechanism and are therefore clearly not restricted to times of lower sea level.

CRITERIA FOR RECOGNIZING THE PETROGENESIS OF SOME TYPES OF MELANGE

Submarine sediment slides are important in understanding melange petrogenesis because sediment slides can result in deposits with a chaotic character typical of melange. However, as discussed in the introduction, melange is thought by many geologists to have formed in accretionary wedges, and the clastic nature of many of these melanges is believed to have been generated by tectonic processes operating there. Because sediment slides and their resulting melange are not restricted to the trench and arc-trench gap, but are common in all submarine tectonic and physiographic provinces, the criteria commonly utilized to determine the petrogenesis and paleotectonic setting of melanges need to be re-evaluated.

Evidence that has been used to support the formation of melange in subduction-zone-related sites includes: (1) the character of the incorporated clasts (e.g., Blake and Jones, 1974; Hsu, 1974; Wood, 1974); (2) "axial-flow" of turbidity currents as indicated by turbidites lying above or below the chaotic units (i.e., turbidity currents flowing down the axis of the paleotrench) (e.g., Piper, 1972; Leggett, 1980; Leggett and others, 1979); (3) regional considerations (e.g., structure and paleontology) (e.g., Bird and Dewey, 1970; Blake and Jones, 1974; Hsu, 1974; Kay, 1976; McKerrow and Cocks, 1978); and (4) the presence of scaly cleavage and fragments sheared by the cleavage (e.g., Bird and Dewey, 1970; Cowan, 1974; Cowan and Page, 1975; Moore and Alwardt, 1980). Considered singly, none of the criteria cited above provide compelling evidence for the subduction-zone-related origin of a melange because most of the features used as supporting evidence also develop in tectonic environments other than accretionary wedges. Figure 6 is a flow chart for melange that displays possible tectonic environments of formation implied by various characteristics of melange. Some of the more important aspects of this chart are discussed below.

Clasts. Blueschist facies clasts in a melange have been cited as compelling evidence for formation of melange in an accretionary wedge. If these clasts were sheared during formation of a tectonic melange (i.e., disruption occurred by tectonic processes), then the clasts may indicate melange formation in an accretionary wedge. However, if the clasts occur in an olistostrome, then the site of formation is equivocal. Other possible sites of deposition include: (1) active forearc basin; (2) forearc which has evolved into a "backarc" because of subduction-polarity reversal; and (3) obducted accretionary wedges.

If clasts derived from an ophiolite are present in the melange (ophiolitic melange), but high P/T minerals are absent, the ophiolitic melange could have formed in several tectonic provinces. If the melange is tectonic, then transform fault zones, accretionary wedges, and zones of backarc thrusting are all possible sites. If the melange is sedimentary, then a great number of generation sites are possible (Fig. 7, Table I). If the ophiolitic sedimentary melange does not have clasts derived from an island arc, then the sites under heading "I" (Table I, Fig. 7) are all possible alternatives. These sites include the base of seamounts and oceanic islands (Type IA), transform faults (Type IB), midocean ridges (Type IC), and accretionary wedges where tectonic dams prevented island-arc derived sedimentation (Type ID). If the ophiolitic sedimentary melange also has clasts from an island arc active at the time of melange generation, then the sites under heading "II" (Table I, Fig. 7) are all possible alternatives. These sites include backarc regions (Type IIA) and forearc regions (Type IIB). In the backarc regions, the clasts with an ophiolitic source could be derived from horsts of ophiolitic basement underlying the island arc (e.g., the model of Mitchell and Reading, 1971) or from seamounts, fracture zones, and scarps of the spreading center in the backarc basin. In the forearc regions, these clasts could be derived primarily from: (1) the oceanic plate forming the base of the island arc or the upper slice of the accretionary wedge, or (2) ophiolite incorporated within the accretionary wedge. From this brief review, it is obvious that the presence of ophiolite-derived clasts in a sedimentary melange does little to limit the possible sites of generation.

If the sedimentary melange contains only clasts derived from a contemporaneous island arc, then the melange formed in the forearc or backarc on the volcaniclastic apron of the island arc. A critical criterion to differentiate forearc melange from backarc melange may be the presence of boninite clasts. The few available studies of the relatively rare boninite suite suggest that these rocks occur only in forearc regions (e.g., Crawford and others, 1981). If this observation continues to be valid as further studies are completed, then the presence of boninite in the melange eliminates consideration of a backarc site of formation (except for subduction-zone polarity reversals).

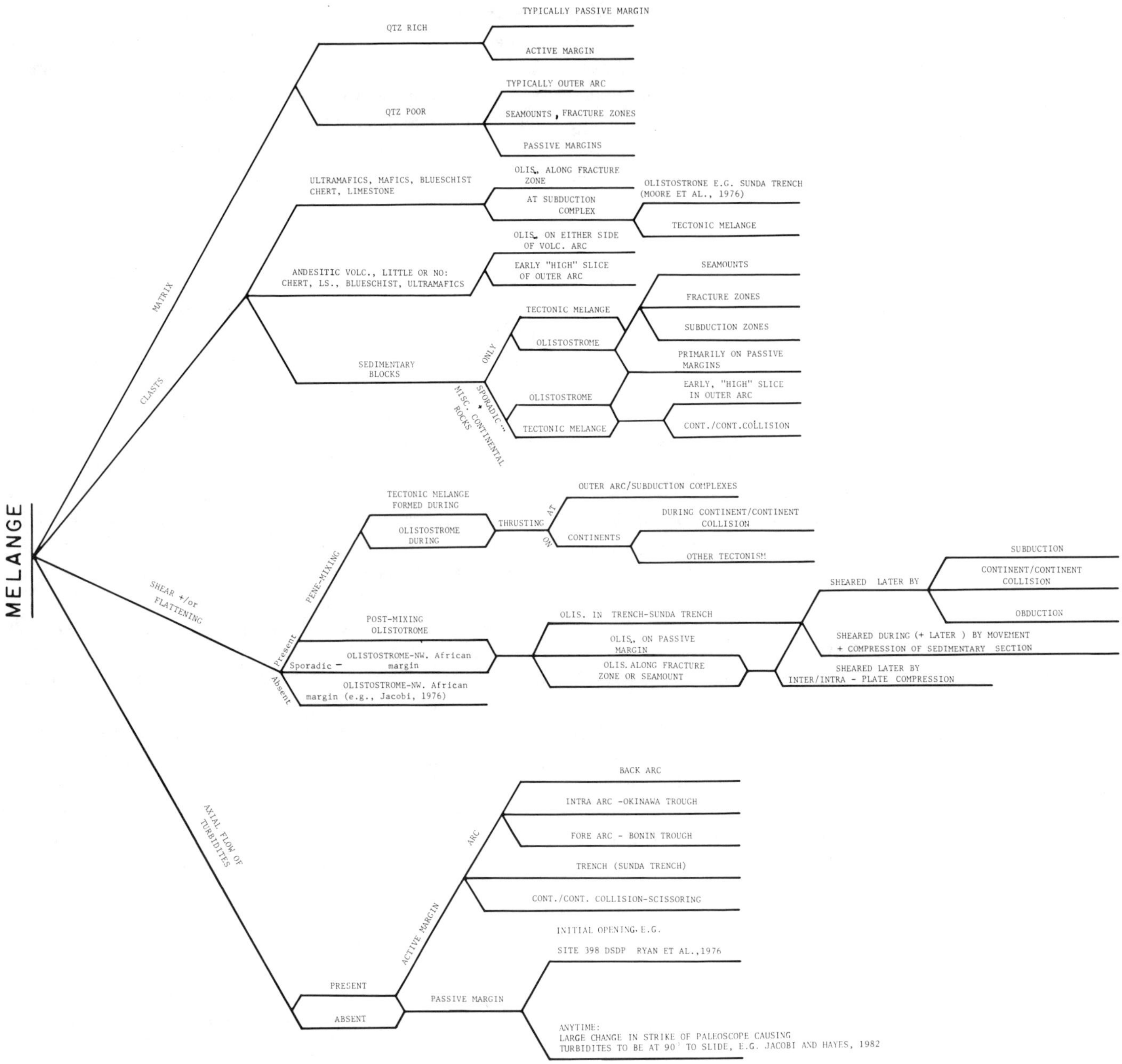

Figure 6. Flow chart displaying the relationship of various melange characteristics to possible sites of generation. Examples in chart are discussed in the text. Olis = olistostrome, qtz = quartz, other tectonism = intraplate tectonism. Figure from Jacobi (1980).

Axial Flow of Turbidity Currents. Axial flow of turbidity currents along trench or terrace basin axes will result in turbidites with current indicators directed perpendicular to the regional paleoslope indicated by slump folds. These turbidites may be observed stratigraphically and/or structurally above or below the melange. This effect commonly is observed in modern convergent margins (e.g., Aleutian Trench, Piper and others, 1973, von Huene, 1974; South China Sea, Damuth, 1979) and thus is a valid criterion for ancient rocks *if* it can be established that: (1) the flow in question was a turbidity flow and not a contour current along a passive margin; and (2) the trough guiding the turbidity flow was indeed a trench or terrace basin near the trench. Although discrimination of distal turbidites from contourites is more difficult than commonly supposed (see Stow, 1979, for a review), the critical problem is the second consideration.

Examples from modern ocean basins show that axial flow of turbidity currents is not unique to trenches (Fig. 6). Along active margins, both interarc basins and abandoned trenches contain

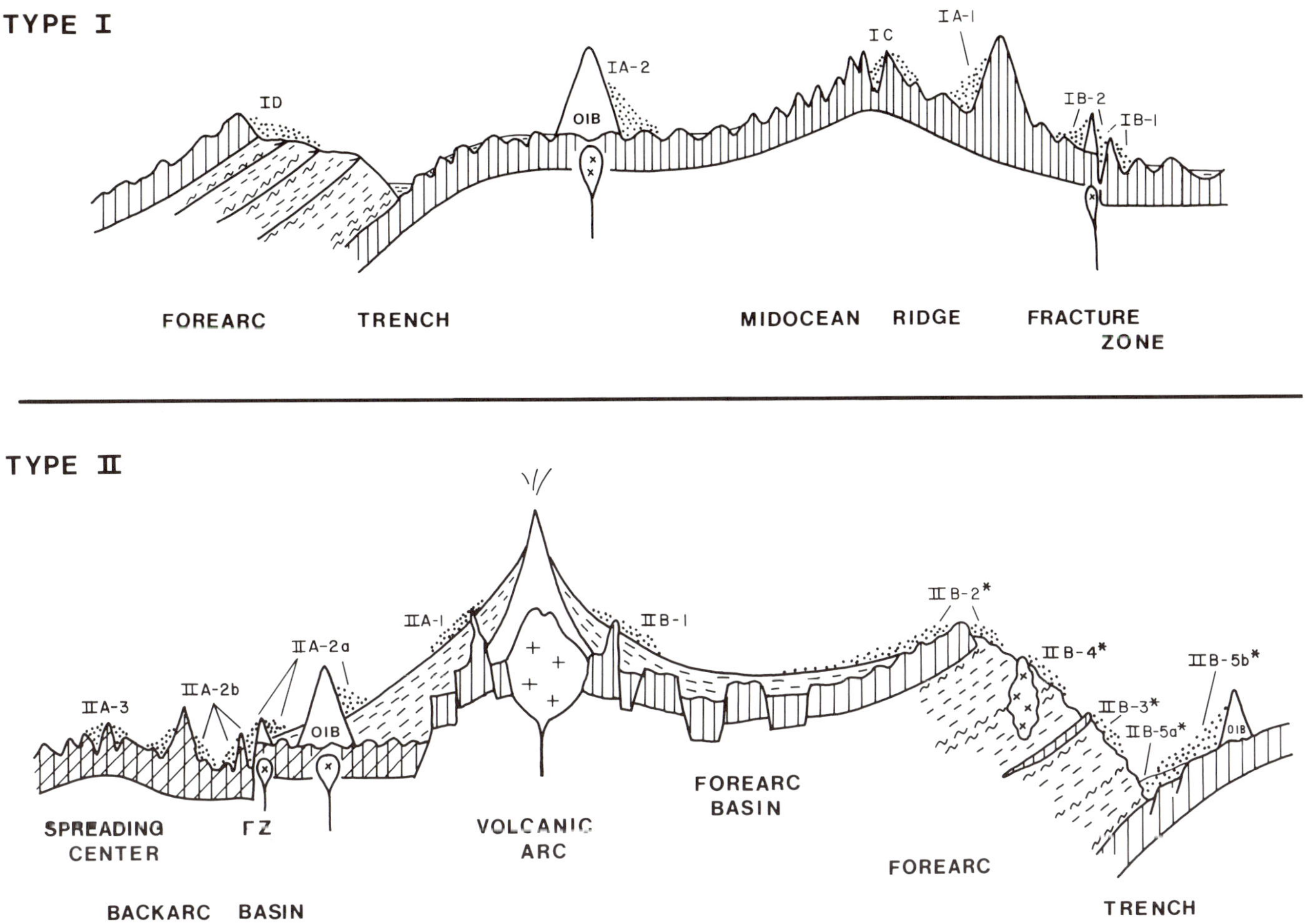

Figure 7. Schematic diagram portraying possible sites of generation of sedimentary melange with clasts derived from oceanic crust (broadly, "ophiolitic melange"). Melange symbols refer to types listed in Table I. MORB = midocean ridge basalt, OIB = ocean-island basalt (off-axis basalt), FZ = fracture zone.

turbidites which have formed from flow parallel to the regional contours. For example, the north-striking Cagayan Trough, Philippines, is an intra-arc basin which has been filled by deposition from north-flowing turbidity currents. Furthermore, trenches abandoned by a reversal in the subduction polarity could have axial flow of turbidity currents, but the tectonic environment of deposited turbidites is now backarc rather than forearc. Finally, passive margins also exhibit axial flow of turbidity currents, both on a local scale (10-50 km) and a regional scale (~200-300+ km). Slump blocks that have slipped significant distances downslope can create along-slope dams, which divert the turbidity flows to along-slope flow (e.g., Wilmington Canyon, McGregor and Bennett, 1979; Stubblefield and others, 1982). Offshore seamounts and island pedestals will also force the turbidity currents to flow parallel to the margin in order to pass around the barrier. For example, in order to skirt the Cape Verde Islands, the Cayar Canyon System flows south 500 km parallel to the northwest African continental margin (Jacobi and Hayes, 1982; Jacobi, 1982). Similarly, "tributaries" of the Agadir Canyon off Morocco parallel the regional slope for a distance of more than 200 km (Jacobi and Hayes, in press). Finally, turbidity currents flowing in mid-ocean canyons and gaps between abyssal plains generally parallel the continental margins. For example, the mid-ocean canyon in the Labrador Sea and North Atlantic follows the continental margin for at least 1500 km (e.g., Drake and others, 1963). It appears that if only local data are utilized in determining the origin of a chaotic unit, axial flow of turbidity currents should be regarded as a suspect criterion until supported by other independent criteria.

Shearing and Tectonic Melange. The identification of tectonic melange is difficult, and the implications of a tectonic origin are far reaching. The presence of scaly cleavage in a melange commonly has been attributed to a tectonic origin (e.g., Cowan, 1974). However, some workers (e.g., Lash, 1980) suggest that scaly cleavage can originate by dewatering of unlithified sediments during sediment sliding. The absence of scaly cleavage and the presence of "soft-sediment" deformational features were thought to imply generation by sediment sliding. However, recent studies suggest that tectonic deformation of unlithified sediments at low temperature (<250°C) may result in uncleaved melange

TABLE 1. POSSIBLE SITES OF FORMATION OF SEDIMENTARY MELANGE THAT CONTAINS CLASTS DERIVED FROM OCEANIC CRUST (Broadly Defined Ophiolitic Melange)

I. Sediment slides without island arc input ± later deformation

- A. Seamounts and islands in main ocean basin
  1. Basalts have mid-ocean ridge geochemical characteristics (MORB's)
  2. Basalts have off-axis geochemical characteristics (OIB's)
- B. Fracture zones
  1. Basalts have mid-ocean ridge geochemical characteristics (MORB's)
  2. Basalts have off-axis geochemical characteristics (OIB's)
- C. Mid-ocean ridges; large slides rarely occur here because of the lack of sufficient fine-grained (pelitic) sediment and long, unbroken slopes

II. Sediment slides with island arc material ± later deformation

- A. Backarc side of island arc
  1. Basalts from oceanic plate forming basement of island arc
  2. Basalts from seamounts or fracture zones in backarc basin
     - a. with OIB characteristics
     - b. with spreading center (~MORB) characteristics
  3. Basalts from spreading center
- B. Forearc side of island arc
  1. Basalts from oceanic plate forming basement of island arc
  2. Basalts from oceanic plate forming upper slice of outer-arc
  3. Basalts from subducting plate incorporated within accretionary wedge
     - a. basalts have MORB geochemical characteristics
     - b. Basalts have OIB geochemical characteristics
  4. Basalts from intrusions into accretionary wedge
  5. Basalts from subducting plate seaward of accretionary wedge
     - a. Basalts have MORB geochemical characteristics
     - b. Basalts have OIB geochemical characteristics

with soft-sediment deformation features (e.g., Cloos, this volume). Clearly, there is little agreement concerning the significance of the presence or absence of scaly cleavage in melange.

Evidence from modern submarine sediment slides and their analogues in the rock record suggests that incipient slaty cleavage forms in zones of high strain in the sediment slide (e.g., Moore and Geigle, 1974). However, this cleavage is not regionally pervasive and does not have a regionally planar orientation. Regionally pervasive, planar slaty cleavage may result from tectonic strain at relatively higher temperatures (e.g., Knipe and White, 1977; Woodland, 1982). In the few available studies of modern slide material, scaly cleavage was not reported. Tectonically-deformed sediments in cores from accretionary prisms generally exhibit spaced fracture cleavages and anastamosing fractures that may develop after dewatering and plastic deformation (e.g., von Huene in Silver and Beutner, 1980; Lundberg and Moore, 1981; Carson and others, 1982). Strain continued beyond the stage observed in the cores could result in the scaly cleavage typical of melanges.

Collectively, the available studies suggest that: (1) scaly cleavage generation synchronous with fragmentation may indicate tectonic melange; (2) local slaty cleavage coeval with fragmentation may indicate a sedimentary melange; (3) slaty cleavage generated at relatively high T may suggest tectonic deformation; and (4) absence of cleavage indicates sedimentary melange or perhaps early stages of tectonic melange. Obviously, the generation of scaly cleavage must be coeval with fragmentation of the sedimentary units in a tectonic melange (i.e., the deformation resulting in the scaly cleavage also caused the fragmentation). However, in most melanges it is extremely difficult to ascertain whether the chaotic nature was caused by the deformation that produced the scaly cleavage or whether the scaly cleavage was imposed long after the initial chaotic nature was formed (e.g., Hsu, 1974). Careful detailed analyses of the relation of clast shape and orientation to the cleavage may reveal the synchronous or diachronous nature of the cleavage/clast formation (e.g., Horne, 1969; Cowan, 1978). Alternatively, it may be possible to detect the cleavage/clast relationship in areas where a transition to an uncleaved chaotic unit occurs (e.g., Dunnage Formation, Kay and Eldredge, 1968; Horne, 1969; Jacobi and Schweickert, 1976; Hibbard and Williams, 1979).

Sediment slides (i.e., olistostromes), overprinted by scaly cleavage after mixing, could form in an almost limitless number of tectonic environments. These environments include: (1) the trench or inner wall of the trench (olistostromes are later sheared during incorporation in the accretionary prism); (2) the backarc volcaniclastic apron (olistostromes are later deformed by backarc thrusting); (3) transform faults (olistostromes are sheared by the transform fault or convergent zone tectonism); (4) a passive margin (olistostromes are later sheared during incorporation in an accretionary wedge or escape later deformation at a convergent margin because of incomplete suturing or obduction, but are then deformed by continental intraplate deformation); and (5) an active margin or on a continental plate (olistostromes are later deformed by intraplate deformation).

Similarly, tectonic melange may form in a number of tectonic settings which include: (1) transform fault; (2) accretionary prism; (3) zones of backarc thrusting; (4) zones of continent-

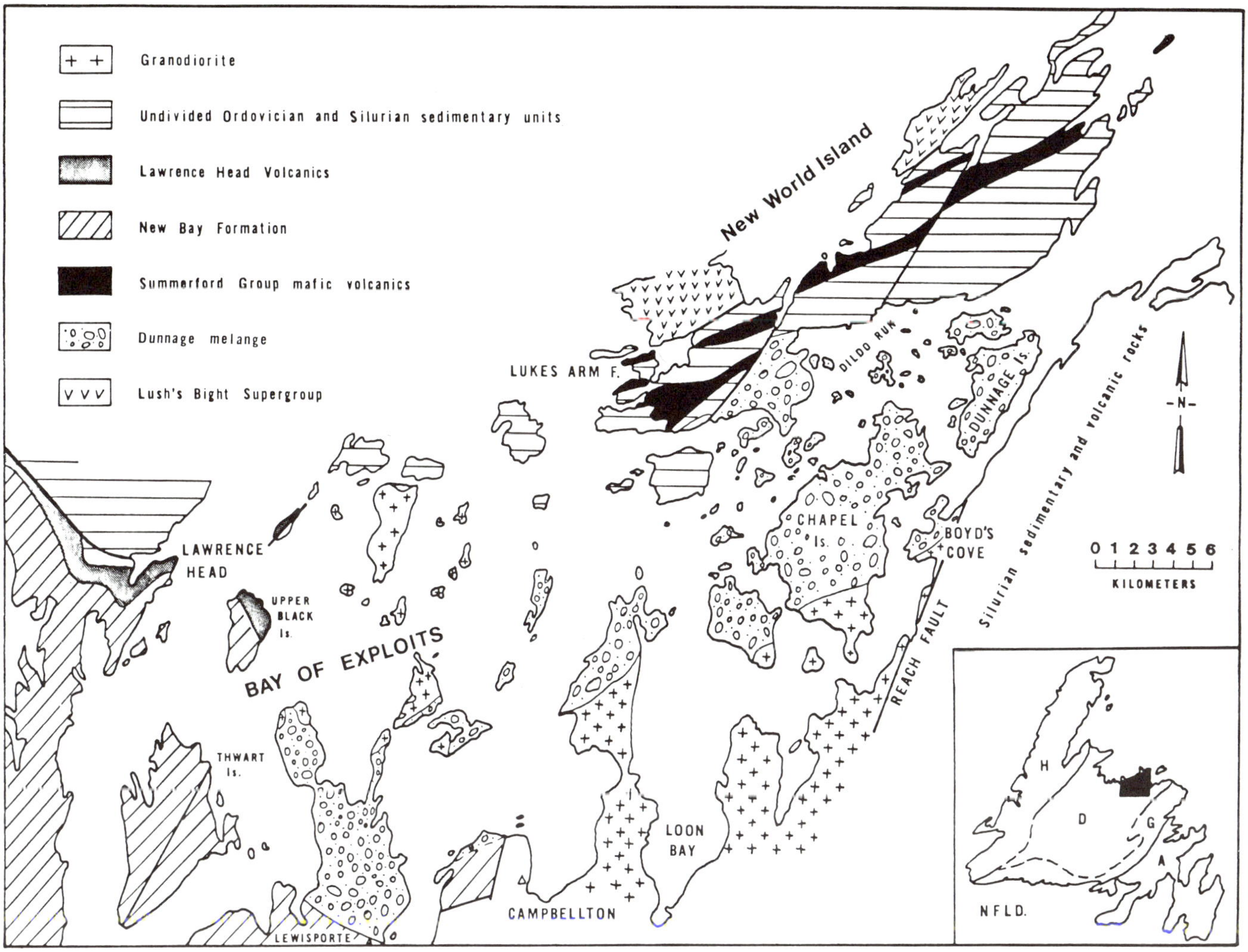

Figure 8. Generalized geology of the Dunnage melange and surrounding area in north-central Newfoundland from Wasowski and Jacobi (1983, 1984) and after Hibbard and Williams (1979). Map location and tectonic-stratigraphic zones of Newfoundland shown on insert (after Williams, 1978). H = Humber zone, D = Dunnage zone, G = Gander zone, A = Avalon zone.

continent and arc-continent collision; and (5) zones of continental intraplate faulting. Discrimination among some of these alternatives is difficult and generally must rely on regional relationships and other criteria. For example, the ophiolitic, tectonic melange at Masirah, Oman (Moseley and Abbotts, 1979), could be considered a typical accretionary prism melange. However, the melange belt is at right angles to sheeted dikes of the associated ophiolite, and the cleavage was formed vertically with respect to the sheeted dikes. These and other considerations suggested to Moseley and Abbotts (1979) that the tectonic melange formed initially in a transform fault.

In summary, the criteria used individually to determine melange origins and sites of formation are largely equivocal. Nevertheless, using a combination of features, it may be possible to identify both the petrogenetic history and the paleotectonic setting for a particular melange. The following section uses this approach to study the petrogenesis of the Dunnage Formation.

PETROGENESIS OF THE DUNNAGE FORMATION

The Dunnage Formation (Kay and Eldredge, 1968) crops out in the Dunnage tectonostratigraphic zone of Williams (1978) along the north-central coast of Newfoundland (Fig. 8). The Dunnage Formation is a melange that illustrates many of the characteristics and problems common to melange. The Dunnage melange displays a wide range of deformational features. Some parts of the melange exhibit soft-sediment deformation with no macroscopic cleavage, whereas other areas are characterized by prominent scaly cleavage. Still other areas exhibit a slaty cleavage (e.g., Kay and Eldredge, 1968; Horne, 1969; Kay, 1976; Jacobi and Schweickert, 1976; Hibbard and Williams, 1979). The origin of these deformational fabrics remains obscure after fifteen years of field study.

The observations and implications detailed below suggest that: initial mixing in the Dunnage Formation was caused by submarine sediment sliding; the sliding was not limited to one episode; local slaty cleavage developed coevally with mixing; the age and origin of the regional slaty and scaly cleavages are unclear, but may be tectonic; the clasts in the Dunnage Formation were derived from both an island arc and the oceanic crust; and the Dunnage Formation deformed via thrusting in a forearc setting.

Pebbly Mudstone and the Pre-Cleavage Character of the Dunnage Formation. In areas where the Dunnage melange lacks scaly cleavage, disrupted bedding, pebbly mudstone (with first and second generation clasts), and "wisps" of argillite and graywacke in an argillite matrix commonly occur (e.g., Kay and Eldredge, 1968; Horne, 1969; Kay, 1970, 1976; Hibbard and Williams, 1979) (Fig. 9). Evidence of large ductility contrasts in the pebbly mudstones is common. For example, angular fragments of graywacke with little or no internal deformation (Figs. 9A and 9B) occur with both graywacke clasts that were rounded by strain (Fig. 9A) and graywacke clasts that are wispy, highly strained, and elongate (Fig. 9B). The short axes of the clasts are generally perpendicular to nearby bedding and to a local, weakly-developed slaty cleavage that generally parallels bedding. In several areas overprinted by later scaly and slaty cleavage, the original parallelism of bedding and clasts can still be observed (e.g., the north shore of Coaker Island and the southwest shore of Dunnage Island).

The disruption of units and consequent lack of marker units cause extreme difficulty in recognizing small-scale pre-cleavage folds. However, tight isoclinal folds (F_1) without axial-planar cleavage are observed in argillites with alternating color banding (e.g., black and green) (Fig. 10). The F_1 folds in Figure 10 clearly predate the prominent cleavage that crosses the F_1 axial surfaces at high angles. Observations of (1) pebbly mudstone with variable ductility contrasts, (2) clasts oriented parallel to nearby bedding, (3) isoclinal folds that lack an axial-planar cleavage, and (4) local weakly developed slaty cleavage generally parallel to bedding are all very similar to characteristics observed in modern submarine sediment slides. However, because tectonic deformation of water-rich sediments also may produce soft-sediment deformational features that lack cleavage, it is not possible to dismiss entirely the hypothesis that low temperature tectonic deformation generated these features observed in the Dunnage melange. Yet, the internally (?) unstrained angular graywacke clasts (which were apparently lithified or semi-lithified at the time of fragmentation) do not exhibit cleavage typical of tectonic deformation of lithified or semi-lithified sediments. Therefore, the very tentative conclusion is that the initial mixing of the Dunnage Formation was caused by submarine sediment sliding.

Several lines of equivocal evidence suggest that the sliding was not limited to one episode; rather, the Dunnage melange represents a series of submarine sediment slides. (1) The composition of clasts does not remain constant throughout the Dunnage. For example, breccia beds with felsic igneous clasts only occur along the southern margin of the Dunnage. (2) Second-cycle pebbly mudstone blocks are common (e.g., Hibbard and

Figure 9. Pebbly mudstone in the Dunnage Formation. In examples A & B (from near Newman's Point, Boyd's Cove), no macroscopic scaly cleavage is present. Note also the ductility contrast between "clasts" of wispy argillite and the discrete clasts forming pebbly mudstone or breccia. Example C illustrates pebbly mudstone with angular clasts locally overprinted by scaly cleavage. Location of example is northwest of Dildo Run Causeway.

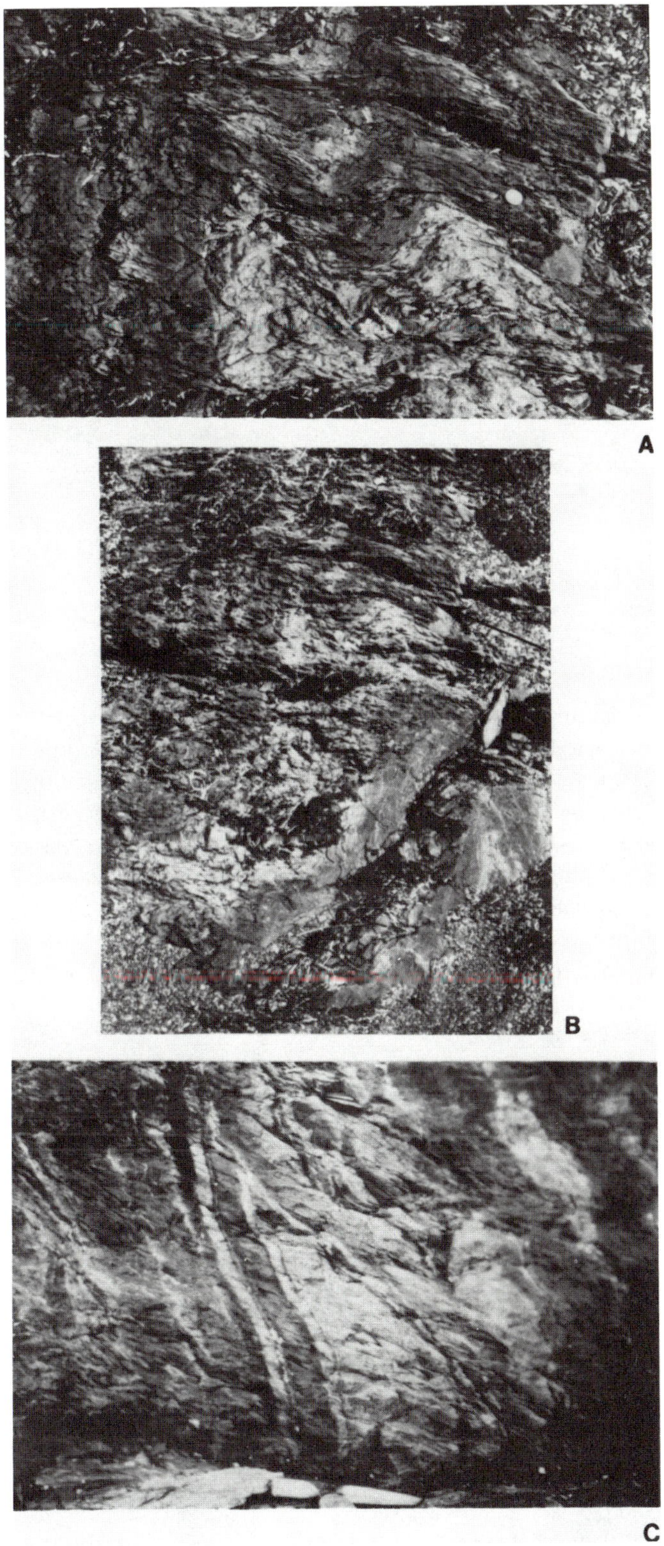

Figure 10. Pre-cleavage folds (F_1) observed in argillites with contrasting color banding. Example A is a more detailed view of example B (arrow in B points to quarter which is in same location in both photographs). Note that the cleavage in all examples is almost at right angles to the axial surfaces of these early folds. All examples are from the island ESE of Black Head in Dildo Run.

Williams, 1979), suggesting multiple stages of sliding and incorporation. (3) Although the Dunnage is probably highly folded, it is abnormally thick compared to single modern submarine sediment slides. (4) The age of clasts in one part of the Dunnage is younger than the matrix in another part of the Dunnage (Hibbard and others, 1977). However, if the initial mixing was tectonically generated, then the above observations are consistent with deformation caused by multiple thrust events.

Scaly Cleavage and Slaty Cleavage. Zones of scaly cleavage are locally to regionally developed. At the margins of scaly-cleavage zones, the scaly cleavage clearly overprints the soft-sediment deformation features (Fig. 9C), indicating that mixing in the scaly cleavage zones occurred before scaly cleavage development. Second-cycle clasts in pebbly mudstone in the Dunnage Formation (Hibbard and Williams, 1979; Wasowski and Jacobi, 1982) invariably exhibit no earlier scaly cleavage, suggesting that the scaly cleavage was not imposed on any part of the Dunnage melange during initial fragmentation and mixing.

A prominent slaty cleavage is observed in the Dunnage melange primarily in argillaceous matrix and blocks that contain few or no clasts (Horne, 1969; Wasowski and Jacobi, 1982). The regional slaty cleavage generally appears to be parallel to the scaly cleavage. The slaty cleavage provides another constraint on the timing and nature of dismemberment during formation of the Dunnage melange. At Boyd's Cove, detailed mapping revealed that bedding/slaty cleavage relationships reversed sense across strike in a sequence of argillite outcrops (Wasowski and Jacobi, 1982). These data might suggest that a fold with axial-planar cleavage exists between the exposures of argillite. However, the facing direction does not reverse across the core of the "fold," implying that the slaty cleavage was imposed on the Dunnage melange after dismemberment and rotation of individual blocks of argillite. Yet, the scaly cleavage and parallel slaty cleavage are axial-planar to rare mesoscopic, asymmetric folds (F_2) throughout the Dunnage melange (e.g., the north shore of Coaker Island, southwest Dunnage Island, and Boyd's Cove). These folds affect both argillite bedding and a parallel, poorly developed slaty cleavage first observed by Horne (1969).

The orientation of many graywacke clasts may suggest that fragmentation continued during cleavage development. The intermediate axes of many graywacke clasts lie within the scaly and slaty cleavage surfaces (Horne, 1969). Where bedding is observed, the plunges of the long axes of the clasts commonly are colinear both with the plunges of the bedding/cleavage intersection (e.g., the southwest shore of Dunnage Island) and with the plunges of axes of tight folds that have axial-planar scaly and slaty cleavage (e.g., north shore of Coaker Island). These clasts oriented parallel to cleavage* may suggest that fragmentation of these clasts was caused by stretching of the units during folding and development of the prominent scaly/slaty cleavage. Thus, although initial mixing probably occurred during sediment slid-

*The cleavage-parallel clasts that exhibit "hooks" (Horne, 1969) perhaps were generated prior to cleavage development and were reoriented during the cleavage development.

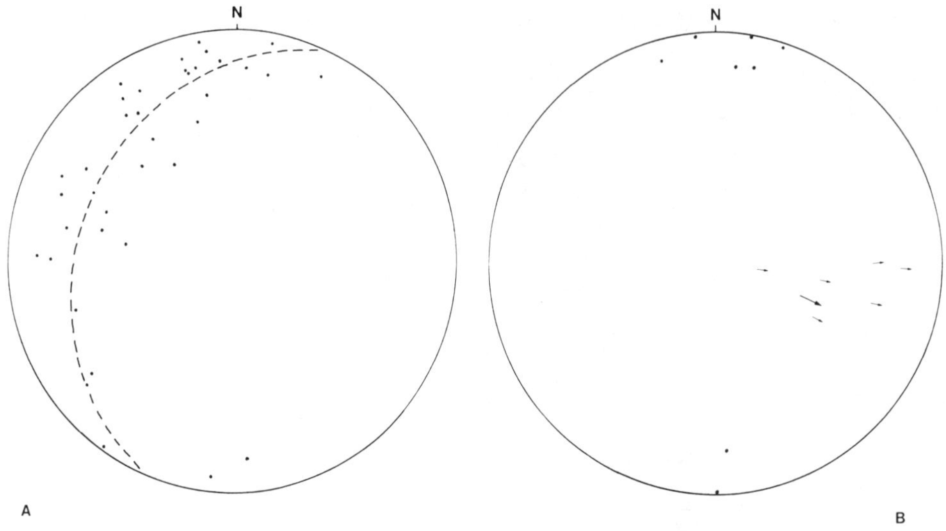

Figure 11. Equal area projection of poles to scaly cleavage on Dunnage Island (Figure 11A) and poles to axial surfaces of folds (F_3) that fold the scaly cleavage on Dunnage Island (Figure 11B). Tails of arrows in Figure 11B are plunges of F_3 axes. Girdle fitted to scaly cleavage poles in Figure 11A implies the F_3 plunge with a large arrow in Figure 11B.

ing, additional fragmentation and mixing of both the slides and remaining intact strata occurred during the generation of the scaly and slaty cleavage.

The trend of the prominent scaly and slaty cleavages is regionally constant, but locally variable. In many of the northeastern exposures of the Dunnage melange, the regional trend is approximately east-west with moderate to steep southerly dips (e.g., Kay, 1976). Along the southeastern-most exposures of the Dunnage melange, the trend is variable, but generally northeast with moderate to steep southeast dips (e.g., Kay, 1976). Some of the local variability in the scaly cleavage trend is an original feature of the scaly cleavage generation. Much of the original variability is a result of structural buttressing in which the cleavage wraps around large blocks of volcanic material and graywacke (e.g., Horne, 1969; Hibbard and Williams, 1979). In other areas, the cause of the original variability is unclear (Fig. 9C). Strain shadows and unrecognized blocks and other inhomogeneities are two of many possibilities. Other local variability is the result of later folding of the scaly cleavage. Along the southern margin of the Dunnage melange at Dunnage Island (Fig. 11), Boyd's Cove (Wasowski and Jacobi, 1982), and western exposures of Dunnage melange (Hibbard, 1976), stereoplots of poles to scaly cleavage suggest that the scaly cleavage is disposed in northeast to southeast plunging folds. On southern Dunnage Island and nearby islands to the west, mesoscopic, upright, open, east-plunging folds with east-west trending axial surfaces (F_3) are observed that fold the scaly cleavage. It is primarily this generation of folds that causes the dispersion of poles to cleavage on the stereoplots (Fig. 11). In some of these folds, slip cleavage and spaced-fracture cleavage are locally developed that are axial-planar to the folds. The orientation of these folds is broadly consistent with the hypothesis that the Dunnage Formation is folded into an east or northeast-plunging antiform (Jacobi and Schweickert, 1976; Karlstrom and others, 1982). If this proposition is valid, the east-west trending scaly cleavage marks the north limb, the northeast-striking scaly cleavage marks the partly-faulted south limb, and the zone of refolded cleavage along the southern margin (especially at the southern tip of Dunnage Island) marks the core of the antiform. The antiform may have formed during the Acadian orogeny, because the F_3 folds are parallel to probable Acadian structures mapped both north and south of the Dunnage Formation (e.g., Karlstrom and others, 1982; Fargo, 1983).

Although the axial surfaces of the F_3 folds are generally oriented east-west, in a few localities the axial surfaces of the F_3 folds are variably oriented. Some of this reorientation could be the result of fault-block reorientations, but in localities where the relationships are clear (e.g., the southwestern side of Dunnage Island), the reorientation was accomplished by folding (F_4) along brittle shear fault zones. The fault zones are oriented approximately parallel to the trace of the Reach Fault (~N35°E, 70°SE), and the F_4 folds suggest that slip along the faults was primarily right-lateral. The age of the F_4 folds and associated shear is probably Carboniferous.

The origin and age of the scaly and slaty cleavages are not entirely clear. Based on the discussions in the preceding sections, tectonic deformation may have generated both the scaly cleavage and the parallel slaty cleavage. The poorly developed slaty cleavage that is folded by the prominent regional scaly and slaty cleavages may be similar to the incipient slaty cleavage observed in submarine sediment slides. The age of the prominent regional cleavage generation is equivocal. Because it is folded by folds thought to be Acadian in age, the cleavage generation must be early Acadian or older (Early Silurian? or Ordovician). The obscure relationship between the cleavage and the poorly-dated Coaker Porphyry (428 m.y. BP to 454 m.y. BP, Kay, 1976) may suggest that the prominent cleavage is Late Ordovician to Early Silurian in age. The local, weakly developed slaty cleavage that parallels bedding is probably coeval with initial mixing and therefore is probably Early-Middle Ordovician.

Clasts in the Dunnage Melange. The source of clasts within the Dunnage melange may provide information concerning the site of generation. The most prevalent clasts in the Dunnage melange are graywacke, argillite, and volcanic clasts (e.g., Kay and Eldredge, 1968; Horne, 1969; Kay, 1970, 1976; Hibbard and Williams, 1979). Lesser amounts of limestone; dolostone; black, white, and red chert; red siliceous argillite, and diabase also occur in the melange. Blocks of breccia beds containing angular felsic igneous clasts and graywacke outcrop along the southern margin at Boyd's Cove (Kay, 1976; Wasowski and Jacobi, 1982).

Andesitic fragments in the graywacke clasts suggested to Horne (1969) that the graywacke clasts have a volcanic arc provenance. It may be argued that sediment slides causing the initial mixing in the Dunnage melange had heads close to a subaerially-exposed portion of an active or remnant volcanic arc (Wasowski and Jacobi, 1982). This hypothesis is based on the following assumptions: (1) the felsic clasts in the igneous breccia blocks at Boyd's Cove were derived from an island arc, and (2) the coarse, angular felsic clasts were transported only minimal distances before they were deposited in the breccia beds. If the felsic clasts were not derived from a volcanic arc (i.e., they have an ophiolitic or continental source), the composition of the graywacke clasts still implies that the Dunnage melange had a volcanic arc source.

Although Horne (1969) observed black and white chert clasts, it was generally believed that little chert existed in the Dunnage melange (Kay, 1976; Jacobi and Schweickert, 1976; Hibbard and Williams, 1979). However, recent mapping on Chapel, Dunnage, and Cross Islands revealed that a ghost-stratigraphy exists with red chert and siliceous argillite as the marker unit. The unit is in depositional contact with tuffaceous, cross-bedded sandstone and is closely associated with blocks of a distinct highly fractured, but intact pillow lava unit. The red siliceous argillite and chert suggest that sedimentary rocks deposited on the ocean floor below the calcite compensation depth (CCD) also supplied detritus to the Dunnage melange.

All pillow lava blocks sampled throughout the Dunnage melange, including those discussed above, appear to have had an oceanic source. Analyses of major elements, trace elements, and rare earths strongly suggest that the pillow lava fragments are alkalic and tholeiitic off-axis basalts (ocean island basalts) (Fig. 12; Wasowski and Jacobi, 1983, 1984). Many of the pillow lava blocks contain limestone and dolostone, while other pillow lava blocks are associated with siliceous argillite and chert. The geochemistry of the basalt and the associated sedimentary rocks suggests that the source for the volcanic rocks was a large seamount or island pedestal that reached above the CCD.

If the felsic igneous clasts and graywacke clasts have a vol-

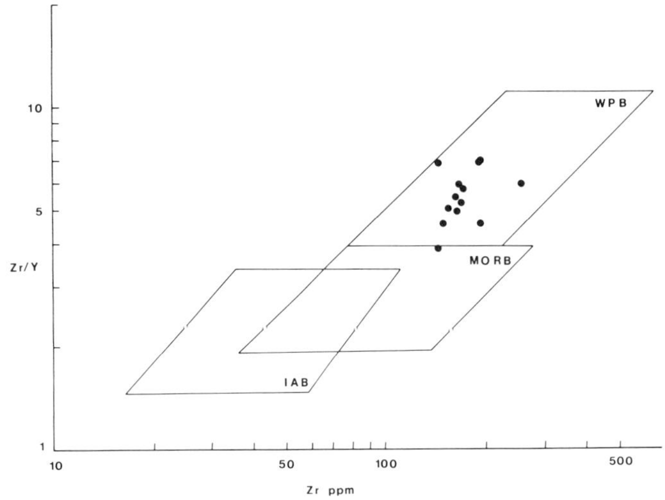

Figure 12. Zr/Y-Zr discrimination plot (Pearce and Norry, 1979) for blocks of mafic volcanics in the Dunnage melange (from Wasowski and Jacobi, 1983, 1984). IAB = island arc basalts, MORB = midocean ridge basalts, WPB = within-plate basalts.

canic arc source, then the blocks of pillow lava derived from oceanic crust suggest that the Dunnage Formation originated as a Type II sedimentary melange (Table I, Fig. 7). At Boyd's Cove, one block of breccia with felsic igneous clasts also contains ophiolitic pillow lavas, whereas another breccia block contains fossiliferous red chert (Wasowski and Jacobi, 1982). If the felsic rocks are volcanic-arc-derived and if the breccias are subaerial, then both oceanic and volcanic arc rocks may have been eroded from the same sourceland. The oceanic crust could be part of the plate upon which the island arc was built (e.g., Mitchell and Reading, 1971) or could be part of an exposed subduction zone complex that was contiguous with the volcanic arc (e.g., Zambales ophiolite, Luzon, Philippines, Schweller and others, 1983). The former case (oceanic basement) does not allow discrimination between a forearc or backarc site of deposition for the Dunnage. The latter case (accretionary wedge) suggests either a forearc site or a recently abandoned forearc site. The second hypothesis is consistent with the suggestion by several researchers that oceanic rises, plateaus, and especially seamounts are preferentially preserved at convergent margins by becoming wedged in the accretionary prism (e.g., Hawkins and others, 1979; Saleeby, 1982).

Regional Considerations. Considerations that could bear on the origin of the Dunnage melange include: (1) possible slide scars; (2) axial flow of turbidites; (3) source of the basaltic blocks in the Dunnage melange; and (4) correlations and regional tectonic models.

The extremely chaotic nature and complex structural style of the Dunnage melange have not yet allowed recognition of internal slide scars. Deposition by axially flowing turbidity currents in stratigraphically equivalent units to the west of the Dunnage outcrops (New Bay Formation) suggested to Horne (1969) that the Dunnage Formation formed in a structural trough (assuming that the present outcrop pattern of the Dunnage Formation was similar to its original distribution). Flute casts both in the Cheneyville Formation (Janish and Jacobi, 1982), which stratigraphically (?) overlies the Dunnage Formation, and in the younger Sansom Formation (Kay, 1967), generally trend south to southeast (restored orientations) near the obscure contact with the Dunnage melange. These directions are somewhat parallel to the present strike of the contacts of the Dunnage melange. However, this parallelism does not necessarily imply axial flow of turbidity currents because the present outcrop pattern of the Dunnage melange is controlled by later folding, not by the original trough or slope along which the Dunnage Formation initially formed. If the Dunnage melange did form along a northeast-southwest trough (e.g., Horne, 1969; Bird and Dewey, 1970), then the turbidites may have an axial-flow relationship with the Dunnage melange.

The submarine sediment slides on the northwest African margin (e.g., Ryan and others, 1976; Lancelot and others, 1977) have implications for stratigraphic correlations involving the Dunnage Formation. A large number of olistostromes with different temporal and/or spatial relationships are present in the African offshore section. These olistostromes all have similar clasts and matrix, making recognition of discrete slides difficult. Failure to recognize the multiple slide nature of a sedimentary melange may result in correlation problems because the outcrops may represent different, but lithologically identical, sediment slides. This consideration may explain the age discrepancy found in the Dunnage melange, in which the matrix in one part of the melange is older than the clasts in another part of the melange (Hibbard and others, 1977). A second explanation is that within the same slide scar the source rock for the matrix was older than the source rock for the clasts (Hibbard and others, 1977). The magnitude of the age discrepancy (early Tremadocian to early or middle Arenigian, Hibbard and others, 1977) suggests that in a conformable sequence a single slide scar would have to be at least 500 m to 1000+ m high in order to tap both source rocks (assuming 10+ m.y. difference and a 5-10 cm/1000 deposition rate). Because this relief is greater than that observed on most present slide scars, explanations other than a single slide and slide scar in a conformable sequence are thought to be more probable. These hypotheses include: (1) multiple slide deposits with or without recycled material, (2) slide(s) in a canyon system, (3) single or multiple slides tapping a terrane of tectonically juxtaposed blocks, (4) single or multiple slides tapping a sequence that contains an unconformity, or (5) tectonic mixing in the Dunnage melange. This discouragingly long list of possible alternatives perhaps can be shortened by facies considerations of the matrix and clasts (argillite and limestone associated with volcanic rocks, respectively). These considerations suggest that alternatives 3, 4, and 5 may be slightly more probable.

Pajari and others (1979) and Karlstrom and others (1982) suggested that the Dunnage melange correlates lithostratigraphically with the ophiolitic Carmanville melange exposed to the east near the Gander zone boundary. Recognition that the volcanic rocks in the Dunnage melange also were derived from oceanic crust supports the proposed correlation. However, as discussed above, the model of modern sediment slides suggests that extreme caution be used in this type of correlation. In fact, Pajari and others (1979) pointed out that the Carmanville melange involves rocks younger than those presently known in the Dunnage melange, implying that the Carmanville sedimentary melange may postdate the initial generation of the Dunnage melange. However, if both melanges were generated in the same paleotectonic framework (Pajari and others, 1979), then the accretionary wedge/obducted margin model proposed for the Carmanville melange (e.g., Pajari and others, 1979) may apply to the Dunnage melange.

What was the paleotectonic setting of Dunnage melange formation? Tectonic and sedimentological data from terranes adjacent to the Dunnage melange help to provide the answer. The upper contact of the Dunnage melange is obscure. It may be a gradational contact (Horne, 1969), an unconformity (e.g., Hibbard and Williams, 1979), or a faulted unconformity (e.g., Kay, 1976). The sedimentary units stratigraphically above the contact show a general coarsening and thickening-upwards sequence (except for the lowermost unit, the Cheneyville Formation). McKerrow and Cocks (1978) interpreted this overlying Ordovician to Lower Silurian (Caradocian to Llandoverian) sequence as a trench-fill deposit. Stratigraphic repetitions in the sequence were thought to represent slide masses (McKerrow and Cocks, 1978), folded thrust faults (e.g., Kay and Williams, 1963; Jacobi and Schweickert, 1976; Dean and Strong, 1977; Karlstrom and others, 1982), or complex dip-slip and transcurrent faults (e.g., Kay, 1969, 1970, 1976). The folded thrust hypothesis is most probable based on structural and stratigraphic relationships along the Cobbs Arm Fault Zone. Here, the Lower Silurian Rogers Cove melange crops out (formerly the Cobbs Arm melange, Jacobi and Schweickert, 1976). Reusch (1982) suggested that this melange represents debris at the toe of thrust slices that repeat the stratigraphic section. If this hypothesis is correct, the "Acadian" overthrusting occurred in Early Silurian times and was directed either toward the southeast (e.g., Dean and Strong, 1977) or the northwest (Karlstrom, personal communication, 1983).

Collectively, the following observations on the terrane north and west of the Dunnage melange suggest an accretionary wedge model (e.g., Seely and others, 1974) or perhaps backarc thrusting (Nelson, 1981). The Summerford Group contains repetitions of section, including dismembered oceanic crust (Jacobi and Wasowski, 1984). Basinal warping occurred synchronously with thrusting (e.g., Helwig, 1970). In addition, possible folded thrusts are present (Jacobi and Schweickert, 1976; Karlstrom and others, 1982), and the section occurs in proximity to an island arc. If Lower Ordovician eastward subduction produced a volcanic arc west of the Dunnage melange (e.g., Strong, 1977; Jacobi, 1981), then the thrusting (if southwestward) was perhaps initially backarc thrusting (Nelson, 1981). In this polarity model, it is difficult to determine whether the "backarc thrusting" developed into full-scale northwestward subduction, or whether the thrusting was actually northwestward subduction approximately synchronous with the eastward subduction. In contrast, if the overthrusting was northwest-directed, then southeastward Silurian subduction is perhaps indicated. In either case, Silurian deformation probably caused the major antiform and the associated refolded scaly cleavage in the Dunnage melange. It is not presently possible to determine whether an earlier phase of this deformation generated the scaly cleavage observed in the Dunnage melange.

CONCLUSIONS

Modern submarine sediment slides provide a model for the non-tectonic generation of melange. Submarine sediment slides are common on both active and passive continental margins. Single sediment slides can affect an area as large as Massachusetts on the continental slope and rise. Internal deformation observed in cores raised from the slide deposits include pebbly mud, isoclinal folds, and faults. Locally developed cleavage is observed in the slide deposits. Because of the ubiquity, immense size, and internal characteristics of modern submarine sediment slides, sediment slides provide an important mechanism for the fragmentation and mixing observed in melange. Because sediment slides are not restricted to convergence zones, the presence of a chaotic unit in the rock record does not necessarily imply a paleosubduction zone.

Criteria that might be utilized to determine the petrogenesis and tectonic setting of a particular melange include: (1) type of structural fabrics present, (2) clast composition and relation to structural fabrics, and (3) regional relationships, including axial flow of turbidity currents and regional tectonic framework. None of the criteria considered singly provides compelling evidence for the petrogenesis and tectonic setting of a particular melange. In fact, melange can form in so many tectonic settings that in some cases it may be fairly impossible to limit the possible tectonic settings of the melange to just one alternative.

The Ordovician Dunnage melange in north-central Newfoundland illustrates the problems inherent in the study of petrogenesis and tectonic setting of melange. Soft-sediment deformational structures in the Dunnage melange include pebbly mudstone, isoclinal folds without axial-planar cleavage, and perhaps a locally and weakly developed slaty cleavage. These characteristics suggest that the chaotic nature originated by sediment sliding, but tectonic deformation of unlithified, water-rich sediment may also result in these same structural features. If the sediment sliding alternative is correct, then the Dunnage melange probably represents several olistostromes. Prominent slaty and scaly cleavages overprint many of the soft-sediment structures, including isoclinal folds and pebbly mudstones. The scaly and slaty cleavage is axial-planar to folds involving argillite and the locally and weakly developed slaty cleavage. Orientations of graywacke clasts suggest that further fragmentation and mixing occurred during this folding and cleavage generation. The origin and age of the prominent, regional, scaly and slaty cleavages are equivocal. The cleavages may be tectonic and are early Acadian (Early-Middle Silurian?) or older. The scaly cleavage is refolded by later Acadian folds that are minor folds of a major antiform that accounts for the present map pattern of the Dunnage melange. Local spaced fracture cleavage and scaly cleavage were generated during this folding phase.

The tectonic setting of the Dunnage melange during its various stages of development is rather unclear. Clasts in the Dunnage melange that were derived from oceanic crust (probably an ocean island) and from a volcanic arc classify the melange as a Type II sedimentary melange, i.e., the olistostromes probably formed on an active (convergent) margin. There is little direct evidence to suggest on which side of the volcanic arc the olistostromes formed. However, by Acadian times, northwestward or southeastward underthrusting associated with an accretionary wedge and plate convergence caused the refolding event in the Dunnage melange. In Ordovician times, two alternative tectonic settings are possible. The Dunnage melange may have been a part of an accretionary wedge and convergence zone that continued into Silurian times (in which case the northwestward subduction was coeval with eastward subduction west of the Dunnage). The other alternative is that the melange was deposited and later thrust on the backarc side of the volcanic arc that was produced by eastward subduction. In the latter case, northwestward subduction either had not yet begun or was located south (?) of its present relative position.

ACKNOWLEDGMENTS

Supported in part by a Petroleum Research Fund Grant (#150-2765A). I thank the officers, crew, and scientists aboard the R/V "Vema" and the R/V "Robert D. Conrad" for their assistance in gathering the deep-sea data, and Tom Fargo, Jerome Janish, and Janusz Wasowski for their assistance in the field. I thank Drs. Flood and Damuth for reviewing this manuscript, and I thank Beth Goldsmith and Doris Millholland for typing and retyping the manuscript.

REFERENCES CITED

Abbott, D., Menke, W., Hobart, M., and Anderson, R., 1981, Evidence for excess pore pressures in southwest Indian Ocean sediments: Journal of Geophysical Research, v. 86, p. 1813–1827.

Almagor, G., and Garfunkel, Z., 1979, Submarine slumping in continental margin of Israel and northern Sinai: American Association of Petroleum Geologists Bulletin, v. 63, p. 324–340.

Alterman, I. B., 1973, Rotation and dewatering during slaty cleavage formation: Some new evidence and interpretations: Geology, v. 1, p. 33–36.

Bentz, F. P., 1974, Marine geology of the southern Lord Howe Rise, southwest Pacific, *in* Burk, C. A., and Drake, C. L., eds., The Geology of Continental Margins: New York, Springer-Verlag, p. 537–547.

Beutner, E. C., 1978, Slaty cleavage and related strain in Martinsburg slate, Delaware Water Gap, New Jersey: American Journal of Science, v. 278, p. 1–23.

Beutner, E. C., Jancin, M. D., and Simon, R. W., 1977, Dewatering origin of cleavage in light of deformed calcite veins and clastic dikes in Martinsburg slate, Delaware Water Gap, New Jersey: Geology, v. 5, p. 118–122.

Bird, J. M., and Dewey, J. F., 1970, Lithosphere plate-continental margin tectonics and the evolution of the Appalachian orogen: Geological Society of America Bulletin, v. 81, p. 1031–1060.

Blake, M. C., Jr., and Jones, D. L., 1974, Origin of Franciscan melanges in northern California, *in* Dott, R. H., Jr., and Shaver, R. H., eds., Modern and Ancient Geosynclinal Sedimentation: Society of Economic Paleontologist and Mineralogists, Special Publication 19, p. 345–357.

Carson, B., von Huene, R., and Arthur, M., 1982, Small-scale deformation structures and physical properties related to convergence in Japan Trench slope sediments: Tectonics, v. 1, p. 277–302.

Chamberlain, T. K., 1964, Mass transport of sediment in the heads of Scripps submarine canyon, California, *in* Miller, R. L., ed., Papers in Marine Geology: New York, The Macmillan Company, p. 42–64.

Corbett, K. D. , 1973, Open-cast slump sheets and their relationship to sandstone beds in an Upper Cambrian flysch sequence, Tasmania: Journal of Sedimentary Petrology, v. 43, p. 147–159.

Coulbourn, W. T., and Moberly, R., 1977, Structural evidence of the evolution of fore-arc basins off South America: Canadian Journal of Earth Science, v. 14, p. 102–106.

Coulter, H. W., and Migliaccio, R. R., 1966, Effects of the earthquake of March 27, 1964, at Valdez, Alaska: United States Geological Survey, Professional Paper 542-C, 36 p.

Cowan, D. S., 1974, Deformation and metamorphism of the Franciscan subduction zone complex northwest of Pacheco Pass, California: Geological Society of America Bulletin, v. 85, p. 1623–1634.

Cowan, D. S., 1978, Origin of blueschist-bearing chaotic rocks in the Franciscan Complex, San Simeon, California: Geological Society of America Bulletin, v. 89, p. 1415–1423.

Cowan, D. S., and Page, B. M., 1975, Recycled Franciscan material in Franciscan Melange west of Paso Robles, California: Geological Society of America Bulletin, v. 86, p. 1089–1095.

Crawford, A. J., Beccaluva, L., and Serri, G., 1981, Tectono-magmatic evolution of the West Philippine-Mariana region and the origin of boninites: Earth and Planetary Science Letters, v. 54, p. 346–356.

Damuth, J. E., 1978, Echo character of the Norwegian-Greenland Sea: Relationship to Quaternary sedimentation: Marine Geology, v. 28, p. 1–36.

Damuth, J. E., 1979, Migrating sediment waves created by turbidity currents in the northern South China Basin: Geology, v. 7, p. 520–523.

Damuth, J. E., 1980a, Quaternary sedimentation processes in the South China Basin as revealed by echo-character mapping and piston core studies, *in* Hayes, D. E., ed., The Tectonic and Geologic Evolution of Southeast Asian Seas and Islands: American Geophysical Union, Mon. 23, p. 105–125.

Damuth, J. E., 1980b, Use of high-frequency (3.5-12 kHz) echograms in the study of near-bottom sedimentation processes in the deep sea: A review: Marine Geology, v. 38, p. 51–75.

Damuth, J. E., Jacobi, R. D., and Hayes, D. E., 1983, Sedimentation processes in the northwest Pacific Basin revealed by echo character mapping studies: Geological Society of America Bulletin, v. 94, p. 381–395.

Dean, P. L., and Strong, D. F., 1977, Folded thrust faults in Notre Dame Bay, central Newfoundland: American Journal of Science, v. 277, p. 97–108.

Diester-Hass, L., 1977, Influence of carbonate dissolution, climate, sea-level changes, and volcanism on Neogene sediments off northwest Africa (Leg 41), *in* Lancelot, Y., Seibold, E., and others, Initial Reports of the Deep Sea Drilling Project, v. 41: Washington, United States Government Printing Office, p. 1033–1047.

Dill, R. F., 1964, Sedimentation and erosion in Scripps submarine canyon head, *in* Miller, R. L., ed., Papers in Marine Geology: New York, The Macmillan Company, p. 23–41.

Dott, R. H., 1963, Dynamics of subaqueous gravity depositional processes: American Association of Petroleum Geologists Bulletin, v. 47, p. 104–108.

Drake, C. L., Campbell, N. J., Sander, G., and Nafe, J. E., 1963, A mid-Labrador sea ridge: Nature, v. 200, p. 1085–1086.

Embley, R. W., 1975, Studies of deep sea sedimentation processes using high frequency seismic data (Ph.D. thesis): New York, Columbia University.

Embley, R. W., 1976, New evidence for occurrence of debris flow deposits in the deep sea: Geology, v. 4, p. 371–374.

Embley, R. W., 1980, The role of mass transport in the distribution and character of deep-ocean sediments with special reference to the North Atlantic: Marine Geology, v. 38, p. 23–50.

Embley, R. W., and Jacobi, R. D., 1977, Distribution and morphology or large submarine slides and slumps on Atlantic continental margins: Marine Geotechnology, v. 2, p. 205–228.

Fargo, T. R., 1983, A re-evaluation of the Loon Harbour Volcanics of the Campbellton sequence, north-central Newfoundland: Geological Society of America, Abstracts with Programs, v. 15, n. 3, p. 189.

Farre, J. A., McGregor, B. A., Ryan, W.B.F., and Robb, J. M., 1983, Breaching the shelf break: Passage from youthful to mature phase in submarine canyon evolution: Society of Economic Mineralogists and Paleontologists, Special Publication 33, p. 25–39.

Ferentinos, G., Brooks, M., and Collins, M., 1981, Gravity-induced deformation on the north flank and floor of the Sporadhes Basin of the North Aegean Sea Trough: Marine Geology, v. 44, p. 289–302.

Flood, R. D., Hollister, C. D., and Lonsdale, P., 1979, Disruption of the Feni sediment drift by debris flows from Rockall Bank: Marine Geology, v. 32, p. 311–343.

Geiser, P. A., 1975, Slaty cleavage and the dewatering hypothesis—an examination of some critical evidence: Geology, v. 3, p. 717–720.

Got, H., Stanley, D. J., and Sorel, D., 1977, Northwestern Hellenic Arc: Concurrent sedimentation and deformation in a compressive setting: Marine Geology, v. 24, p. 21–36.

Gray, D. R., 1979, Microstructure of crenulation cleavages: An indicator of cleavage origin: American Journal of Science, v. 279, p. 97–128.

Groshong, R. H., Jr., 1976, Strain and pressure solution in the Martinsburg Slate, Delaware Water Gap, New Jersey: American Journal of Science, v. 276, p. 1131–1146.

Hampton, M. A., 1972, The role of subaqueous debris flow in generating turbidity currents: Journal of Sedimentary Petrology, v. 42, p. 775–793.

Hawkins, J., Bloomer, S., Evans, C., and Melchior, J., 1979, Mariana arc-trench system: Petrology of the Inner Trench Wall: EOS (American Geophysical Union Transactions), v. 60, no. 46, p. 968.

Heezen, B. C., and Ewing, M., 1962, Turbidity currents and submarine slumps, and the 1929 Grand Banks earthquake: American Journal of Science, v. 250, p. 849–873.

Helwig, J., 1979, Slump folds and early structures, northeastern Newfoundland Appalachians: Journal of Geology, v. 78, p. 172–187.

Hibbard, J. P., 1976, The southwest portion of the Dunnage Melange and its relationships to nearby groups (M.S. thesis): St. John's, Newfoundland,

Memorial University of Newfoundland, 131 p.

Hibbard, J. P., Stouge, S., and Skevington, D., 1977, Fossils from the Dunnage Melange, north-central Newfoundland: Canadian Journal of Earth Science, v. 14, p. 1176–1178.

Hibbard, J., and Williams, H., 1979, Regional setting of the Dunnage melange in the Newfoundland Appalachians: American Journal of Science, v. 279, p. 993–1021.

Horne, G. S., 1969, Early Ordovician chaotic deposits in the central volcanic belt of northeastern Newfoundland: Geological Society of America Bulletin, v. 80, p. 2451–2464.

Horne, G. S., 1970, Complex volcanic-sedimentary patterns in the Magog Belt of northeastern Newfoundland: Geological Society of America Bulletin, v. 81, p. 1767 1788.

Houtz, R. E., 1962, The 1953 Suva earthquake and tsunami: Bulletin of the Seismological Society of America, v. 52, p. 1–12.

Hsu, K. J., 1968, Principles of melanges and their bearing on the Franciscan-Knoxville paradox: Geological Society of America Bulletin, v. 79, p. 1063–1074.

Hsu, K. J., 1969, Preliminary report and geologic guide to the Franciscan melanges of the Morro Bay-San Simeon area, California: California Division of Mines and Geology, Special Publication 35, 46 p.

Hsu, K. J., 1971, Franciscan melanges as a model for eugeosynclinal sedimentation and underthrusting tectonics: Journal of Geophysical Research, v. 76, p. 1162–1170.

Hsu, K. J., 1973, Mesozoic evolution of the California Coast Ranges: A second look, *in* DeJong, K. A., and Scholten, R., eds., Gravity and Tectonics: New York, John Wiley & Sons, p. 379–396.

Hsu, K. J., 1974, Melanges and their distinction from olistostromes, *in* Dott, R. H., and Shaver, R. H., eds., Modern and Ancient Geosynclinal Sedimentation: Society of Economic Paleontologists and Mineralogists, Special Publication 19, p. 321–333.

Jacobi, R. D., 1980, Geology of part of the terrane north of Lukes Arm fault, north-central Newfoundland (Part I): Modern submarine slides and their geological implications (Part II) (Ph.D. thesis): New York, Columbia University, 422 p.

Jacobi, R. D., 1982, Microphysiography of the southeastern North Atlantic and its implications for the distribution of near-bottom processes and related facies: Bordeaux, Bulletin Institute Geologic Bassin d'Aquitaine, n. 31, p. 31–46.

Jacobi, R. D., Damuth, J. E., and Hayes, D. E., 1980, Evaluation of the North Pacific subseabed for disposal of high level nuclear waste: II, Analysis of geologic data in the Pac 1 area, northwest Pacific to locate potential candidate disposal sites [abs.]: EOS (American Geophysical Union Transactions), v. 61, p. 988.

Jacobi, R. D. and Hayes, D. E., 1982, Bathymetry, microphysiograhy and reflectivity characteristics of the West African margin between Sierra Leone and Mauritania, *in* vonRad, U., Hinz, K., Sarnthein, M. and Seibold, E., eds., Geology of the Northwest African Continental Margin: Berlin, Springer-Verlag, p. 182–212.

Jacobi, R. D. and Hayes, D. E., 1984, Echo character, microphysiography and geologic hazards, *in* Rabinowitz and Hayes, eds., OMD Regional Data Synthesis Series Atlas XI (Morocco margin), in press.

Jacobi, R. D. and Mrozowski, C. L., 1979, Sediment slides and sediment waves in the Bonin Trough, western Pacific: Marine Geology, v 29, p. M1-M9.

Jacobi, R. D. and Schweickert, R. A., 1976, Implications of new data on stratigraphic and structural relations of Ordovician rocks on New World Island, north-central Newfoundland [abs.]: Geological Society of America Abstracts with Programs, v. 8, no. 2, p. 206.

Jacobi, R. D. and Wasowski, J. J., 1984, Geochemistry and plate-tectonic significance of the volcanic rocks of the Sommerford Group, north-central Newfoundland: Geology, in press.

Janish, J. K., and Jacobi, R. D., 1982, Nature, origin, and significance of the Cheneyville Formation, south coast of New World Island, Newfoundland: Geological Society of America, Abstracts with Programs, v. 14, p. 29.

Karlstrom, K. E., van der Pluijm, B. A., and Williams, P. F., 1982, Structural interpretation of the eastern Notre Dame Bay area, Newfoundland: Regional post-Middle Silurian thrusting and asymmetrical folding: Canadian Journal of Earth Science, v. 19, p. 2325–2341.

Kastens, K. A., 1981, Structural causes and sedimentological effects of "cobblestone topography" in the eastern Mediterranean (Ph.D. thesis): San Diego, University of California at San Diego, 207 p.

Kay, G. M., 1967, Stratigraphy and structure of northeastern Newfoundland bearing on drift in North Atlantic: America Association of Petroleum Geologists Bulletin, v. 51, p. 579–600.

Kay, M., 1970, Flysch and bouldery mudstone in northeast Newfoundland: Geological Association of Canada, Special Paper No. 7, p. 155–164.

Kay, M., 1976, Dunnage melange and subduction of Protacadic Ocean, northeast Newfoundland: Geological Society of America, Special Paper 175, 49 p.

Kay, M., and Eldredge, N., 1968, Cambrian trilobites in central Newfoundland volcanic belt: Geological Magazine, v. 105, p. 373–377.

Kay, M., and Williams, H., 1963, Ordovician-Silurian relations on New World Island, Notre Dame Bay, northeast Newfoundland: Geological Society of America Bulletin, v. 74, p. 807–808.

Kenyon, N. H., Belderson, R. H., and Stride, A. H., 1978, Channels, canyons, and slump folds on the continental slope between south-west Ireland and Spain: Oceanologica Acta, v. 1, p. 369–380.

Knipe, R. J., and White, S. H., 1977, Microstructural variation of an axial plane cleavage around a fold—a H.V.E.M. study: Tectonophysics, v. 39, p. 355–380.

Lancelot, Y., Seibold, E., Dean, W. E., Jansa, L. F., Eremeev, V., Gardner, J., Cepek, P., Krasheninnikov, V. A., Pflavmann, V., Johnson, D., Rankin, J. G., Trabant, P. et al., 1977, Initial Reports of the Deep Sea Drilling Project, Volume 41: Washington, United States Government Printing Office, 1259 p.

Lash, G. G., 1980, Structural geology and stratigraphy of the allochthonous and autochthonous rocks of the Kutztown and Hamburg 7½ minute quadrangles, eastern Pennsylvania (unpublished Ph.D. thesis): Bethlehem, PA, Lehigh University, 240 p.

Leggett, J. K., 1980, The sedimentological evolution of a Lower Palaeozoic accretionary fore-arc in the Southern Uplands of Scotland: Sedimentology, v. 27, p. 401–417.

Leggett, J. K., McKerrow, W. S., and Eales, M. H., 1979, The Southern Uplands of Scotland. A Lower Palaeozoic accretionary prism: Journal of the Geological Society of London, v. 136, p. 755–770.

Lewis, K. B., 1971, Slumping on a continental slope inclined at 1°-4°: Sedimentology, v. 16, p. 97–110.

Lonsdale, P., 1975, Sedimentation and tectonic modification of Samoan Archipelagic apron: American Association of Petroleum Geologists Bulletin, v. 59, p. 780–798.

Lundberg, N., and Moore, J. C., 1981, Structural features of the Middle American Trench slope off southern Mexico, Deep Sea Drilling Project Leg 66, *in* Watkins, J. S., Moore, J. C., et al., Initial Reports DSDP, 66: Washington, United States Government Printing Office, p. 793–814.

McGregor, B. A., and Bennett, R. H., 1979, Mass movement of sediment on the continental slope seaward of the Baltimore Canyon Trough: Marine Geology, v. 33, p. 163–174.

McKerrow, W. S., and Cocks, L.R.M., 1978, A lower Paleozoic trench-fill sequence, New World Island, Newfoundland: Geological Society of America Bulletin, v. 89, p. 1121–1132.

Maxwell, J. C., 1962, Origin of slaty and fracture cleavage in the Delaware Water Gap area, New Jersey and Pennsylvania, *in* Engle, A.E.J., James, H. J., and Leonard, B. F., eds., Petrologic Studies: A Volume in Honor of A. F. Buddington: Geological Society of America, New York, p. 281–311.

Middleton, G. V., and Hampton, M. A., 1973, Sediment gravity flows: Mechanics of flow and deposition, *in* Middleton, G., and Brown, A., eds., Turbidites and Deep Water Sedimentation: Society of Economic Paleontologists and Mineralogists Short Course, p. 1–38.

Mitchell, A. H., and Reading, H. G., 1971, Evolution of island arcs: Journal of Geology, v. 79, p. 253–284.

Moore, D. G., Curray, J. R., and Emmel, J. F., 1976, Large submarine slide (olistostrome) associated with Sunda Arc subduction zone, northeast Indian Ocean: Marine Geology, v. 21, p. 211–226.

Moore, J. C., and Allwardt, A., 1980, Progressive deformation of a Tertiary trench slope, Kodiak Islands, Alaska: Journal of Geophysical Research, v. 85, p. 4741–4756.

Moore, J. C., and Geigle, J. E., 1974, Slaty cleavage: Incipient occurrences in the deep sea: Science, v. 183, p. 509–510.

Morgenstern, N. R., 1967, Submarine slumping and the initiation of turbidity currents, *in* Richards, A. F. ed., Marine Geotechnique: Urbana, Illinois, University of Illinois, p. 189–220.

Moseley, F., and Abbotts, I. L., 1979, The ophiolite melange of Masirah, Oman: Journal of Geological Society of London, v. 136, p. 713–724.

Nelson, K. D., 1981, Melange development in the Boones Point Complex, north-central Newfoundland: Canadian Journal of Earth Science, v. 18, p. 433–442.

Normark, W. R., 1974, Ranger submarine slide, northern Sebastian Vizcaino Bay, Baja California, Mexico: Geological Society of America Bulletin, v. 85, p. 781–784.

Pearce, J. A., and Norry, M. J., 1979, Petrogenic implications of Ti, Zr, Y, and Nb variations in volcanic rocks: Contributions to Mineralogy and Petrology, v. 69, p. 33–47.

Piper, D.J.W., 1972, Trench sedimentation in the Lower Palaeozoic of the Southern Uplands: Scottish Journal of Geology, v. 8, p. 289–291.

Piper, D.J.W., von Huene, R., Duncan, J. R., 1973, Late Quaternary sedimentation in the active eastern Aleutian Trench: Geology, v. 1, p. 19–22.

Popenoe, P., Coward, E. L., and Cashman, K. V., 1982, A regional assessment of potential environmental hazards to and limitations on petroleum development of the southeastern United States Atlantic continental shelf, slope, and rise, offshore North Carolina: United States Geological Survey Open File Report 82-136, 67 p.

Raymond, L. A., Classification of melanges (this volume).

Reusch, D., 1982, Structural and stratigraphic relationships along the Cobbs Arm volcanic belt, New World Island [abs.]: Geological Society of America Abstracts with Programs, v. 14, p. 76–77.

Ross, D. A., and Shor, G. G., Jr., 1965, Reflection profiles across the Middle America Trench: Journal of Geophysical Research, v. 70, p. 5551–5572.

Rowley, D. B., 1982, Nature of pre-Caradocian unconformity on the Cambro-Ordovician shelf [abs.]: Geological Society of America Abstracts with Programs, v. 14, p. 78.

Ryan, W.B.F., von Rad, U., Sibuet, J. C., and scientific party, 1976, Passive continental margin: Geotimes, v. 21, p. 21–24.

Saleeby, J. B., 1982, Polygenetic ophiolite belt of the California Sierra Nevada; Geochronological and tectonostratigraphic development: Journal of Geophysical Research, v. 87, p. 1803–1824.

Schweller, W. J., Karig, D. E., and Backman, S. B., 1983, Original setting and emplacement history of the Zambales ophiolite, Luzon, Philippines, from stratigraphic evidence, *in* Hayes, D. E., ed., The Tectonic and Geologic Evolution of Southeast Asian Seas and Islands: American Geophysical Union, Geophysical monograph 27, p. 95–138.

Seely, D. R., Vail, P. R., and Walton, G. G., 1974, Trench slope model, *in* Burk, C. A., and Drake, C. L., eds., The Geology of Continental Margins: New York, Springer-Verlag, p. 249–260.

Sheridan, R. E., Pastovret, L., and Mosditchian, G., 1978, Seismic stratigraphy and related lithofacies of the Blake-Bahama Basin, *in* Bensen, W. E., Sheridan, R. E., et al., eds., Initial Reports of the Deep Sea Drilling Project, v. 44: Washington, United States Government Printing Office, p. 529–546.

Silver, E. A., and Beutner, E. C., 1980, Melanges: Geology, v. 8, p. 32–34.

Sorby, H. C., 1853, On the origin of slaty cleavage: New Philosophic Journal of Edinburgh, v. 55, p. 137–148.

Stanley, D. J., 1974, Pebbly mud transport in the head of Wilmington Canyon: Marine Geology, v. 16, p. M1-M8.

Stow, D.A.V., 1979, Distinguishing between fine-grained turbidites and contourites on the Nova Scotian deep water margin: Sedimentology, v. 26, p. 371–387.

Strong, D. F., 1977, Volcanic regimes of the Newfoundland Appalachians, *in* Barager, W.R.A., Coleman, L. C., and Hall, J. M., eds., Volcanic Regimes in Canada: Geological Association of Canada, Special Paper 16, p. 61–90.

Stubblefield, W. L., McGregor, B. A., Forde, E. B., Lambert, D. N., and Merrill, G. F., 1982, Reconnaissance in DSRV *Alvin* of a "fluvial-like" meander system in Wilmington Canyon and slump features in South Wilmington Canyon: Geology, v. 10, p. 31–36.

Tyler, J. H., 1972, Pigeon Point Formation: An Upper Cretaceous shoreline succession, central California Coast: Journal of Sedimentary Petrology, v. 42, p. 537–557.

Vassallo, K., Jacobi, R. D., and Shor, A., 1983, Echo character, microphysiography and geologic hazards, *in* Ewing, J. I. and Rabinowitz, P. O., eds., OMD Regional Data Synthesis Series Atlas 1 (Eastern North American Continental Margin and Adjacent Ocean Floor, 28° to 36°N and 70° to 82°W): Washington, JOI, Inc., map #40.

von Huene, R., 1974, Modern trench sediments, *in* Burk, C. A., and Drake, C. L., eds., The Geology of Continental Margins: New York, Springer-Verlag, p. 207–211.

Wasowski, J., and Jacobi, R. D., 1982, Nature and significance of igneous-clast breccias and adjacent units in the Dunnage melange, north-central Newfoundland [abs.]: Geological Society of America Abstracts with Programs, v. 14, p. 94.

Wasowski, J., and Jacobi, R. D., 1983, Geochemistry and tectonic significance of mafic volcanic blocks in the Dunnage melange, north-central Newfoundland [abs.]: Geological Society of America Abstracts with Programs, v. 15, p. 189.

Wasowski, J., and Jacobi, R. D., 1984, Geochemistry and tectonic significance of mafic volcanic blocks in the Dunnage melange, north-central Newfoundland: Canadian Journal of Earth Science, in press.

Williams, H., 1978, Geological development of the northern Appalachians: Its bearing on the evolution of the British Isles, *in* Bowes, D. R., and Leake, B. E., eds., Crustal Evolution in Northwestern Britain and Adjacent Regions: Liverpool, United Kingdom, Seal House Press, Geological Journal Special Issue, no. 10, p. 1–22.

Williams, P. F., Collins, A. R., and Wiltshire, R. G., 1969, Cleavage and pencontemporaneous deformation structures in sedimentary rocks: Journal of Geology, v. 77, p. 415–425.

Wood, D. S., 1974, Ophiolites, melanges, blueschists, and ignimbrites: early Caledonian subduction in Wales? *in* Dott, R. H., and Shaver, R. H., eds., Modern and Ancient Geosynoclinal Sedimentation: Society of Economic Paleontologists and Mineralogists, Special Publication 19, p. 334–344.

Woodcock, N. H., 1979, Sizes of submarine slides and their significance: Journal of Structural Geology, v. 1, p. 137–142.

Woodland, B. G., 1982, Gradational development of domainal slaty cleavage, its origin and relation to chlorite porphyroblasts in the Martinsburg Formation, eastern Pennsylvania: Tectonophysics, v. 82, p. 89–124.

Manuscript Accepted by the Society July 12, 1984

Printed in U.S.A.

Geological Society of America
Special Paper 198
1984

Ophiolitic olistostromes in the basal Great Valley sequence, Napa County, northern California Coast Ranges

*Stephen Paul Phipps**
Department of Geological and Geophysical Sciences
Princeton University
Princeton, New Jersey 08544

ABSTRACT

The basal Great Valley sequence in Napa and southern Lake Counties, California, is a mappable chaotic unit composed largely of ophiolitic debris. Serpentinite flows and breccias, mafic breccias and associated finer-grained clastic rocks, and blocks of extrusive greenstone, mafic breccia, chert, bedded and unbedded clastic sedimentary rocks, phyllites, actinolitic greenschists, and hornblende amphibolites are mixed with Great Valley sequence mudstone and serpentinous mudstone. The chaotic unit extends along strike for at least 50 km. Cross-sections indicate that it extends for at least 20 km across strike and is up to 1 km thick. It is involved in complex folds caused by imbricate thrust faulting. The unit lies directly above the serpentinite that represents the Coast Range Ophiolite within the study area and below the well-bedded Great Valley sequence of Upper Jurassic and Cretaceous age. Its lower contact is enigmatic but is probably depositional; the upper contact is sheared and gradational. Locally the unit represents the entire Tithonian Stage.

Ophiolitic detritus in the lower Great Valley sequence is also found elsewhere in the Northern California Coast Ranges—near the Geysers, in Rice Valley, near Wilbur Springs, along the Bartlett Springs Road near Walker Ridge, near Cooks Springs, and at Crowfoot Point west of Paskenta. Other accumulations of ophiolitic debris are interlayered in the Great Valley sequence at various stratigraphic levels in and near the study area. This detritus takes four forms, which may be mixed together: (1) sedimentary serpentinite debris flows; (2) mafic breccias; (3) basaltic sandstones; and (4) polymict, polymorphous chaotic units with blocks-in-matrix texture, like the rocks in Napa County described here.

Widespread detrital textures and the common occurrence of spaced, rather than penetrative, shear foliation in its matrix demonstrate that the chaotic unit in Napa County is not a tectonic melange, and I interpret it to be an amalgam of olistostromes. These ophiolitic olistostromes are a facies distinct from the overlying turbidites. Thus, the basal Great Valley sequence in this area is composed of two different rock types: very proximal ophiolitic debris flows, and substantially more distal subsea-fan rocks derived from a volcanic arc.

Ophiolites may form at mid-ocean ridges, in back-arc or forearc basins, or in island arcs. Ophiolitic detritus may be eroded and deposited on ophiolitic basement in any environment in which the oceanic crust is deformed. The geology of surrounding terranes and the petrologic features of the ophiolitic basement below the Great Valley sequence suggest that the basement was formed in a back-arc basin. The stratigraphy of

*Present address: Geology Department, Hayden Hall, The University of Pennsylvania, Philadelphia, Pennsylvania 19104.

the chaotic rocks that overlie the basement in Napa County suggests that they were deposited on deeply eroded basement in a tectonically active forearc basin. Large volumes of rock stuffed under the hanging-wall slab after the onset of subduction may have uplifted the forearc basin and subjected its basement to erosion. A wave of uplift may have passed across the basin, so that debris shed from eroding oceanic basement was deposited directly on freshly exposed harzburgite tectonite. Some blocks may have been carried completely across the forearc basin and into the trench, and incorporated into the Franciscan melange wedge, which is also rich in ophiolitic blocks. The change from back-arc to forearc basin was probably caused by collisional tectonics and the establishment of a new subduction zone off the western coast of California during the Late Jurassic Nevadan orogeny.

Stratigraphic relationships in and above the Coast Range Ophiolite are unusual through much of the Northern Coast Ranges. Nearly complete ophiolites are the exception rather than the rule, and in many areas only serpentinite is present. In some areas, ophiolitic debris different from that described here overlies the serpentinite. In other areas, arc-derived submarine fan rocks of the Great Valley sequence directly overlie serpentinized harzburgite tectonite. The relationships described here suggest that many of these contacts are not tectonic, and that the Coast Range Ophiolite does not owe its fragmentary nature to tectonic dismemberment. Rather, it is likely that the ophiolitic basement below the Great Valley sequence was deeply eroded during Mesozoic time. Many of the contacts throughout the Coast Ranges along which sedimentary rocks overlie serpentinite—which must represent deep layers of the oceanic crust or the upper mantle—are in the main nonconformities.

INTRODUCTION

This paper describes an extensive, mappable chaotic unit, composed largely of ophiolitic debris, which lies in the basal Great Valley sequence in Napa and southern Lake Counties in the Northern California Coast Ranges (map, Figure 1). It then discusses a number of environments in which ophiolites may be formed and in which ophiolitic detritus may be formed overlying the oceanic crust. Finally, it interprets the Napa County rocks in terms of these environments.

This study is an outgrowth of a continuing regional structural mapping project begun in 1976. In the course of the work, an area of 500 square kilometers was mapped at a scale of 1:24,000. This area includes the following 7½′ quadrangles: all of Aetna Springs and Walter Springs, most of Knoxville and Jericho Valley, and the northern portions of St. Helena and Chiles Valley.

PREVIOUS STUDIES

Portions of the study area and adjacent areas have been mapped by Averitt (1945), Taliaferro and his students (Taliaferro, 1943; Angel, 1948; Borglin, 1949; Clark, 1948; Conrey, 1948; Shouldice, 1947), Brice (1953), Lawton (1956), Swe and Dickinson (1970), and Wagner (1975). Several of these workers observed the chaotic rocks described here but misinterpreted them. In particular, Taliaferro (1943) and his students believed that igneous rocks within the chaotic unit were intrusions and interbedded lava flows in a stratigraphic unit transitional between the Franciscan Complex and the Great Valley sequence.

Previous workers have noted the existence of ophiolitic debris of various sorts elsewhere in the basal Great Valley sequence (Brice, 1953; Lawton, 1956; Moiseyev, 1966, 1970; Rich, 1968; Bailey and others, 1970; Maxwell, 1974; Raney, 1976; McLaughlin, 1977, 1981; Etter, 1979; Carlson, 1981a, 1981b). Although many of these other occurrences are similar in one aspect or another, the rocks I describe here are unique because they combine all of the following features: great lithologic heterogeneity; large extent along strike (50 km); and substantial thickness (1 km or more).

ROCK UNITS

Three rock units dominate the outcrop pattern of the study area, as well as of the Northern Coast Ranges as a whole. The Great Valley sequence, largely a subsea-fan complex of Late Jurassic and Cretaceous age derived from the Sierran volcanic and plutonic arc, lies on ophiolitic basement, which consists almost entirely of serpentinite in most localities. These two rock bodies have been superposed along a regional thrust fault over the coeval Franciscan Complex, an extremely heterogeneous mass of rocks that formed in and near a subduction zone. Within the study area, this regional thrust fault has been repeatedly cut and folded by a complex system of much younger thrusts, resulting in an outcrop pattern of anastomosing thrust faults and plunging anticlines and synclines strongly reminiscent of the foreland overthrust belts of compressional orogens (Suppe and Foland, 1978; Suppe, 1979a, 1979b; Phipps, 1981, 1983). The area is now beginning to undergo strike-slip deformation on faults of the San Andreas system.

In the Coast Ranges generally, the Franciscan Complex con-

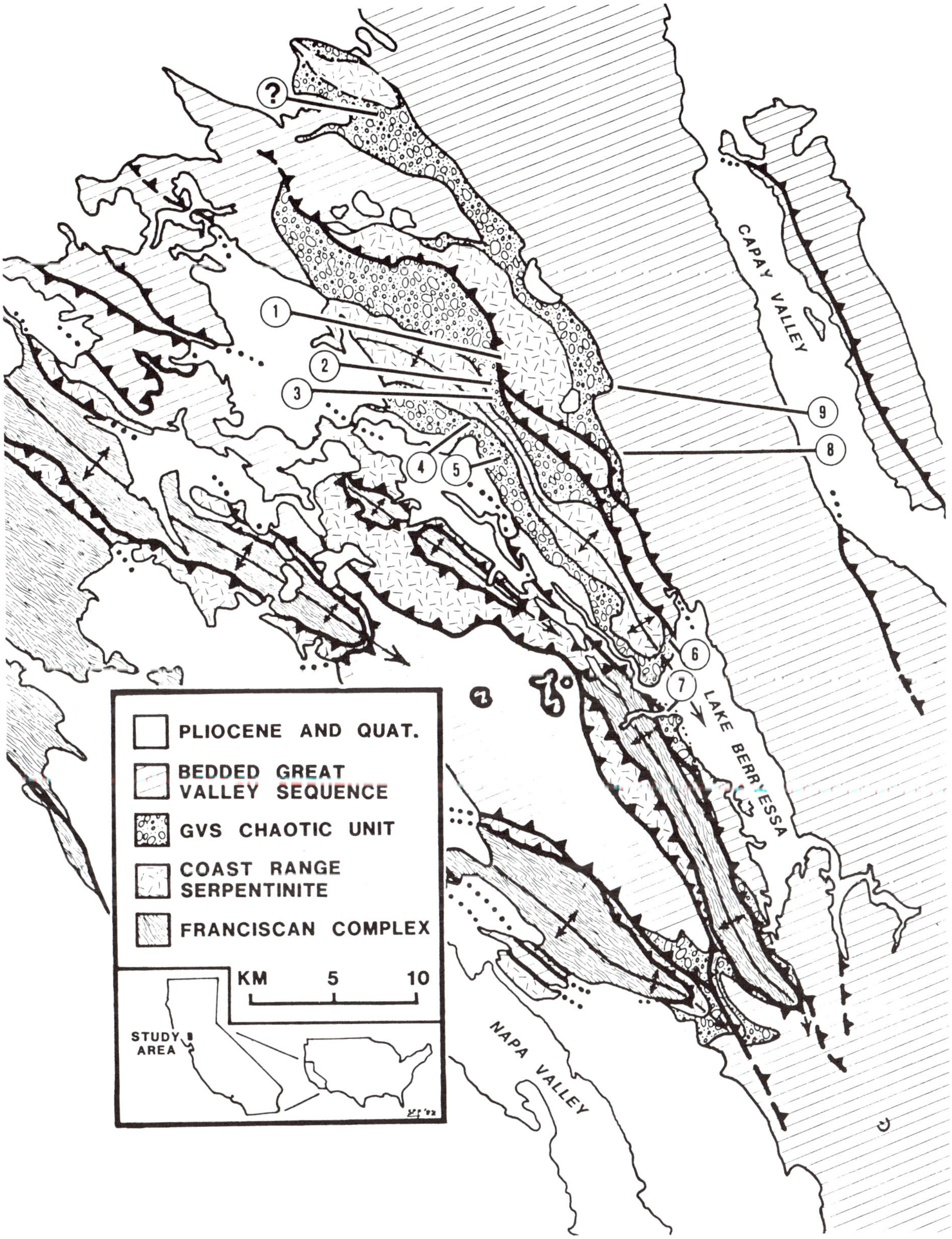

Figure 1. Geologic sketch map of Napa County study area. Circled numbers mark localities referred to in text; detailed locations are given in Appendix 1. Light contacts are depositional or intrusive; heavy contacts are faults; faults with sawteeth (on hanging wall block) are thrusts. Query denotes reconnaissance identification. North is at top of map.

sists of four rock types: (1) coherent, well-recrystallized blueschist formations; (2) coherent metagraywacke units, often poorly recrystallized, with well-preserved sedimentary textures; (3) serpentinite- and mudstone-matrix melanges, containing blueschist and eclogite blocks and showing pervasive intense shearing; and (4) younger (Early and mid-Cenozoic) and less metamorphosed broken formations (the Coastal Belt) (Bailey and others, 1964; Maxwell, 1974). These categories are not clear-cut; where metamorphic grade decreases and shearing and stratal disruption increase, coherent metagraywacke grades into broken formation; where igneous and metamorphic blocks become more common, broken formation grades into melange. Where recrystallization increases, coherent metagraywacke may grade into coherent blueschist.

In the study area, the Franciscan includes all types except the fourth (Coastal Belt). In particular, both serpentinite- and mudstone-matrix melanges are present. Franciscan rocks crop out principally in the cores of anticlines, where they appear repeatedly in windows through folded thrust faults previously grouped together as the "Coast Range Thrust" (Figure 1).

The serpentinite that represents the Coast Range Ophiolite in the study area consists of moderately- to highly-serpentinized (50-100%) peridotite, which is largely harzburgite tectonite, and dunite. It is similar to that found elsewhere in the Coast Ranges (Loney and others, 1971). The serpentinite is cut by local mafic dikes and sills (for instance, at Locality 1, Figure 1 and Appendix 1), and by stock-like and plug-like diabase intrusions, typically a few meters across. These intrusions do not cut the overlying chaotic rocks or the bedded Great Valley sequence.

Serpentinite crops out in several thrust sheets. The easternmost is a sheet that, with the overlying Great Valley sequence, dips eastward beneath the Sacramento Valley, forming the Great Valley sequence homocline. Other sheets are folded over the crests of Franciscan-cored anticlines and are thrust out over both Great Valley sequence and Franciscan rocks (Figure 1).

In general, the Coast Range Ophiolite is characterized by incompleteness and extreme lateral variability. Its mafic part is notably underrepresented, and the "ophiolite" in many places consists merely of serpentinite, as in the study area. Elsewhere in the Northern Coast Ranges, the ophiolite varies from a thin curtain of serpentinite directly overlain by bedded, arc-derived Great Valley sequence sedimentary rocks, to a fairly well developed ophiolite with a somewhat thin plutonic section (locally quite felsic), to an ophiolitic tectonic melange with a serpentinite matrix (Bailey and others, 1970; Hopson and others, 1981; and my own reconnaissance study).

A widely distributed, mappable ophiolitic chaotic unit overlies the serpentinite of the Coast Range Ophiolite and lies within the basal Great Valley sequence in the study area. The ophiolitic chaotic rocks are directly overlain by well bedded Upper Jurassic or basal Lower Cretaceous Great Valley sequence rocks. Most of these are fine-grained, thin-bedded mudstones and massive chert-pebble sandstones and conglomerates. The brown to gray mudstones locally contain a few laminated sandstone interbeds up to 10 cm thick, with sharp tops and bottoms. Higher in the Lower Cretaceous section, the mudstone contains interbeds to 1 m thick, which show Bouma sequences with base cut-out. A Lower Cretaceous boulder- and cobble- conglomerate rich in felsic plutonic and feldspar porphyry clasts is locally present. Upper Jurassic and Lower Cretaceous sandstones are lithic arenites. Still higher in the section, probable Upper Cretaceous sandstones are mostly arkosic arenites.

The Great Valley sequence rocks are dominantly turbidites and can be related well to sedimentologic models of subsea fans (Ingersoll, 1978a). Upper Jurassic mudstones, conglomerates, and laminated sandstones probably represent interchannel, channel, and levee deposits, respectively, on the slope and upper fan. Lower Cretaceous mudstones and sandstones were laid down after rapid fan regression and represent distal turbidites of the basin plain and lower fan. This Great Valley sequence stratigraphy is similar to that elsewhere in the Northern Coast Ranges (Ingersoll, 1976, 1978a, 1978b, 1979; Ingersoll and Dickinson, 1981). Bedded Great Valley sequence rocks crop out in the homocline along the eastern edge of the study area and in several structurally deeper thrust sheets to the west (Figure 1).

GENERAL FEATURES OF THE CHAOTIC UNIT

The chaotic unit in the basal Great Valley sequence is a polymict, polymorphous accumulation of ophiolitic detritus. It consists of ophiolitic blocks, flows, and slide masses ranging from sand-sized grains to blocks and elongate masses as much as 2 km long, set in a fine-grained (mud-sized) matrix, usually sheared, consisting of serpentinite and clay.

The most striking facts about the chaotic unit are as follows.

1. It extends for more than 50 km along strike and makes up much of the Upper Jurassic Great Valley sequence throughout Napa County and southern Lake County (Figure 1). Cross-sections indicate that it is typically about 1 km thick and that its lateral extent across strike before the late thrust-faulting was at least 20 km. Laterally, it is very continuous.

2. Block lithologies are dominantly ophiolitic, with greenstone, mafic breccia, and chert blocks found in approximately a 3:1:1 ratio, respectively, based on a count of 167 blocks. All other blocks except serpentinite, when summed, would add to approximately 1 on this scale. Sandstone blocks are rare.

3. The unit contains samples from all parts of an ophiolite. All rock types are represented: ultramafics (serpentinite); gabbros (rare blocks of gabbro and gabbro breccia); sheeted dikes (diabase-rich mafic breccias?); pillow basalts; and pelagic sediments (chert blocks).

4. The mafic part of the ophiolite is badly underrepresented within the chaotic unit, as it is within the underlying basement. In particular, gabbro blocks are conspicuously rare.

5. The unit contains blocks of metamorphic rocks that represent the greenschist and amphibolite facies. These blocks show an extremely broad spectrum of textures, ranging from well-

recrystallized polygonal granular rocks to well-foliated amphibolites to amphibolite tectonites to mylonites. Most rocks are metabasites, but one is ultramafic. One enigmatic metasedimentary block was found.

6. Much of the mafic debris—and virtually all the material that might represent a sheeted dike complex—is second-generation detrital material, that is, it occurs as blocks of mafic breccia that are themselves composed of clasts.

7. Much of the ophiolitic debris is very coarse-grained (1-300 m).

8. Although the unit is locally homogeneous (i.e., almost all the larger blocks in a given square kilometer may be greenstone, or serpentinite), it is regionally quite heterogeneous.

9. Matrix consists of clay mudstone and small, thin flakes and scales of sheared serpentinite. It is apparently continuous from mudstone indistinguishable from that in the bedded Great Valley sequence (except for its lack of bedding), through serpentinous mudstone, to sedimentary serpentinite flows.

10. Detrital textures are common in the matrix of the unit.

11. Although the matrix is commonly sheared, especially near mapped faults, it never contains a penetrative shear foliation or cleavage like that of the *argille scagliose* matrix of Franciscan melanges. Locally, quite unsheared matrix may preserve silty laminae or textures showing serpentine detritus in mudstone.

12. Everywhere the chaotic rocks occur, they lie on serpentinized ultramafic rocks, frequently harzburgite tectonite—never on pillow basalts, sheeted dikes, gabbros, or ultramafic cumulates.

13. The chaotic rocks are immediately overlain by bedded subsea-fan rocks of Tithonian and Valanginian ages. These Great Valley sequence rocks apparently represent slope and inner fan facies. The sedimentary rocks were derived from the Sierra Nevadan/Klamath volcanic and plutonic arc.

14. The chaotic unit is not cut by mafic dikes or "necks" like those cutting the serpentinite, nor are there any lava flows interlayered within it.

15. The chaotic unit is Tithonian (late Late Jurassic) in age, and locally represents the entire stage.

Because it has been folded and repeated by thrust-faulting, the chaotic unit occurs in several outcrop belts (Figure 1). The easternmost of these, containing Localities 8 and 9 on Figure 1, is the Knoxville belt. The next belt to the west, containing Localities 2 and 3, is the Mysterious Valley belt. Farther to the west, containing Localities 4 and 5, is the Putah Creek belt. The Lake Berryessa belt lies to the south, along the west shore of the lake.

The outcrop belts have distinctive hummocky topography, with resistant blocks protruding above grassy swales developed on the matrix. Thick soil forms on the matrix (3 m is not unusual). The matrix is weak and landslides massively, and the enclosed blocks tend to creep downhill. As a result, it is rare to find a block that is clearly in place, and block-matrix contact relations are often difficult to determine.

Hundreds of blocks protrude above the surface throughout the terrane. Where streams have cut deeply into it, blocks in profusion crown knobs high above the stream valleys, jut out from valley walls, and stick up from stream bottoms, commonly deflecting the streams around them. Blocks range from pebble-sized or even smaller to mountain-sized. Very small pieces of block mateial—fragmented either during the disintegration that originally formed the blocks or during transport of those blocks—may be the same size as matrix grains. In this sense, the distinction between blocks and matrix is arbitrary—the two are gradational. This is especially true of serpentinite within the unit, which occurs as unsheared blocks, foliate sedimentary serpentinite flows, serpentinite-mudstones containing serpentinite pebbles, and serpentinous mudstone matrix in which clay and serpentinite are intimately intermixed.

BLOCKS

The chaotic unit contains a wide variety of block types-igneous, sedimentary, and metamorphic. Blocks of greenstone are by far the most common. Many of these show subophitic textures, and many are pillowed, but most appear altered, some very badly. These basalts are not vesicular and rarely have phenocrysts, although a few are plagioclase-phyric. Glassy rinds usually occur only on fresher outcrops, for instance in stream beds. Some pillow blocks contain breccia zones, which may be interlayered with pillows within a single block. The igneous textures of breccia clasts are generally indistinguishable from those of intact pillows. Locally, it is obvious that many of the clasts are pillow fragments.

Although some greenstone blocks may have been derived from hypabyssal intrusions within an ophiolite, greenstone has never been observed intruding the chaotic unit. I found no chilled margins in blocks or baked contacts in matrix.

Greenstone blocks range in size from cobbles to small mountains (over 1 km across). Where several pillowed greenstone blocks are found in close proximity (at the mouth of Pope Canyon, for instance—Figure 1, Locality 7), the younging directions shown by the pillows are random.

Coarse-grained mafic plutonic blocks are very rare. Only two examples of blocks as coarse as fine-grained gabbro and no blocks of coarse-grained gabbro are known.

Many different sorts of mafic breccias occur as blocks. These blocks range in size from 1 m to 500 m across. The most common breccias are clast-supported and consist largely of angular diabase clasts, but some breccias are matrix-supported, some contain subangular to subrounded clasts, and some contain clasts of gabbro, plagiogranite, or basalt. Clasts of medium- to coarse-grained gabbro are rare. Clast sphericity is low, and the breccias are poorly sorted. The angularity and size and, to a lesser extent, the composition and igneous grain size of most clasts in a single breccia block are similar. The matrix between larger clasts consists merely of sand- and grit-sized grains that are smaller versions of the large clasts. Therefore, it may be said that clasts range in size from microscopic to 1 m across. Clasts larger than 10 cm are rare, and 2-5 cm clasts are characteristic.

Hopson and others (1981) have described mafic breccias,

found as *in situ* accumulations lying depositionally on the Coast Range Ophiolite, from several localities, including Mt. St. Helena and the South Fork of Elder Creek (Crowfoot Point). I have visited most of these localities. The breccias are similar in texture and in clast composition, size, density, sorting, and angularity to those in the blocks described here. Wagner (1975) has also described mafic breccias, which are not blocks in a larger chaotic unit, within the present study area (his Pope Creek Breccia). These rocks are found in the basal Great Valley sequence in Pope Valley at the entrance to Pope Canyon, about 5 km northwest of Locality 7, Figure 1 (at Locality 18, Appendix 1). The Pope Creek breccias are bedded on a scale of tens of meters. Other breccia localities exist; it is evident that mafic breccias are a common feature of the Coast Range Ophiolite, at least in Northern California. However, the breccia within blocks in the chaotic unit I describe here, and that in *in situ* accumulations such as those at Crowfoot Point and Mt. St. Helena, may have different origins.

Similar ophiolitic breccias are found in modern ocean basins. Piles of angular, unsorted breccia are common in the Mid-Atlantic Ridge rift valley, where they are talus breccias accumulating at the bases of submarine escarpments (ARCYANA, 1975; Ballard and others, 1975; Ballard and Moore, 1977). It is notable that in two of the California occurrences, at Mt. St. Helena and Pope Creek, the breccias are themselves intruded by dikes. In the Pope Creek breccias, these intrusions are clearly dikes, not sills, and they show dike-in-dike textures (Wagner, 1975; and my own observations). It seems likely that many masses of mafic breccia in California, including the blocks described here, accumulated as talus piles at the bases of submarine scarps. At least some of them formed in zones of mafic igneous activity.

Within blocks, and within single outcrops, mafic breccias grade into a distinctive sandstone that macroscopically has a speckled, mottled, or flecked appearance. Where weathered, this sandstone is tan to brown; where fresh, it is gray. Thin section studies reveal it to be composed entirely of basaltic and diabasic rock fragments, angular pieces of plagioclase crystals, and occasional detrital clinopyroxene grains. No orthopyroxene grains have been found. This sandstone is essentially a fine-grained equivalent of the mafic breccias; it is like the sandy matrix found between larger clasts in many blocks of mafic breccia. In general, the sandstone is massive and structureless, but in one exposure (in the bed of Pocock Creek—Figure 1, Locality 3), it shows clear planar bedding on a scale of 1-10 cm. I have not observed grading, cross-bedding, ripple marks, or burrowing in these rocks, nor have I seen pelagic sediment or more typical Great Valley sequence sandstones or mudstones interbedded with them.

Associated with the mafic breccias and speckled sandstones is a distinctive olive-drab to gray mudstone that is much better indurated than chaotic unit matrix or well bedded Great Valley sequence mudstone. The mudstone typically displays pervasive fractures lacking slickensides.

The diabase-rich mafic breccias, speckled sandstones, and related mudstones form a distinctive and recurring association within the chaotic unit, an association I call the *diabase breccia assemblage.* Several outcrops demonstrate that the three rock types of the assemblage grade into each other. For example, at Locality 3, Figure 1, the transition between mafic breccia, speckled sandstone containing a few isolated clasts of diabase, and pure speckled sandstone occurs within a meter or two, both across and along bedding. The mudstone and the speckled sandstones probably represent reworking of the finer-grained fraction of the mafic breccias by submarine currents. Such current-transported ophiolitic sand grains have been observed in the Quebrada Transform on the East Pacific Rise (Lonsdale, 1978). Isolated larger clasts found in the speckled sandstone may have rolled or fallen onto the surface of the sand deposits.

In addition to diabase-rich breccias, mafic breccia blocks include abundant aphanitic greenstone breccias and very rare gabbro and plagiogranite breccias. Equally rare are blocks of massive gabbro.

Chert blocks are common. Bedded cherts are typically green to greenish-white; many are red, and some are gray to black. All include thin shale interbeds. Most bedded chert blocks contain only chert and shale, but one block contains chert interbedded with pillow lavas. This block lies in the stream bottom of Mysterious Creek just below the point where the creek exits from the gorge it has cut where it leaves Mysterious Valley (Locality 19, Appendix 1, near Locality 5, Figure 1). In this block, somewhat less than 100 meters of beautifully preserved pillow lavas are overlain by about 2 m of bedded, manganiferous, red radiolarian chert. Locally about 1 m of mafic breccia intervenes between pillow lavas and chert. Overlying the bedded chert are a few more meters of pillow lavas; above this, the block is covered by Quaternary alluvium. Some bedded chert blocks show alteration colors—commonly pasty yellow—that may be caused by hydrothermal circulation. This alteration does not extend into the surrounding matrix. Some cherts, like those described above, are manganiferous.

Other chert bodies are amorphous, unbedded masses of red chert, with vuggy centers and outward-tapering, branching, wedge-shaped, quartz-filled cavities radiating from the central vugs. (Occasionally, bedded cherts show similar fractures.) Some amorphous chert masses show altered color similar to that described in bedded cherts above. Bright orange-yellow is a typical color.

Rare blocks of chert breccia occur that contain only chert clasts. Clasts, which are usually less than a centimeter across, are significantly smaller than those in typical mafic breccia blocks. Chert clasts are extremely rare in mafic breccias.

An unusual, 15-m long block of bedded sandstone consisting of well-indurated, gray, planar-laminated lithic arenite occurs at Locality 2, Figure 1. This block contained many bright green grains that appear to be celadonite. In hand sample, it appears to have significantly lower content of lithic grains than typical basal Great Valley sequence sandstones, and is much poorer in volcanic lithic grains than the "speckled" sandstones.

TABLE 1. METAMORPHIC BLOCK TYPES

Type	Number of Localities	Texture	Mineralogy	Protolith
I. Phyllite	1	tightly folded, quartz-veined	unidentified	mudstone
II. Weakly foliated schist	1	fine-grained, cleaved, relict spherulites, badly weathered	clays, possible pumpellyite	volcanic, probably basaltic
III. Gabbroic amphibolite	1	(1) gabbroic; (2) sheared amphibolite, minerals show pinch-and-swell, lineation	act + grn hbl + brn hbl + cloudy plag + sph + opaque	gabbro
IV. Amphibolites		strong pinch-and-swell fabric,	rock: hbl + plag + sph	
A. Amphibolite tectonites	(3)	hornblende necked, fragments rotated	veins: clinozoisite,	basalt or gabbro
1. amphibolite mylonites	1	with crush trails, gran. plag matrix	pumpellyite, thomsonite	
2. "felted" amphibolites	2	amphiboles w/und. ext. woven into felted mats, anastomose around plag w/pinch-and-swell	act + brn hbl + cloudy plag + clinozoisite or zois + altered titanomagnetite	basalt or gabbro
B. Highly foliated amphibolites	4	Hand sample: fine-med-grained, strong amphibole lineation Thin sect.: strong foliation, little und. ext. or mylon.	grn to bl-grn amphib. (hbl) ± cloudy plag + sph + clinozois; plag is ab; cloudiness is clinozois. No opaque	basalt or gabbro; ultramafic
C. Massive amphibolite	2	Hand sample: dark grn to black, dense unfol. Thin sect.: well reXized, gran. polygonal, unfol., equant hbl.	hbl + plag + sph + diops; relect ign plag; diopside cores, probably relict igneous, in hornblende	basalt?

Metamorphic Blocks

The chaotic unit contains blocks of metamorphic rock of several types, ranging from phyllites to amphibolites (Table 1). These blocks were first reported by Phipps and others (1979). Subsequently, another occurrence of metamorphic blocks in chaotic rocks in the lower Great Valley sequence was reported by Carlson (1981a, 1981b) from the Wilbur Springs area.

The occurrences—resistant masses of schist lying on grassy hillsides—are like those of other blocks, with the same difficulties attendant in finding blocks that are definitely in place. However, one phyllite (metamorphosed mudstone) block that is clearly in place is well exposed along Mysterious Creek approximately 1 km southwest of where it leaves the gorge it cut at the edge of Mysterious Valley (Figure 1, Locality 5). This block is entirely surrounded by sheared mudstone matrix.

About two dozen occurrences of metamorphic blocks are known in all. In no instance are they found downslope from Franciscan outcrops, and cross-sections indicate that Franciscan thrust slices never overlay outcrop belts containing the metamorphic blocks. Therefore, they are not float blocks from the Franciscan Complex. Blocks are typically 10 cm to 20 m across. They are commonly found in clusters of from six to several dozen blocks. In most occurrences, they are found on hillsides or hilltops and appear to be weathering out of the matrix. Matrix surrounding metamorphic blocks is similar to matrix elsewhere in the chaotic unit. Occurrences of schist blocks are no different from those of any other block lithology, and there seems to be no reason to exclude them from the chaotic assemblage. This, combined with the existence of the phyllite block that is demonstrably in place, indicates that blocks of metamorphic rocks are present in small numbers within the chaotic unit.

Chemically, the blocks are metabasites with the exception of one ultramafic rock and one phyllite. They contain greenschist and amphibolite facies mineral assemblages, and their textures are highly variable.

The metamorphic rocks can be placed in four groups on the basis of texture: phyllites, weakly foliated schists, gabbroic amphibolites, and amphibolites. Within the groups, the variety of textures is extreme, so great that virtually every locality is unique. Table 1 describes the mineralogies and textures of 14 of the metamorphic blocks.

Many of these rocks are similar to metamorphic rocks dredged from the mid-ocean ridges and transform faults. Rocks from these dredge hauls contain greenschist and amphibolite facies mineral assemblages, and some possibly transitional to granulite facies. The mineral assemblages are much like those reported here. The rocks show a wide variety of textures, many of which closely correspond to those of metamorphic rocks from the Great Valley sequence chaotic unit. The rocks include metamorphosed diabases, plagioclase-phyric diabases, and gabbros, all preserving their igneous textures, banded greenschists and amphibolites (not cataclastic), intergranular amphibolites, banded cataclastic to mylonitic granular amphibolites, and gabbro mylonites (Melson and Van Andel, 1966; Bogdanov and Ploshko, 1967; Ploshko and others, 1970; Chernysheva, 1970; Bonatti and others, 1971; Miyashiro and others, 1971; Cann, 1971; Engel and Fisher, 1975; Fox and others, 1976; Helmstaedt and Allen, 1977). In particular, gabbroic amphibolites from Napa County are texturally very similar to rocks dredged from the Cayman Trough, the axial valley of the slow-spreading, short Mid-Cayman Rise (Malcolm, 1981). They are texturally and mineralogically similar to rocks from the axial rift valley of the mid-Atlantic Ridge (Bonatti and others, 1975) (Sample # P6903-26, illustrated in their Figures 2 and 6).

Metamorphic rocks similar to those described here have been reported from several other ophiolites, among them the Del Puerto occurrence of the Coast Range Ophiolite (Evarts and Schiffman, 1979, 1983), the Zambales Ophiolite in the Philippines (Geary and Kay, 1983), and the East Taiwan Ophiolite (Liou and Ernst, 1979). Seafloor metamorphic rocks are probably a common, if minor, member of the ophiolite assemblage.

Carlson (1981a, 1981b) has found rare metamorphic blocks in a similar terrane in the Great Valley sequence in the Wilbur Springs area 75 km to the north. These rocks also apparently display great diversity within a relatively small number of localities. The metamorphic blocks from the Wilbur Springs area are amphibolites and greenschists displaying metamorphic assemblages much like those reported here. However, Carlson (1981b) also reports blueschist blocks (lawsonite- and glaucophane- bearing). The blueschist rocks are mostly overprinted hornblende amphibolites, but include one specimen of metachert.

No high-pressure minerals have been found to date in the Napa County rocks. There is no known source within the Coast Range Ophiolite for blocks of high P/T facies. Therefore, blueschist blocks within the Great Valley sequence were probably derived from the Franciscan Complex. Their presence at Wilbur Springs in itself is a problem; so is their presence at Wilbur Springs and their absence in the Napa County chaotic unit, if the origins of the two occurrences of ophiolitic chaos are similar.

However, although there are similarities between the ophiolitic detrital rocks in Napa County and those at Wilbur Springs—sedimentary debris flows of foliate serpentinite (discussed below) and polymict, polymorphous chaotic rocks are found at both sites—there are also differences. In the rocks studied by Carlson (1981a, 1981b), the blocks are found entirely enclosed within sedimentary serpentinite flows, whereas those in the Napa County chaotic unit occur within mixed clay mudstone and serpentinite matrix. The Wilbur Springs serpentinite flows are thicker and make up more of the Upper Jurassic/Lower Cretaceous section than those in the Mysterious Valley area. Finally, the ophiolite debris at Wilbur Springs studied by Carlson is Hauterivian to Barremian (middle Early Cretaceous) and late Aptian (late Early Cretaceous) in age and is thus substantially younger than the rocks described here.

The mechanisms and/or paleogeography that emplaced blueschist blocks within ophiolitic olistostromes in the Great Valley sequence during the later Cretaceous may have been somewhat different from those that formed the ophiolitic chaos in Napa County. The blueschist blocks might have been shed into the Great Valley sequence from an eroding outer-arc high that contained Franciscan material. This high might have contained fault slices of Franciscan melange bearing blueschist blocks (Maxwell, 1974), or might have been an uplifted blueschist-bearing Franciscan melange wedge (Cloos, 1982).

Serpentinite within the Chaotic Unit

Serpentinite is among the most abundant rock types in the Napa County chaotic unit. It occurs as: (1) angular pebbles in serpentinite-pebble mudstones and other detrital serpentinites; (2) blocks up to 100 m in diameter; and (3) sedimentary serpentinite flows consisting of sheared serpentinite with a strong, anastomosing foliation formed by phacoids of sheared serpentinite, with entrained blocks of unsheared serpentinite from 1 cm to 10 m in diameter.

1. Serpentinite-pebble Mudstones. Serpentinite-pebble mudstones consist of subrounded to angular serpentinite pebbles set in a very fine-grained (mud or silt sized) matrix. The matrix is usually serpentinite but may be clay shale. The rocks are matrix-supported. Clasts are angular to subrounded. In any one sample, the larger clasts are poorly size-sorted but are usually of about the same angularity. Matrix is typically pastel green, light green, or light gray; pebbles are usually darker and range from blue-green to dark gray to black. The pervasive shear foliation characteristic of sedimentary serpentinite flows is usually absent or poorly developed. Serpentinite-pebble mudstones locally exhibit crude layering on a scale of 1-10 cm, with different layers varying in pebble concentration. In layers that contain many pebbles, the pebbles may be aligned in trains extending for a meter or more.

Serpentinite-pebble mudstones weather deeply, and good exposures are rare. However, excellent outcrops are exposed at low water below the north abutment of the bridge over the Putah Creek arm of Lake Berryessa on the Berryessa-Knoxville Road, and nearby at Locality 10 (Appendix 1). During the unusually wet winter of 1978, good exposures were also found in the fresh headwall scarps of very recent landslides (Locality 11, Appendix 1).

By analogy to more conventional pebbly sandstones and mudstones common in submarine fans, serpentinite-pebble mudstones probably represent debris flows (in the strict sense of Blatt and others, 1972). In these flows, the larger clasts (serpentinite pebbles) are supported by the finite yield strength of the matrix. Therefore, the serpentinite-pebble mudstones represent disturbance, remobilization, and resedimentation of pre-existing serpentinite detritus. The extreme angularity of many pebbles suggests that they are very proximal, and is consistent with their production in a submarine environment.

*2. **Serpentinite Blocks.*** Serpentinite blocks, such as those exposed at Locality 13 (Appendix 1), are characterized principally by their large aspect ratios and by relatively little shearing. It is unlikely that there is a minimum block size, as there are isolated serpentinite pebbles in the chaotic unit matrix, but blocks range up to 100 m across. Large blocks are commonly cut by sheared zones, but these do not make up most of the serpentinite mass and do not have a single dominant orientation as they do in sedimentary serpentinite flows.

The large, unsheared serpentinite blocks probably represent slump or slide blocks that broke off or slid down from serpentinite escarpments. They probably rolled or glided downhill into the accumulation of ophiolitic debris.

*3. **Sedimentary Serpentinite Flows.*** Serpentinite debris flows are commonly lobate in plan view and lens-shaped in cross-section. Within the study area, they range in size from small, perhaps 10 m thick and 100 m long, to enormous 300-500 m thick and 2 km long. The largest flows may be amalgamations of smaller ones. The sheared serpentinite within the flows is best described as a foliate breccia, as it consists largely of aligned phacoidal serpentinite scales. Most phacoids range in length from 0.5 cm to 10 cm in long dimension, though they may be larger or smaller, and their aspect ratios vary from 1:3 to 1:5. The phacoids are bounded by polished and slickensided surfaces and are the typical blue-green color of sheared serpentinite. The foliation formed by the phacoidal scales anastomoses around unsheared serpentinite blocks entrained within the flows. These blocks range in size from a few centimeters to several meters across.

Both modern and ancient serpentinite flows exist; the presence of actualistic examples has been a great help in understanding the ancient flows. Presently active flows are common in the California Coast ranges today (Dickinson, 1966; Cowan and Mansfield, 1970). Many occur within the study area.

These modern serpentinite flows commonly develop wherever a sheared serpentinite body is exposed at the ground surface above a slope. The sheared serpentinite flows out onto the slope and downhill. It commonly retains a remarkable coherence, and its anastomosing shear foliation is not disrupted. Within the study area, the sources of modern flows are typically the highly sheared lower or outer margins of large serpentinite bodies. These bodies have been emplaced at their present structural level by thrust faults and have then been exposed by erosion. Excellent examples of modern serpentinite flows occur on the northeast sides of Mysterious and Pope Valleys (Localities 14 and 15, Appendix 1). In Pope Valley, one flow is 700 m long, 250 m wide, and 30 m or more thick.

Mesozoic serpentinite flows very similar to modern, subaerial flows are interbedded in the Great Valley sequence at many localities: near Cooks Springs (Phipps and others, 1979), near Wilbur Springs (Moiseyev, 1966; Rich, 1971; Carlson, 1981a, 1981b), at the southeast end of Little Blue Ridge above the Reid Mine (Averitt, 1945), a mile northeast of Knoxville (Borglin, 1949), near Middletown (Brice, 1953), at Fir Creek (Moiseyev, 1970), and in southern Pope Valley (Shouldice, 1947), as well as throughout the chaotic unit I describe here. Within the study area, particularly well developed Mesozoic serpentinite flows occur within the Putah Creek outcrop belt of the chaotic unit, in the west central portion of the Walter Springs 7½′ Quadrangle, near Locality 5, Figure 1 (Locality 16, Appendix 1).

As most outcrops of the chaotic unit dip moderately to steeply and most serpentinite flows within it are seen in oblique cross-section, Mesozoic serpentinite flows in map view appear as cross-sections of double-convex lenses: that is, they are elliptical with pointed ends, with their long axes parallel to regional strike. Their phacoid foliation also parallels regional strike and dips at approximately the dip of the chaotic unit. The flows are typically 100-200 m long and 10-40 m thick; thus, they have aspect ratios in the range 1:10 to 1:5. Some very large Mesozoic serpentinite flows exist; one, in the Mysterious Valley outcrop belt in the northeastern quarter of the Walter Springs 7½′ Quadrangle, is at least 300-500 m thick and 2 km long. This may be an amalgamation of flows—the quality of exposure does not permit differentiation into separate, smaller flow units. Coherent flows of foliate serpentinite breccia are resistant to erosion and are commonly found capping knolls and hummocks in the chaotic unit.

The internal structure of Mesozoic flows is not distinguishable from that of modern ones. I have not found blocks of other rock types within serpentinite flows.

No high-quality exposures of the contact relations between serpentinite flows and chaotic unit matrix have been observed. Where exposed by digging, the contacts appear gradational, with the serpentinite fraction of mixed sheared mudstone and serpentinite matrix increasing rapidly over a distance of no more than a few meters until, at the margin of the flow, serpentinite comprises 100% of the material.

Serpentinite Emplacement by Faulting and Solid-state Extrusion

Various workers have attempted to explain serpentinite-rich

ophiolitic debris at several localities, including the blueschist-bearing debris at Wilbur Springs, as derived from diapirically emplaced serpentinite "protrusions" (for instance, Lockwood, 1972; Carlson, 1981a, 1981b). I believe, however, that other processes have emplaced the serpentinite-rich debris, including blockfalls from serpentinite escarpments, debris flows of piles of serpentinite detritus at the bases of such escarpments, and downslope creep of foliate serpentinite flows derived from sheared serpentinite bodies faulted up to the seafloor. These means of serpentinite emplacement may render serpentinite diapirism an unnecessary hypothesis.

Furthermore, I would argue that several lines of evidence indicate that serpentinite diapirism is *not* responsible for the ophiolitic debris rich in serpentinite that is so common in the Coast Ranges. Firstly, much of the ophiolitic detritus at Wilbur Springs, particularly the Upper Jurassic material, is similar in many ways to that described here. The blueschist-bearing olistostromes in the Wilbur Springs area represent only one episode in the longest history of ophiolitic submarine slide and debris flow activity in the northern Coast Ranges, and it seems likely that the rocks there were formed by mechanisms generally similar to those proposed here.

Secondly, at many localities in the Coast Ranges, serpentinites are observed that represent the probable sources of ophiolitic debris within the Great Valley sequence. In all cases, the serpentinites are not diapiric intrusions through the basement but rather *are* the basement, the Coast Range "Ophiolite."

Finally, I consider diapiric emplacement to be mechanically improbable because the density inversion necessary to produce diapirs probably did not exist. Densities of typical California Coast Range serpentinite at Burro Mountain range from 2.52 (almost 100% serpentinized) to 3.14 gm/cm^3 (lightly serpentinized) (Christensen, 1966). On the other hand, densities of Great Valley sequence sediments average 2.59 (Upper Jurassic), 2.57 (Lower Cretaceous), and 2.55 gm/cm^3 (Upper Cretaceous) (Bailey and others, 1964, p. 141). Thus, assuming the extreme values—a density of 2.52 gm/cm^3 for 100% serpentinite and a density of 2.59 for a well compacted modern Upper Jurassic Great Valley sequence sedimentary rock—the density contrast will only be –0.07 gm/cm^3. This value is based on the density of present-day, well lithified and compacted Upper Jurassic Great Valley sequence rocks, which have been buried beneath 11 km of sediment and mildly metamorphosed (Dickinson and others, 1969). The hypothesized serpentinite diapirism took place in the Late Jurassic and Early Cretaceous. In a geologically more realistic situation, with moderately serpentinized ultramafic rock and sediments that had not been deeply buried, and may not have been well lithified and compacted, or even completely dewatered, the density contrast would be zero or positive. This is a stable configuration. No matter *how* weak the serpentinite is, which is not clear (Raleigh and Paterson, 1965), diapirism will not occur. Thus, seemingly diapiric serpentinites—for instance, at Table Mountain in the southern Coast Ranges (Dickinson, 1966)—may have been faulted to the surface. Demonstrable ongoing upward movement of sheared serpentinite may be caused by lateral pressure exerted by the enclosing rocks on the plastic serpentinite body. Such lateral pressure is a product of overburden stress and the incompressibility of the surrounding rocks, and may be supplemented by stresses applied by larger-scale tectonics. Therefore, I suggest that serpentinite upwellings are solid-state extrusions rather than *diapiric* protrusions. The term "protrusion," as originally used by Lockwood (1971, p. 920) to mean "masses of any rock which are believed to have reached their present positions through forceful injection in the solid state," if divorced from its diapiric connotations, is probably still valid for many Coast Range serpentinite bodies. These protrusions (*sensu strictu*) or solid-state extrusions give rise to serpentinite flows.

Serpentinite Flows and Serpentinite Matrix

Within the Napa County chaotic unit there is a continuous range of serpentinite bodies that includes: (1) coherent foliate breccia flows with well-preserved, coarse phacoid foliation; (2) flows with more finely comminuted, looser, and less coherent foliation that forms a flaky or scaly mass; (3) serpentinite-pebble sandstones and mudstones in which the sand- or mud-sized fraction is entirely serpentinite; (4) serpentinite-pebble sandstones and mudstones in which the fine-grained fraction contains considerable bluish-gray clay similar to that in the bedded Great Valley sequence; and (5) serpentinous matrix consisting of a mixture of finely comminuted serpentinite scales and sheared mudstone. It seems likely that this range of rock types represents a gradational series.

MATRIX

The matrix of the chaotic unit consists of comminuted flakes of serpentinite and of unsheared to highly sheared bluish-gray to gray to black clay mudstone. The serpentinite is finely sheared and usually has the same light green to blue-green color as the larger, phacoidal scales of the coherent foliate breccias. Mudstone matrix is frequently quite sheared. However, sheared phacoids of mudstone matrix, when broken open, commonly reveal unsheared mudstone indistinguishable in color and texture from mudstone of the bedded Great Valley sequence, except that bedding is lacking. Locally, mudstone matrix may be unsheared but highly fractured (*i.e.,* fracture surfaces are not polished and slickensided). Sometimes, mudstone matrix is unsheared, unfractured, and identical to the mudstone in the Great Valley sequence, except that bedding is lacking. Disturbed silty laminae may be present. Pebbles of mudstone are sometimes found in serpentinous matrix, but I have not seen larger blocks of mudstone. Especially toward the top of the chaotic unit, mudstone matrix preserves the limestone concretions characteristic of many Great Valley sequence mudstones. Thus, "normal" Great Valley sequence sedimentary rock—clay mudstone—is found with the ophiolitic debris in many places.

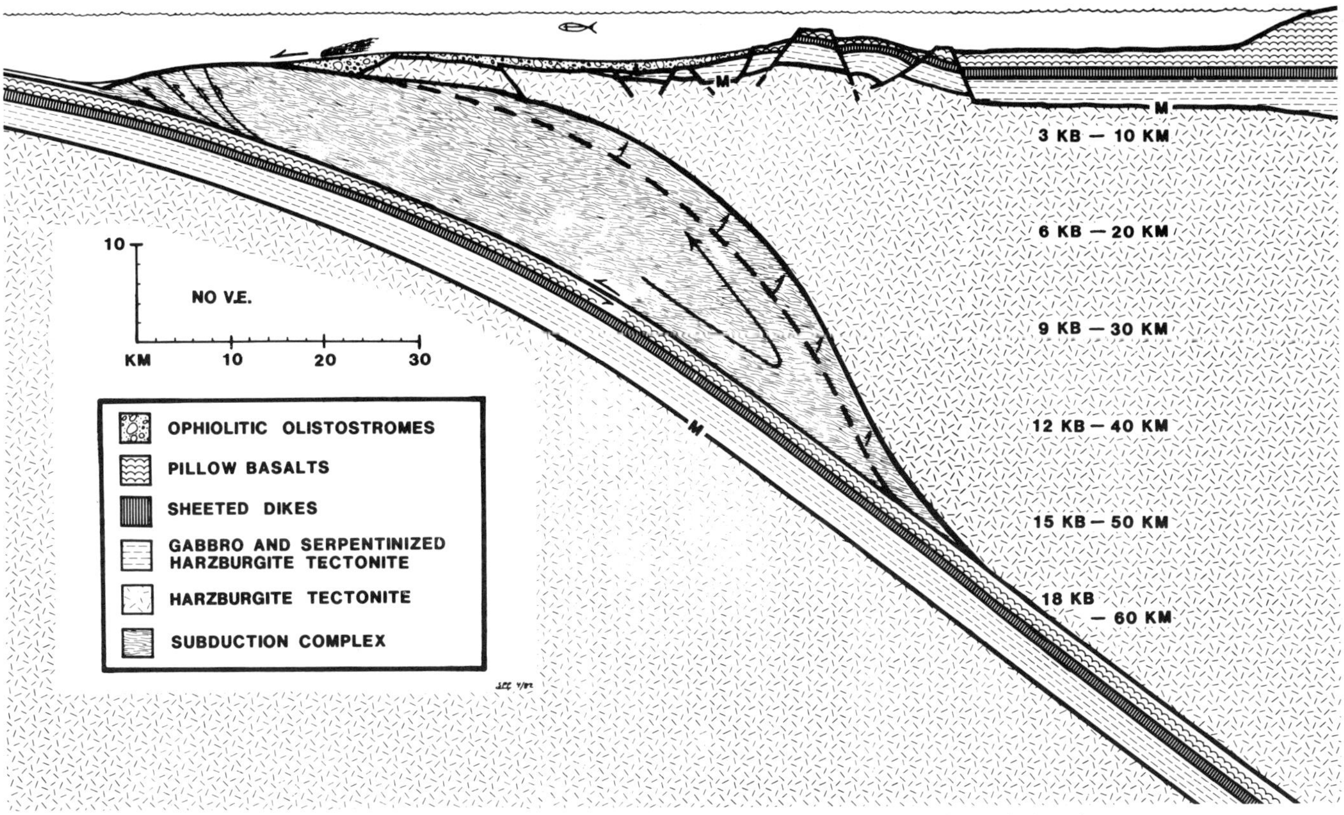

Figure 2. Schematic diagram of structure, stratigraphy, and processes which might exist in a forearc basin underlain by an actively forming melange wedge. "M" is Moho. Some concepts from Cloos (1982).

In some areas, serpentinite-rich zones in the matrix alternate on a scale of 1 to 10 meters with mudstone-rich zones. In many outcrops, however, serpentinite and mudstone matrix are intimately intermixed at millimeter scale. Here, the mixing of mudstone and serpentinite has given the matrix a marbled, mottled, or swirled appearance. This is the case at several localities 300 to 500 meters southeast of Hunting Creek (the Napa/Lake County line) in the Putah Creek outcrop belt of the chaotic unit, where matrix was sampled extensively (Localities 11 and 12, Appendix 1, which are near Locality 4, Figure 1). Here light pastel-green, thin, flaky serpentinite scales only one or a few mm long form an anastomosing foliation that wraps around angular serpentinite and mudstone pebbles, internally unsheared mudstone phacoids 0.5 to 5 mm thick and 1 to 10 mm long, and mudstone scales and flakes the same size as serpentinite flakes. Overall, serpentinite and clay mudstone appear to make up roughly subequal portions of the rock. These rocks are very poorly lithified, although individual pebbles and flakes of serpentinite and mudstone may be well lithified. When transported or handled, the matrix material quickly crumbles into piles of serpentinous dirt.

The matrix appears to be more serpentinous at the base of the chaotic unit and more mudstone-rich at the top, toward the upper contact with the bedded Great Valley sequence. In addition, some outcrop belts are richer in mudstone matrix (Mysterious Valley, containing Localities 2 and 3, Figure 1) and some in serpentinite matrix (Putah Creek, containing Locality 5, Figure 1).

The mudstone matrix of the Great Valley sequence chaotic unit is distinctly different from the scaly-clay (*argille scagliose*) matrix of the Franciscan melanges, which is blacker, scalier, and generally more pervasively sheared. The shearing in the Great Valley sequence chaotic unit is greatest near major faults, and much of it, especially in the mudstone, was probably caused by later thrust-faulting.

CONTACTS

The nature of both the upper and lower contacts of the chaotic unit is critical to its interpretation. The best exposure of the lower contact, at the mouth of the lower gorge of Hunting Creek on the Napa-Lake County boundary (Figure 1, Locality 4), shows the massive, slabby serpentinite of Mysterious Ridge grading into moderately sheared serpentinite. Above this is a covered interval 2 m wide, succeeded by serpentinite breccia and then by mudstone containing angular cobbles and pebbles. Both clasts and mud-sized matrix are serpentinite. At no other locality is the contact so well bracketed. However, where the chaotic unit occurs, it always directly overlies massive serpentinite, and its matrix is always quite serpentinous. The matrix of the basal part of the unit frequently shows detrital textures and contains serpentinite clasts. In general, the lower contact can be located within a few meters or tens of meters.

Wherever the chaotic unit is found, it lies directly on the serpentinite that represents the Coast Range Ophiolite; no intervening rocks occur. In particular, the "ophiolite" contains no mafic section anywhere in the study area, except for that contained as fragments within the chaotic unit.

The chaotic unit's upper contact with the bedded Great Valley sequence is well exposed at Locality 8, Figure 1, where it is a transition zone a few meters wide between mudstone-matrix chaotic unit and bedded mudstone and sandstone. The contact corresponds to an abrupt decrease in the intensity of shearing. A major thrust fault (the Cedar Creek Thrust) passes very near these outcrops, and other bedding-plane thrusts probably lie nearby. The greater shearing of the chaotic unit may reflect its role as a preferred decollement horizon for the late thrust-faulting.

In the Putah Creek outcrop belt, the contact is also sharp. Here the deformation in both the chaotic unit and the bedded Great Valley sequence is even more severe because of the proximity of the Putah Creek Thrust.

The best exposures of the upper contact are found along the Knoxville-Bereryessa Road on the nose of Adams Ridge (Figure 1, Locality 6). Here the contact is more broadly gradational, and olistostromes of the chaotic unit interfinger with the subsea-fan rocks of the bedded Great Valley sequence. In a series of roadcuts, a fine-grained, serpentinous phase of the chaotic unit is succeeded stratigraphically upward by lithic sandstone and clay mudstone with badly disrupted and almost unrecognizable bedding. These are overlain by more fine-grained, serpentinous material of the chaotic unit, containing greenstone blocks. Above this, well bedded, fossiliferous Great Valley sequence rocks crop out. No major faults pass near this area.

During Mesozoic time, there was some topographic relief on the lower contact, that between the serpentinite and the chaotic unit, and there was substantial relief on the upper contact. This is evident on the geologic map, as the moderately to steeply dipping stratigraphy is seen in oblique cross-section. Unevenness in the trace of the contacts on a scale of a kilometer or less, where not caused by present-day topography, is probably due to structural relief on the contacts rather than to local reversals of fold plunge. Topography on the chaotic unit's lower contact can be seen at the southeast end of the Mysterious Ridge Anticline serpentinite body, where the serpentinite core of the anticline plunges southeast (near Locality 6, Figure 1). Here, on the southwest limb of the anticline, the serpentinite is thicker and the chaotic unit noticeably thinner. Relief on the upper contact is dramatic south of Knoxville (Figure 1, Locality 9). Here the chaotic unit thickens, and its top rises through the section to the base of the Lower Cretaceous. The chaotic unit breaches an otherwise continuous strike ridge held up by chert-pebble sandstone and conglomerate, representing a filled submarine fan channel, which lies on the Upper Jurassic-Lower Cretaceous boundary through much of the area. This must represent a topographic high on the Late Mesozoic seafloor, which divided the channel. The turbidites of the bedded Great Valley sequence apparently lap onto the chaotic unit.

It is particularly significant that this high occurs in the Knoxville area, for a large serpentinite flow is found in the Great Valley sequence near the Upper Jurassic/Lower Cretaceous boundary at the southeast end of Little Blue Ridge, a few kilometers to the north. Another is mapped well up in the Lower Cretaceous section to the northeast of Knoxville (Averitt, 1945; Borglin, 1949). It is possible that these flows were derived from the Knoxville high. Thus, in this area, the processes shedding ophiolitic debris into the Great Valley sequence basin were active into the Early Cretaceous.

AGE OF THE CHAOTIC UNIT

Rare fossils found within the chaotic unit are *Buchia piocchii,* pelecypods of Tithonian (Late Jurassic) age. Within the study area, the chaotic unit is bounded almost continuously above by bedded rocks that also contain *Buchia piochii.* Locally in the Putah Creek outcrop belt, the first overlying fossil locality is Valanginian (Early Cretaceous), on the basis of the occurrence of *Buchia pacifica.* The Coast Range Ophiolite is Middle Jurassic in age. Radiometric ages of mafic intrusions in the serpentinite of the ophiolite fall into two groups: those in the early 150 m.y. range (Callovian) and those in the early to mid-160 m.y. range (Bathonian) (Lanphere, 1971; Fritz, 1975, cited by Hopson and others, 1981; Evarts, 1978; Hopson and others, 1981). The age of the chaotic unit therefore falls in the range Callovian to Tithonian (Middle to Late Jurassic), is probably Tithonian, and locally may represent the entire Tithonian stage.

NATURE AND ORIGIN OF THE BASEMENT BELOW THE CHAOTIC UNIT

The nature and origin of the basement below the chaotic unit are important in any discussion of the origin of the chaotic rocks themselves. Blocks within the chaotic unit should reflect the origin of the Coast Range Ophiolite, if they were derived from it. Furthermore, any environment suggested for the formation of the ophiolite should be compatible with its later history—its fragmentation and the formation of the chaotic unit. If the environment in which the ophiolitic basement and the chaotic unit formed can be deduced, this information may place significant constraints on Californian paleogeography during Late Mesozoic time.

Hopson and others (1981) have described 23 occurrences of the Coast Range Ophiolite throughout the Coast Ranges, emphasizing common aspects of the occurrences. They have argued that the Coast Range Ophiolite was formed at a slow-spreading mid-ocean ridge, citing chemical arguments, among others, to support this conclusion.

Unfortunately, major- and perhaps even trace-element

chemistry may be an ambiguous tool in inferring ophiolite origin. In particular, source materials and magma-generating processes may be quite similar in back-arc basins and at mid-ocean ridges, even though the fundamental processes driving seafloor spreading may be quite different in the two environments. Some ophiolites may never be definitively assigned to one environment or another solely on the basis of their chemistry. More powerful constraints on the origin of most ophiolites may be those from their stratigraphy, structural settings and styles, and from their relationships to the surrounding rocks, especially sedimentary ones. Relationships between ophiolites and ophiolitic detritus may be particularly informative.

The data presented by Hopson and others (1981) include interesting stratigraphic features of the Coast Range Ophiolite that suggest to me that much of it was *not* formed at a mid-ocean ridge. Firstly, the ophiolite is usually quite thin, its mafic portion being especially so. Excluding the Stonyford high-TiO_2 basalt pile, which is possibly a seamount and may be part of the Franciscan Complex, and the metamorphosed and laterally far-traveled Santa Cruz Island remnant, only four ophiolite occurrences aggregate more than 1 km of mafic section or 2 km of total exposed ophiolite. Secondly, the stratigraphy of the Coast Range Ophiolite at many localities is quite unusual. Occurrences fall into three geographical groups, each with distinctive stratigraphy. Only in the first of these is the stratigraphy like that suggested by models of ophiolite formation at mid-ocean ridges (Dewey and Kidd, 1977).

Ophiolite occurrences west of the San Andreas Fault—Point Sal and Cuesta Ridge—are characterized by substantial sheeted sill complexes and by well developed piles of pillow basalt. These are overlain conformably by tuffaceous radiolarian cherts, followed by Great Valley sequence subsea-fan rocks. Neither large polymict olistostromes with blocks-in-matrix textures nor sedimentary serpentinite flows are reported at these localities, although mafic breccia is common below the chert at Cuesta Ridge (Page, 1972) and occurs locally within the volcanic section at Point Sal (Hopson and Frano, 1977, p. 166).

At two localities in the Diablo Range—Llanada and Del Puerto—ophiolitic volcanic rocks are overlain by thick (1.2 and 2.5 km, respectively) sequences of volcanic breccia, conglomerate, and tuff. At Llanada, the overlying rocks are siliceous volcanic conglomerates, breccias, and pyroclastic rocks. At Del Puerto Creek, the overlying rocks, which have basaltic to intermediate compositions, are interlayered with pillow basalts and cut by dikes and sills toward the base of the section (Hopson and others, 1981; Evarts, 1977, 1978).

In general, Northern Coast Range ophiolite occurrences are characterized by accumulations of ophiolitic debris of various descriptions directly overlying the ophiolite and/or interbedded with the Great Valley sequence. The ophiolitic debris ranges in thickness from tens to hundreds of meters and extends as high in the section as the upper Lower Cretaceous. Perhaps most importantly, in those areas where higher levels of the ophiolite are preserved, the debris is relatively thin and consists mainly of mafic breccia. In contrast, in those areas in which only the serpentinized harzburgite tectonite is preserved, the ophiolitic debris is often much thicker and typically contains abundant sedimentary serpentinite flows. In some areas such as the study area, these flows are part of a larger ophiolitic chaotic unit that includes blocks of diabase-rich mafic breccias and other ophiolitic rock types (Brice, 1953; Lawton, 1956; Moiseyev, 1966; Rich, 1968, 1971; Bailey and others, 1970; Maxwell, 1974; Raney, 1976; McLaughlin, 1977; Etter, 1979; Phipps and others, 1979; Carlson, 1981a, 1981b).

The described relationships suggest that at least some parts of the Coast Range Ophiolite (e.g., in the Diablo Range) formed near or even within a volcanic arc. Evarts (1977, 1978) has contended on stratigraphic and geochemical grounds that the Del Puerto ophiolite was formed near or in a back-arc basin, but could not have formed at a mid-ocean ridge. Furthermore, many features of the rocks representing the Coast Range Ophiolite within the area of the present study and elsewhere in the Northern Coast Ranges are consistent with its formation within a back-arc basin. Notably, the serpentinitized dunite and harzburgite tectonite of the Coast Range Ophiolite are cut by mafic dikes, sills, and stocks, both in the area of this study and at many other localities (for example, see Page, 1972, pp. 960–961). These intrusions into deep levels of the oceanic crust or upper mantle would not be expected in an ophiolite formed at a mid-ocean ridge and may represent back-arc intrusions.

Hopson and others (1981) have cited an apparent hiatus of 15 ± 5 m.y. between U/Pb ages of zircons from felsic plutonic rocks within the Coast Range Ophiolite and the fossil ages of radiolarians from the first overlying tuffaceous cherts. They have argued that this indicates a sustained period of non-deposition, which would not be expected in a near-arc environment. The length of the hiatus falls within the error range of fossil and radiometric dates used in the evaluation, making it suspect. Furthermore, even if the hiatus exists in the well developed ophiolite fragments to the west of the San Andreas Fault, I believe that the data cited here indicate a near-arc environment for other occurrences.

CHAOTIC ROCKS—TECTONIC SETTING AND DEPOSITIONAL ENVIRONMENT

Although stratigraphic relationships demonstrate that the chaotic unit formed within 20 m.y. or so after the ophiolite, it is possible that the processes that formed the two units are unrelated. Clearly, the Coast Range Ophiolite as it now exists is a mere remnant of a true ophiolite in many areas. This is especially true in the study area, where it consists of 2 km of serpentinite with scattered mafic intrusions. It is unlikely that the ophiolite was formed as we now see it—that harzburgite tectonite was formed on what was once the seafloor and was blanketed by later sediment. The mafic section that must have overlain the ultramafic rocks at some time in the past either must have been removed tectonically in Mesozoic or Cenozoic time or must have been stripped off by erosion during Mesozoic time. Because the mafic part of the ophiolite is not preserved in any thrust slice within or near the field area, it seems unlikely that tectonic dismemberment of the ophiolite is responsible for its present state. Tectonic dismemberment would require that the mafic section of the ophiolite has been selectively hidden. Rather, the abundance of mafic blocks in the chaotic unit and the common presence of the chaotic unit between the serpentinite and the overlying bedded Great Valley sequence suggest that the ophiolite was eroded during Mesozoic time and its erosional products redeposited on top of it. Both were then buried by Great Valley sequence submarine fans. Thus, at least in this region, contacts on which sedimentary rocks, whether ophiolitic or not, overlie serpentinite or other deep levels of the ophiolite may be mainly nonconformities, some of which may have been modified by later faulting.

Among the environments in which oceanic crust, once formed, is tectonized and may shed ophiolitic debris into nearby sedimentary basins are: the axial valleys and swells of mid-ocean ridges; major transform fault zones (whether on mid-ocean ridges or in back-arc basins); back-arc basins in general; forearc basins; and trenches (where the deep layers of tectonized ophiolite are exposed on inner trench walls). In each of these environments, peculiarities of the interactions between deformation, igneous activity, and erosion and sedimentation should be reflected in the stratigraphic relations of the ophiolitic chaotic rocks. I now examine each of these environments actualistically, attempt to infer the stratigraphy expected within each, and assess whether or not any of these match the stratigraphy of the California rocks as described above.

Mid-ocean Ridges

Stratigraphic relationships between oceanic crust and ophiolitic debris formed at a mid-ocean ridge are similar in some ways to those found in the Napa County rocks, although they are very dissimilar in other ways. Detritus formed at a mid-ocean ridge should be thin and comprised mainly of mafic rocks and should overlie complete sections of oceanic crust. None of these is true of the Napa County rocks.

Seismic activity and the presence of young faults demonstrate that the axial valleys and swells of mid-ocean ridges are tectonically active. Furthermore, axial valleys are the sites of widespread, major accumulations of talus breccias (Ballard and Moore, 1977; ARCYANA, 1975). Because similar tectonic and sedimentary processes should act all along the strike of the axial valley, basaltic and perhaps diabasic breccias should be widespread as a thin, discontinuous blanket overlying basalt and underlying the bulk of the pelagic sediment accumulations.

However, the topographic relief within and flanking axial valleys is rather low (300 m in the FAMOUS area), and individual fault scarps are usually less than 50 m high (ARCYANA, 1975; Ballard and Moore, 1977). Relief near axial swells is similar (Rea, 1975). Thus, deep levels of the oceanic crust and mantle are generally not exposed along these scarps. Although serpentinite is widely sampled in the crestal mountains of mid-ocean ridges, it is less common on the walls of axial valleys and is absent from their floors (Hekinian, 1982). Even where serpentinite is found, sampling indicates that it makes up only a minority of the outcrops, which are dominated by basaltic rocks (Hekinian, 1982, pp. 351-2). Therefore, serpentinite-rich debris accumulations like those in Napa County are not expected in or near the axial valley.

Even if serpentinite "protrusion" or solid-state extrusion is an important process in the crestal mountains away from the axial valley, as envisioned by Bonatti (1976) and Bonatti and Honnorez (1976), the stratigraphy would not be like that present in Napa County. In a mid-ocean ridge ophiolite, the basement below the ophiolitic debris should be normal, well-layered oceanic crust. Thus, ophiolitic debris formed at mid-ocean ridges should consist largely of a thin, discontinuous layer, rich in mafic breccia, overlying well-layered oceanic crust with a well developed mafic section, and in turn overlain by bedded sedimentary rocks. Considering that the Napa County chaotic rocks are thick and serpentinite-rich and lie directly on deep levels of the oceanic crust or on the upper mantle, a mid-ocean ridge origin seems unlikely.

Transform Faults

The unique history of oceanic crust formed in and very near fracture zones should result in distinctive stratigraphy (Karson and Dewey, 1978; Saleeby, this volume). Some elements of this stratigraphy are much like those seen in the Napa County rocks, but others are very different. In particular, ophiolitic chaotic rocks formed in a fracture zone should display a history of later igneous activity, which the chaotic rocks in Napa County lack.

Major transform fault zones are the sites of substantial relief on the seafloor, and large, rugged escarpments are found within them (Lonsdale, 1978; Bonatti and Honnorez, 1976; Engel and Fisher, 1975; Melson and Thompson, 1971). Topographic relief is commonly 4 km in great fracture zones such as the Romanche; average topographic slopes of 20° are common over distances of 10 km. Furthermore, deep levels of the oceanic crust and perhaps the mantle are exposed. Gabbros are frequently dredged from these fracture zones. Major ridges held up by serpentinite extend along much of their length (Bonatti and Honnorez, 1976; Engel and Fisher, 1975; Hekinian and Aumento, 1973; Melson and Thompson, 1971). Evidently, the rock types found within the Napa County chaotic unit can be found in surface exposures in major fracture zones. In contrast, in smaller transform fault zones, large serpentinite ridges do not seem to occur. During the FAMOUS dives to Transform A, a small-offset transform on the Mid-Atlantic Ridge, no large ridges were seen (ARCYANA, 1975). Small serpentinite ridges may exist in the small-offset Tamayo fracture zone in the Gulf of California (Macdonald and others, 1979), although they were not found in Transform A (ARCYANA, 1975).

In crust formed along, but not in, major transform fault

zones, ophiolitic debris should be found in two different stratigraphic situations. Crust that is formed flanking a transform fault on the side of the ridge inboard of the ridge-ridge offset, and which therefore must move along the active portion of the fault between the offset ridge segments, should have the following stratigraphy: normal ophiolite—ophiolitic debris—pillow or flow basalt—ophiolitic debris. The ophiolitic debris is shed from escarpments across the transform during times when crust on the far side is higher because of its different subsidence history. Pillow or flow basalts may be deposited above the early sediment accumulations by "spillover" when the crust passes by the offset ridge segment on the opposite side of the transform. Shallow mafic intrusions may also be emplaced at this time. Overprinting this stratigraphy will be ongoing deformation, expressed as brittle fracturing in very shallow layers and as ductile shearing in moderately deeper ones (Karson and Dewey, 1978). Thus, this crust will be characterized by a normal ophiolite sequence overlain by ophiolitic detritus, then possibly by more pillow basalt, and by more ophiolitic detritus. Complex interlayering of ophiolitic debris and pillow lavas is also possible, with both cut by shallow mafic intrusions.

Rocks formed flanking the transform fault, but outboard of the ridge-ridge offset, will not be subjected to the intense strike-slip deformation that affects crust that passes through the active ridge-ridge fault segment. It will also not be exposed to basalt spillover or diabase intrusion from the offset ridge segment. It is unlikely to accumulate large thicknesses of breccia, because it will be younger and, in general, bathymetrically higher than the crust immediately across the transform fault. However, it will undergo vertical tectonism along the fault, as it will be cooling and subsiding faster than the already cool crust across the fault. The crust outboard of the active ridge segment, then, will look substantially similar to that formed at axial valleys well away from transform faults, except that it may show evidence of high-angle dip-slip faulting.

We know little about the stratigraphy of crust within, as opposed to adjoining, modern transform zones, although several studies suggest transform origins for on-land ophiolites (Saleeby, 1978, 1979, 1982, and this volume; Karson and Dewey, 1978; Spray and Roddick, 1981, among others). The stratigraphy will probably be similar to that described above for crust formed outside, but near, the transform zone and that passes along the ridge-ridge offset. The bathymetry of the transform zone—a trough bounded by steep, high escarpments—should make it the site of thick breccia accumulations. Lithologies represented in the detritus may range from pelagic sedimentary rocks and basalt to serpentinite. These may be jumbled together in piles of debris. Some blocks may be very large. Dredge hauls, cores, and deep-tow sonar and camera studies in the Quebrada transform on the East Pacific Rise revealed abundant basaltic breccia within and flanking the transform valley. They also showed that current-transported basaltic sand is commonly intermixed with pelagic sediment (Lonsdale, 1978). The close proximity of subcrustal magmas, especially at the cut-off ridge segments, probably makes the trough the site of greater late igneous activity than any other area in or near the transform zone. Lonsdale (1978) reported that the floor of the Quebrada transform valley opposite the offset ridgecrest was "flooded with young pahoehoe flows." At such a site, then, mafic igneous rocks probably are erupted above the ophiolitic detritus and intruded into it.

Crust formed within the transform zone will show a tectonic overprint that reflects the tectonic environment of the zone. Observation of continental transform faults, of other fault types, and of suspected onland oceanic transform faults (Spray and Roddick, 1981) suggests that the zone will consist of an anastomosing network of fault strands, with major movement switching from one strand to another in space and time. A small component of vertical movement on one of these strike-slip faults may result in substantial vertical offsets, and rocks formed at distinctly different levels may be juxtaposed across these faults. The faults should cut rock layering at high angles. At very shallow levels, the fault zones should reveal brittle behavior. Because of high heat flow, zones of brittle behavior will pass rapidly downward into ductile shear zones. Therefore, crust formed within the transform should display a stratigraphy of interlayered ophiolitic detritus and pillow basalts, cut by discrete mafic intrusions, and transected by a network of faults at a high angle to layering. The faults may display brittle or ductile features, and substantial dynamothermal metamorphism of both sedimentary and igneous rocks may occur along them, especially at deeper levels.

In summary, crust formed in and near major transform fault zones should consist of normal, well-layered ophiolite overlain by interlayered basalt flows and ophiolitic debris, much of which may have been derived from the deep layers of the oceanic crust or even from the upper mantle. All rocks may be cut by discrete mafic intrusions. In addition, crust formed within the fault zone itself should be highly tectonically disrupted.

The Napa County chaotic rocks probably were not formed in a transform zone, since they lack intrusions or interlayered lava flows. Further, they lie directly on serpentinite. It might be supposed that they accumulated in the bathymetric trench along a fault zone that was underlain by serpentinite, in an area where there has been no basalt eruption. However, even in great fracture zones such as the Romanche and Vema, the transform valleys are quite narrow, about 20 km wide (Bonatti and Honnorez, 1976; Melson and Thompson, 1971). Relatively normal oceanic crust flanks the fracture zone and should be found in close proximity to hypothesized transform fault deposits. But normal oceanic crust is not found in or near the study area, and the chaotic unit outcrops in an area 50 km long by 20 km across strike. If the Napa County chaotic rocks were formed in a transform zone, that zone must have been almost parallel to the regional strike of the Great Valley sequence homocline. This would be a substantial coincidence.

Back-arc Basins

Ophiolitic detritus is likely to be formed in a back-arc

basin, and its stratigraphic relationship to the underlying basement is similar in some ways to that found in the Napa County rocks. Yet significant differences exist. In particular, detritus formed in a back-arc basin should lie on more varied basement than the Napa County chaotic unit does, and it should be interlayered with or intruded by later igneous rocks, which the Napa County chaotic unit is not.

Back-arc basins are tectonically very active and are characterized by substantial escarpments and topographic relief that commonly attains 1 km (Karig and others, 1978; Mrozowski and Hayes, 1979; Taylor, 1979). This is less than the relief in major transform fault zones, but substantially more than that in the axial valleys or highs of mid-ocean ridges. It may be enough to expose serpentinite to erosion. Even if not, ongoing tectonism within an active back-arc basin might cause a crustal block to be uplifted, eroded, and covered, and uplifted and eroded again, this time to a deeper level. By intermittent uplift, deep levels of the oceanic crust or upper mantle might be exposed over large areas without relief exceeding 1 km at any one time. Ongoing back-arc tectonism might also result in solid-state extrusion of serpentinites. Because back-arc basins can be several hundreds of kilometers long and wide, extensive layers of ophiolitic detritus, much of it consisting of deep-seated rocks, might be laid down in them.

Back-arc basins are probably the sites of substantial mafic igneous activity. The distribution of the igneous rocks is poorly known, but will probably be irregular. Diffuse spreading (Lawver and Hawkins, 1978; Dick, 1982) and ongoing tectonism within the basin probably mean that faulting, ophiolite debris deposition, and intrusion and extrusion will alternate repeatedly. This should form a stratigraphy of ophiolitic detritus interlayered with basalts. Feeders for the basalts and other intrusions should also cut these interlayered rocks, including the ophiolitic detritus.

Igneous breccias are indeed common in present-day back-arc basins. Drilling at Deep Sea Drilling Project (DSDP) Site 447 on Leg 59 in the West Philippine Basin encountered coarse basaltic sedimentary breccia (Scott and Kroenke, 1980). The drillhole at Site 453 on Leg 60, in the Mariana Trough, bottomed in 150 m of mafic breccia. Gabbro breccia overlies basaltic breccia there, and both were subjected to greenschist facies metamorphism (Hussong and others, 1978). DSDP boreholes have also penetrated large numbers of basaltic intrusions, mostly sills, in several back-arc basins (Scott and Kroenke, 1980; Dick, 1982).

The Napa County chaotic unit lacks crosscutting intrusives or interlayered lavas of any composition: its tectonic, erosional, and depositional history was not followed by a later igneous history. Hence it seems unlikely that it was formed in a back-arc basin.

Island Arcs and Trenches

Oceanic island arcs may be tectonically active, and deep levels of the oceanic crust may be exposed within them. Ophiolitic debris may accumulate. For example, at DSDP Site 448 (Leg 59) on the Palau-Kyushu Ridge, a remnant volcanic arc, the basement consists of pillowed island-arc tholeiites interbedded with volcaniclastic breccias (Scott and Kroenke, 1980). The presence of tuffs and tuffaceous cherts in the basal Great Valley sequence (Evarts, 1977, 1978) and the island arc source of the main part of the Great Valley sequence imply that an arc was nearby. However, the general absence of calc-alkaline lavas (but see Evarts, 1978) and nearby intrusives of island-arc affinity argues against an interpretation of the Napa County rocks as an arc terrane.

Trenches are sites of active tectonism. Some trenches exhibit exposures of deep layers of oceanic crust or upper mantle, especially on their inner walls (Fisher and Engel, 1969; Evans and Hawkins, 1979; Hawkins and others, 1979; Lawver and Hawkins, 1978; Bloomer and Hawkins, 1983). However, regional relationships argue against Great Valley rocks being deposited in a trench environment. The Great Valley sequence chaotic rocks lie in a basin interpreted to be a forearc basin (Ingersoll, 1978b) and are separated from the subduction complex—the Franciscan Complex—by the ophiolite on the hanging-wall of the thrust that separates the two rock units. Furthermore, Great Valley sequence rocks do not show the characteristics of a subduction complex, such as pervasive shearing or high P/T metamorphism.

Forearc Basins

Relatively little is known about the crust beneath modern forearc basins, most attention having been devoted to the sedimentary accumulations within them and scraped off beneath them (Dickinson and Seely, 1979). Forearc crust is unlikely to show significant igneous history after the onset of subduction; the forearc is generally an area of low heat flow and little or no magmatic activty. Amphibole-bearing metabasites exist in forearc basins, as breccias containing them have been drilled at DSDP Sites 460 and 461 (Leg 60) on the inner trench wall of the Mariana Trench (Hussong and others, 1978).

Forearc basins may be tectonically active, because accretionary prisms form in front of and below them (Karig and Sharman, 1975; Snavely and others, 1980; Speed and Larue, 1982). Faults exist that presumably cut the forearc basement, exposing mafic crust at the surface (Karig, 1970; Coulbourn and Moberly, 1977). In some cases, basement rocks are uplifted above sea level, for instance, on Eua in the Tonga Islands (Ewart and Bryan, 1972). The known basement in oceanic forearcs is mafic volcanic rock (Ewart and Bryan, 1972; Hussong and others, 1978). Whether this is old oceanic crust trapped in the arc-trench gap or crust created by arc magmatic processes soon after the onset of subduction, it was formed in deep water, and its present exposure near sea level, as on Eua, represents vertical uplift of the order of several km. Mafic volcanic rocks on Eua, representing oceanic basement, are presently eroding and providing detritus to the forearc basin (Ewart and Bryan, 1972).

Uplift of forearc basin crust is also observed in the Sunda arc (Karig and others, 1980) and the East Luzon arc (Karig and Wageman, 1975). In the Sunda arc, uplift of several km during the Late Oligocene and earliest Miocene exposed radiolarian cherts and basalts to erosion; these are found as conglomerate clasts in later forearc basin sediments (Karig and others, 1980, p. 84). Exposure of the forearc was widespread, and it probably served as a major sediment source. Uplift was restricted to the forearc region (Karig and others, 1980, p. 85).

The Mariana forearc basin is cut by extensive high-angle normal faults (Mrozowski and Hayes, 1980). Most of the faults strike parallel to the trend of the arc and trench. Horst blocks are common; basement is exposed on the seafloor in an outer-arc high. Many faults cut the sedimentary rocks ponded behind this high. Fault scarps range in height from 10 to 250 m. Extensive vertical tectonism persists all the way to the trench inner wall. Exposures of deep-seated rocks—serpentinite with some cumulus gabbro—are common in the Marianas forearc (Bloomer, 1982). It is probably significant that the accretionary wedge here is poorly developed. Sediments on the subducting plate are not being scraped off in significant quantities at shallow levels (Mrozowski and Hayes, 1980), but may be detached from the subducting slab at depth.

The causes of forearc uplift and vertical tectonism are not well known. Karig and others (1980) suggest that it may be due to elastic upbowing of the hanging-wall slab caused by compressive stresses at the onset of subduction. However, they point out that such uplift should not reverse quickly, but should subside slowly. Such an explanation is not consistent with situations such as the Sunda arc, where uplift is rapidly followed by subsidence. Karig and others (1980) also feel that known examples of ongoing uplift during continuing subduction are difficult to explain in this way, presumably since stresses in the hanging-wall slab caused by the initiation of subduction should relax with time.

An intuitive model for the vertical uplift of the forearc basin crust, suggested by the presence of subduction complexes beneath some forearcs, is illustrated in Figure 2. Most subduction complex material is sediment scraped off the footwall slab. This material may be pelagic sediment deposited on the footwall slab or arc-derived sediment carried through or over the forearc basin and into the trench. Some sediment may be scraped off at fairly shallow levels, forming an accretionary prism (Karig and Sharman, 1975). Such prisms are characterized by discrete, radiating imbricate thrust faults, general younging of the thrust slices downwards and toward the trench, and very low metamorphic grade. However, much of the sediment may be carried to deeper levels (Scholl and others, 1980), where it may become detached from the subducting slab and accreted to the hanging-wall block at depth (underplating). Underplating of significant volumes of low-density material should result in isostatic uplift of the overlying crust. Such uplift has apparently occurred in the South Aegean region, in the hanging-wall block above the subduction zone marked by the Hellenic Trench, where a zone averaging 70 km wide has been uplifted 2.5 to 4 km (Le Pichon and Angelier, 1981; Angelier and others, 1982).

If the thickness of sediment subducted and detached at depth is substantial, a wedge-shaped mass of underplated water-charged sediments will be trapped between the downgoing slab and the lower edge of the hanging-wall plate. Such a wedge may quickly grow to extend tens of km across strike and hundreds of km along strike and will reach to metamorphic depths. A wedge of the size shown in Figure 2 will grow in about 12 m.y. if a layer of sediment 1 km thick is subducted at a subduction rate of 6 cm/yr. If the mechanical properties of the underplated material are right, a forced corner-flow convection cell may be set up within this melange wedge (Silling and Cowan, 1977; Cowan and Silling, 1978; Cloos, 1982). A melange wedge will be characterized by continuous shear caused by approximately laminar Newtonian viscous flow of the water-charged sediment, by the absence of discrete faults, by the lack of an ordered stratigraphy, and by high P/T metamorphism (Cloos, 1982, 1983).

The incorporation of large volumes of partially-subducted material into a growing melange wedge should affect both the subducting plate and the hanging-wall plate. The former should re deflected downward and backward, causing a retreat of the bathymetric trench and a widening of the arc-trench gap. Widening has been noted in a systematic study relating the width of arc-trench gaps to the duration of the magmatism in the associated volcanic arc (Dickinson, 1973). Furthermore, accretion seems to depress the upper portion of the subducting slab (Karig and others, 1976).

The hanging-wall slab should be uplifted isostatically. A small fractional growth in the size of the melange wedge may cause several kilometers of uplift (Fig. 2). I suggest that it is this uplift that causes exposure of the forearc basement. At any one time, the zone of fastest uplift of the forearc basin will lie above the zone of most rapid thickening of the melange wedge.

Forearc uplift will expose the oceanic crust to erosion and landsliding processes. Ophiolitic olistostromes may be shed from the bathymetric high backward and forward into the forearc basin. Some of them may reach the trench, contributing ophiolitic debris derived from the hanging-wall slab to the growing subduction complex (Fig. 2). Continued uplift and erosion in such a situation may in time expose deep layers of the oceanic crust or even the upper mantle of the hanging-wall plate. If the hanging-wall crust is complex, a wide variety of ophiolitic rocks may be exposed to subaerial or submarine erosion and to landsliding.

The location of the bathymetric high will migrate across the forearc basin toward the arc as the melange wedge grows. A wave of uplift and erosion will pass across the basin, and each area of the ophiolitic basement will be eroded and then buried by debris shed from the adjoining area. In this way an extensive, thick carpet of ophiolitic debris may be formed overlying deeply-eroded oceanic crust or upper mantle. Because this occurs in the non-magmatic forearc environment, there will be no later igneous history of crosscutting intrusions or interlayered lavas. The eroded basement and ophiolitic debris could then be buried by prograding submarine fan deposits derived from the arc.

The stratigraphy suggested by this model fits that of the Napa County ophiolitic and chaotic rocks admirably. It matches all three of the salient characteristics of these rocks: the deep level of exposure of the basement, the wide variety of blocks in the chaotic unit, and the lack of igneous activity after the deposition of the chaotic unit. Because of the similarity between the stratigraphy suggested by the model and that observed in the Napa County rocks, and because of paleogeographic considerations (e.g., Ingersoll, 1978b), I conclude that the Great Valley sequence chaotic rocks were formed in a forearc basin.

Neither the corner-flow melange wedge nor its Newtonian viscous behavior is *required* to produce the forearc uplift suggested here, although textural and metamorphic evidence from Franciscan melanges suggests that it may be the correct model (Cloos, 1982, 1983). Underplating of subducted material is the critical factor, not its mechanical behavior. We still do not know how material in subduction complexes behaves; brittle Coulomb, plastic, and Newtonian viscous models have been suggested (Davis and others, 1983; Stockmal, 1983; Emerman and Turcotte, 1983), and different models may be correct in different subduction complexes, or at different places or times in any one subduction complex.

Uplift of the forearc by sediment being stuffed into the melange wedge may solve both problems noted by Karig and others (1980) in theories explaining forearc uplift as an elastic slab phenomenon. Uplift followed by subsidence may be explained by the passage of the wave of uplift across the area, and ongoing uplift may be explained merely by ongoing addition of material to the melange wedge.

Furthermore, the process of forearc uplift and erosion suggests a source for many of the ophiolitic blocks found within subduction melanges of the Franciscan Complex. These blocks may have been eroded from the Coast Range Ophiolite and carried by debris flows across the forearc basin and into the trench, where they were incorporated into the Franciscan melange wedge. Blueschist metamorphism of some melange blocks may have occurred when circulation within the wedge carried them to metamorphic depths. Debris shed from uplifted and eroded forearcs may be a significant component of other subduction complexes.

TECTONIC HISTORY

The history I suggest for the study area fits well into the larger paleotectonic scenario suggested by Schweickert and Cowan (1975). Significant petrologic and stratigraphic evidence suggests that the Coast Range Ophiolite formed in a back-arc basin. Timing considerations suggest that this may have occurred shortly before an arc-continent collision in the Sierra Nevada (the classic Nevadan orogeny) (Schweickert and Cowan, 1975). A new east-dipping subduction zone formed, trapping the back-arc-basin-generated oceanic crust in its forearc basin. As subduction continued, the addition of offscraped sediment to the melange wedge caused uplift of the forearc basin crust. This crust was exposed to erosion and then covered by the Great Valley sequence chaotic unit. Finally, the bedded Great Valley sequence, a subsea-fan complex derived from the Sierran/Klamath volcanic arc, transgressed across the area, burying the chaotic unit.

CONCLUSIONS AND SUMMARY

The basal Great Valley sequence in Napa and southern Lake Counties, California, consists of an ophiolitic chaotic unit up to 1 km thick. This chaotic unit lies above the Coast Range Serpentinite and below the bedded, arc-derived subsea-fan rocks of the Great Valley sequence. It appears repeatedly in a series of fault slices formed by Cenozoic imbricate thrust-faulting and extends for at least 50 km along strike and 20 km across strike. The unit is characterized by ophiolitic blocks set in a mudstone, serpentinous mudstone, and serpentinite matrix.

Blocks in the chaotic unit include: bedded cherts and unbedded, amorphous cherts; pillow basalts and other greenstones; unsorted aphanitic and diabasic mafic breccias, many of which probably represent submarine talus deposits; rare gabbro and plagiogranite breccias and massive gabbros; metamorphic rocks; and serpentinite. Metamorphic blocks include one phyllite and a wide variety of mafic greenschists and amphibolites. Some of these rocks show relict igneous textures. Many metamorphic textures are represented, indicating a wide range of differential stresses and degrees of syn- and post-recrystallization deformation. Mineral assemblages and textures are much like those of dredge hauls from the mid-ocean ridges, and I infer that the rocks formed by seafloor metamorphism in a spreading environment. No high-pressure phases have been found. The blueschist-bearing ophiolitic olistostromes at Wilbur Springs (Carlson, 1981a, 1981b) are younger and must have had a somewhat different origin. Serpentinite in the chaotic unit ranges from blocks to lobate, lensoid foliate breccia flows to serpentinite-pebble sandstone and mudstone. These serpentinite bodies probably represent blockfalls, landslides, and debris flows from serpentinite escarpments and detritus piles and submarine foliate serpentinite flows similar to modern subaerial ones. It is unlikely that serpentinite-rich ophiolitic debris interbedded in the Great Valley sequence was derived from diapirically emplaced serpentinites. The field evidence suggests that serpentinite was exposed on the seafloor by faulting and erosion of the Coast Range Ophiolite. Furthermore, serpentinite diapirism is mechanically implausible because the requisite density inversions probably did not occur.

Chaotic unit matrix is serpentinite, unbedded clay mudstone, or an intimate admixture of the two. Matrix has been sheared by late tectonism, but relatively unsheared mudstone matrix is common and is similar to mudstone in the bedded Great Valley sequence. Mudstone matrix differs from the blacker, scalier, and generally more pervasively sheared scaly clay matrix of the Franciscan melanges.

The chaotic unit is an accumulation of olistostromes; its initial mixing was sedimentary rather than tectonic. The ophiolitic debris is very proximal, much more so than the subsea-fan deposits with which it is to some degree mixed and which overlie it.

Throughout the study area, the chaotic unit lies directly above serpentinite that represents the Coast Range Ophiolite. No mafic section ever intervenes. The lower contact is badly exposed but is probably a nonconformity, possibly disturbed by later faulting. The processes that formed the chaotic rocks were the successive uplift, erosion, and burial by ophiolitic debris of each piece of crust. Ophiolitic debris never overlies well-formed oceanic crust. The chaotic unit is bounded above by the arc-derived subsea-fan rocks of the bedded Great Valley sequence. The contact is gradational and interfingering. Mafic intrusions that cut the serpentinite never cut the chaotic unit or the overlying Great Valley sequence.

The chaotic unit was deposited shortly after formation of the ophiolite and then was rapidly covered by the transgressing submarine fan complex. However, bedded chert blocks entrained within the chaotic unit suggest that some time elapsed between the formation of the ophiolite and the deposition of the chaotic unit.

Relief on both the lower and upper contacts of the chaotic unit was as much as 1 km. Locally within the area, basement was exposed on the seafloor, and topographic highs shed ophiolitic debris into the Great Valley sequence until at least Valanginian time (early Early Cretaceous). North of the study area, ophiolitic debris extends as high in the section as the lower Albian (late Early Cretaceous).

The Great Valley sequence chaotic unit is Late Jurassic and probably Tithonian in age, and locally makes up the entire Tithonian stage. The underlying Coast Range Ophiolite is Middle Jurassic (Bathonian and Callovian). Fossils from the immediately overlying bedded Great Valley sequence are Late Jurassic (Tithonian) and, in one case, Early Cretaceous (Valanginian). Sparse fossils from the chaotic unit itself are Tithonian.

The Coast Range Ophiolite is thin, fragmentary, and blanketed by ophiolitic detritus in the study area, as it is in much of the Northern Coast Ranges (Hopson and others, 1981). Stratigraphic, petrologic, and geochemical considerations together suggest that the Coast Range Ophiolite in the study area formed in a back-arc basin.

Although ophiolitic debris may be shed onto the oceanic crust in mid-ocean ridge axial valleys or highs, transform faults, trenches, back-arc basins, and forearc basins, study of present-day examples and their stratigraphies indicates that the Great Valley sequence chaotic rocks probably formed in a tectonically active forearc basin. An important constraint is that there was no igneous activity at the site during or after formation of the chaotic unit. Uplift and erosion of the forearc basin crust may have been caused by the rapid growth of the underlying Franciscan melange wedge shortly after subduction began. Growth of the melange wedge laterally across strike would have caused a wave of uplift to pass across the forearc basin, first uplifting the basement and exposing it to erosion, then allowing it to be buried by debris shed from the adjoining uplift. This would yield a thick blanket of ophiolitic detritus directly overlying deep levels of the ophiolite.

Blocks from the eroding Coast Range Ophiolite may also have been transported across the forearc basin and incorporated into the Franciscan melange wedge. Uplifted and eroded forearc basins represent a potential source of ophiolitic debris found in other subduction complexes as well.

The origins of the Coast Range Ophiolite and chaotic unit I propose here are compatible with the Mesozoic tectonic history of California proposed by Schweickert and Cowan (1975). The Coast Range Ophiolite probably formed by back-arc spreading behind a Middle Jurassic west-dipping subduction zone. The westward-subducting oceanic plate was also being subducted eastward beneath a proto-Sierran continental arc. When this plate was consumed, the oceanic arc to the west collided with this continental arc (Nevadan Orogeny). The oceanic arc and the suture are preserved in the Sierra Nevada foothills (Western Terrane and Melones Fault Zone). Eastward subduction beneath the new continental margin started shortly after this, during the Late Jurassic, trapping the Great Valley sequence crust, which had been formed in a back-arc basin, in a new forearc basin. The Franciscan subduction complex grew quickly after subduction restarted, and soon a growing melange wedge lifted the basement of the forearc basin and exposed it to erosion. The wave of uplift passed, and the remnant of the Coast Range Ophiolite and its blanket of ophiolitic debris were buried by the Great Valley sequence subsea fans.

ACKNOWLEDGMENTS

This work is part of the writer's doctoral thesis at Princeton University. It was funded largely by National Science Foundation grants to John Suppe, supplemented by a G.S.A. Penrose Research Grant and in the final stages by funds from the Department of Geology and the College of Arts and Sciences at Oklahoma State University and from the University of Pennsylvania. I am grateful for all this support.

John Suppe identified the chaotic unit, provided invaluable discussions as the study progressed, and carefully edited another version of the manuscript. John Maxwell has shown continuing and kindly interest in the work. His emphasis on the importance of ophiolitic debris in the Great Valley sequence has been seminal. Glenn MacPherson provided vital help in petrographic and petrologic questions. David L. Jones and Jack Miller identified the megafossils. I thank Mark Cloos, Ray Ingersoll, and Darrell Cowan for preprints; Enrico Bonatti, Christine Carlson, Mark Cloos, R. G. Coleman, R. Nowell Donovan, Karen Kleinspehn, Bob McLaughlin, Ray Ingersoll, Charles Stone, Don Videgar, David Wagner, and Mel Erskine, Jim Wolleben, and the other geologists of Eureka Resource Associates for helpful discussions both in and out of the field; and Dana Murphy, Barbara Hill, Dick George, Carol Faul, and C.M.C. Phipps for their help at various stages of the project. My special thanks to James O. Berkland and two anonymous reviewers for their painstaking and constructive criticism of the original manuscript, and to Loren Raymond for his careful editing of a later version. Their efforts to improve this work do not constitute endorsement of my views; I am solely responsible for any errors of observation and interpretation.

Finally, I thank the many landowners of the study area for their cooperation, especially Norman Pease of Mysterious Valley Ranch.

APPENDIX 1. LOCATIONS OF LOCALITIES CITED

Locality Number	Location (including 7 1/2' quadrangle)	Comments
1	Jericho Valley Quadrangle: NW corner, NE 1/4 SW 1/4, Sec. 14, T 11 N, R 5 W, Hunting Cr. gorge wall	Dike and sill complex in serpentinite
2	Jericho Valley Quadrangle: SE corner, NW 1/4 SE 1/4, Sec. 23, T 11 N, R 5 W, on Paradise Valley Rd.	Enigmatic lithic sandstone block in chaotic unit
3	Jericho Valley Quadrangle: SE corner, SW 1/4 SE 1/4, Sec. 22, T 11 N, R 5 W, bed of Pocock Cr.	Mafic breccia grading into bedded "speckled" sandstone
4	Jericho Valley Quadrangle: center N edge, SW 1/4 SW 1/4, Sec. 27, T 11 N, R 5 W, Hunting Cr. gorge exit	Probable depositional contact, chaotic unit on serpentinite
5	Jericho Valley Quadrangle: center N edge, NW 1/4 NE 1/4, Sec. 3, T 10 N, R 5 W, N bank Mysterious Cr.	Quartz-veined phyllite block in chaotic unit
6	Walter Springs Quadrangle: 200 m E of E termination of T 10 N, T 11 N bdry., Berryessa-Knoxville Rd.	Interfingering contact, chaotic unit and bedded Great Valley Sequence
7	Chiles Valley Quadrangle: general vicinity 100-400 m W of bridge over Pope Cr. arm, Lake Berryessa	Large blocks of pillow lavas with different "up" directions
8	Knoxville Quadrangle: SE corner, NW 1/4 SW 1/4, Sec. 33, T 11 N, R 4 W, 100 m NE of Nevada Cr.	Gradational contact, chaotic unit and overlying bedded Great Valley Sequence
9	Knoxville Quadrangle: Sections 20 and 21 and vicinity, T 11 N, R 4 W	Relief on contact, chaotic unit and overlying bedded Great Valley Sequence
10	Walter Springs Quadrangle: at waterline 400 m E SE corner, Sec. 3, T 9 N, R 4 W, E side headland 520'+	Serpentinite-pebble mudstones in chaotic unit; 1.5 km S of Locality 6, Fig. 1
11	Jericho Valley Quadrangle: NE corner SW 1/4 SW 1/4, Sec. 27, T 11 N, R 5 W	Serpentinite-pebble mudstones in chaotic unit; near locality 4, Fig. 1
12	Jericho Valley Quadrangle: center SW 1/2 SW 1/4, Sec. 27, T 11 N, R 5 W	Several good outcrops of of chaotic unit matrix; near Locality 4, Fig. 1
13	Walter Springs Quadrangle: 650 m N and 200 m E of SE corner, Sec. 3, T 9 N, R 4 W, on Knoxville-Berryessa Rd.	Serpentinite block in chaotic unit 1 km S of Locality 6, Fig. 1
14	Knoxville Quadrangle: NW 1/4, Sec. 36, T 11 N, R 5 W, NE side of Mysterious Valley	Modern sedimentary flows of foliate serpentinite breccia, derived from sheared serpentinite thrust sheet
15	Aetna Springs Quadrangle: Sec. 3 and E 1/2 Sec. 4, T 9 N, R 5 W, and area 500 m to N, NE side Pope Valley	(As Locality 14)
16	Walter Springs Quadrangle: E 1/2 sec. 11 and SW 1/2 SW 1/4 Sec. 12, T 10 N, R 5 W, esp. vicinity Hill 1287'	Mesozoic sed. serp. flows, elongate lenses in map view; near Locality 5, Fig. 1
17	Walter Springs Quadrangle: NE corner, SE 1/4 SE 1/4, Sec. 3, T 9 N, R 4 W, Mile 15.50, Knoxville-Berryessa Rd.	Well-preserved pillow lavas in block in chaotic unit 1.3 km S of Locality 6, Fig. 1
18	Walter Springs Quadrangle: Center E 1/2 Sec. 12, T 9 N, R 5 W, round hill on stream near gravel pit	Pope Creek Breccia (Wagner, 1975): bedded mafic breccia cut by dikes w/dike-in-dike
19	Jericho Valley Quadrangle: Center SW 1/4 SW 1/4 Sec. 35, T 11 N, R 5 W, mouth of Mysterious Cr. gorge	Block of pillow basalt overlain by mafic breccia, red chert, and more pillow basalt

REFERENCES CITED

Angel, L. H., 1948, Geology of a portion of the St. Helena quadrangle, California [M.A. thesis]: University of California at Berkeley.

Angelier, J., Lyberis, N., Le Pichon, X., Barrier, E., and Huchon, P., 1982, The tectonic development of the Hellenic Arc and the Sea of Crete: A synthesis: Tectonophysics, v. 86, p. 159–196.

ARCYANA, 1975, Transform fault and rift valley from bathyscaph and diving saucer: Science, v. 190, p. 108–116.

Averitt, P., 1945, Quicksilver deposits of the Knoxville district, Napa, Yolo, and Lake Counties, California: California Journal of Mines and Geology, v. 41, p. 65–89.

Bailey, E. H., Blake, M. C., Jr., and Jones, D. L., 1970, On-land Mesozoic oceanic crust in California Coast Ranges: U.S. Geological Survey Professional Paper 700-C, p. C70–C81.

Bailey, E. H., Irwin, W. P., and Jones, D. L., 1964, Franciscan and related rocks and their significance in the geology of western California: California Division of Mines and Geology Bulletin 183, 177 p.

Ballard, R. D., and Moore, J. G., 1977, Photographic atlas of the Mid-Atlantic Ridge Rift Valley: New York, Springer-Verlag, 114 p.

Ballard, R. D., Bryan, W. B., Heirtzler, J. R., Keller, G., Moore, J. G., and van Andel, Tj., 1975, Manned submersible observations in the FAMOUS area: Mid-Atlantic Ridge: Science, v. 190, p. 103–108.

Blatt, H., Middleton, G., and Murray, R., 1972, Origin of Sedimentary Rocks: Englewood Cliffs, N.J., Prentice-Hall, 634 p.

Bloomer, S. H., 1982, Structural and magmatic evolution of the Mariana forearc [abs.]: Geological Society of America Abstracts with Programs, v. 14, p. 445.

Bloomer, S. H., and Hawkins, J. W., 1983, Gabbroic and ultramafic rocks from the Mariana Trench: An island arc ophiolite, *in* Hayes, D., ed., The tectonic and geologic evolution of southeast Asian seas and islands, Part 2: American Geophysical Union Geophysical Monograph 27, p. 294–317.

Bogdanov, Yu.V., and Ploshko, V. V., 1967, Igneous and metamorphic rocks from the abyssal Romanche Depression: Doklady Akademii Nauk SSSR, v. 177, p. 173–176 (translation).

Bonatti, E., 1976, Serpentinite protrusions in the oceanic crust: Earth and Planetary Science Letters, v. 32, p. 107–113.

Bonatti, E., and Honnorez, J., 1976, Sections of the Earth's crust in the equatorial Atlantic: Journal of Geophysical Research, v. 81, p. 4104–4116.

Bonatti, E., Honnorez, J., and Ferrara, G., 1971, Peridotite-gabbro-basalt complex from the equatorial Mid-Atlantic Ridge: Philosophical Transactions of the Royal Society of London, Series A, v. 268, p. 385–402.

Bonatti, E., Honnorez, J., Kirst, P., and Radicati, F., 1975, Metagabbros from the Mid-Atlantic Ridge at 06°N: Contact-hydrothermal-dynamic metamorphism beneath the axial valley: Journal of Geology, v. 83, p. 61–78.

Borglin, E. K., 1949, The geology of a part of the Morgan Valley quadrangle, California [M.S. thesis]: University of California at Berkeley, 85 p.

Brice, J. C., 1953, Geology of the Lower Lake quadrangle, California: California Division of Mines and Geology Bulletin 166, 72 p. (includes 1: 62,500 map of 15′ quadrangle).

Cann, J. R., 1971, Petrology of basement rocks from Palmer Ridge, NE Atlantic: Philosophical Transactions of the Royal Society of London, Series A, v. 268, p. 605–617.

Carlson, C., 1981a, Upwardly mobile melanges, serpentinite protrusions, and transport of tectonic blocks in accretionary prisms [abs.]: Geological Society of America Cordilleran Section, Abstracts with Programs, v. 13, p. 48.

Carlson, C., 1981b, Sedimentary serpentinites of the Wilbur Springs area—a possible Early Cretaceous structural and stratigraphic link between the Franciscan Complex and the Great Valley Sequence [M.S. thesis]: Palo Alto, California, Stanford University, 105 p.

Chernysheva, V. I., 1970, Greenstone-altered rocks of rift zones in median ridges of Indian Ocean: International Geology Review, v. 13, p. 903–913.

Christensen, N. I., 1966, Elasticity of ultrabasic rocks: Journal of Geophysical Research, v. 71, p. 5921–5931.

Clark, A. W., 1948, Geology of a portion of the St. Helena quadrangle, California [M.A. thesis]: University of California at Berkeley.

Cloos, M., 1982, Flow melanges: Numerical modelling and geological constraints on their origin in the Franciscan subduction complex, California: Geological Society of America Bulletin, v. 93, p. 330–345.

Cloos, M., 1983, Comparative study of melange matrix and metashales from the Franciscan subduction complex with the basal Great Valley Sequence, California: Journal of Geology, v. 91, p. 291–306.

Conrey, B. L., 1948, Geology of the southern portion of the Morgan Valley quadrangle, California [M.A. thesis]: University of California at Berkeley, 88 p.

Coulbourn, W. T., and Moberly, R., 1977, Structural evidence of the evolution of fore-arc basins off South America: Canadian Journal of Earth Science, v. 14, p. 102–116.

Cowan, D. S., and Mansfield, C. F., 1970, Serpentinite flows on Joaquin Ridge, Southern Coast Ranges, California: Geological Society of America Bulletin, v. 81, p. 2615–2628.

Cowan, D. S., and Silling, R. M., 1978, A dynamic, scaled model of accretion at trenches and its implications for the tectonic evolution of subduction complexes: Journal of Geophysical Research, v. 83, p. 5389–5396.

Davis, D., Suppe, J., and Dahlen, F. A., 1983, Mechanics of fold-and-thrust belts and accretionary wedges: Journal of Geophysical Reseach, v. 88, p. 1153–1172.

Dewey, J. F., and Kidd, W.S.F., 1977, Geometry of plate accretion: Geological Society of America Bulletin, v. 88, p. 960–968.

Dick, H.J.B., 1982, The petrology of two back-arc basins of the northern Philippine Sea: American Journal of Science, v. 282, p. 644–700.

Dickinson, W. R., 1966, Table Mountain serpentinite extrusion in California Coast Ranges: Geological Society of America Bulletin, v. 77, p. 451–472.

Dickinson, W. R., 1973, Width of modern arc-trench gaps proportional to past duration of igneous activity in associated magmatic arcs: Journal of Geophysical Research, v. 78, p. 3376–3389.

Dickinson, W. R., and Seely, D. R., 1979, Structure and stratigraphy of forearc regions: American Association of Petroleum Geologists Bulletin, v. 63, p. 2–31.

Dickinson, W. R., Ojakangas, R. W., and Stewart, R. J., 1969, Burial metamorphism of the late Mesozoic Great Valley sequence, Cache Creek, California: Geological Society of America Bulletin, v. 80, p. 519–526.

Emerman, S. H., and Turcotte, D. L., 1983, A fluid model for the shape of accretionary wedges: Earth and Planetary Science Letters, v. 63, p. 379–384.

Engel, C. G., and Fisher, R. L., 1975, Granitic to ultramafic rock complexes of the Indian Ocean ridge system, western Indian Ocean: Geological Society of America Bulletin, v. 86, p. 1553–1578.

Etter, S. D., 1979, Geology of the Lake Pillsbury area, Northern Coast Ranges, California [Ph.D. thesis]: Austin, Texas, The University of Texas.

Evans, C., and Hawkins, J., 1979, Mariana arc-trench system: Petrology of "Seamounts" on the trench-slope break: EOS, v. 60, p. 968.

Evarts, R. C., 1977, The geology and petrology of the Del Puerto ophiolite, Diablo Range, Central California Coast Ranges, *in* Coleman, R. G., and Irwin, W. P., eds., North American Ophiolites: Oregon Department of Geology and Mineral Industries Bulletin 95, p. 121–139.

Evarts, R. C., 1978, The Del Puerto ophiolite complex, California: a structural and petrologic investigation [Ph.D. thesis], Palo Alto, Calif., Stanford University, 409 p.

Evarts, R. C., and Schiffman, P., 1979, Low-pressure hydrothermal metamorphism of the Del Puerto Ophiolite, Central California: Geological Society of America Abstracts with Programs, v. 11, p. 422.

Evarts, R. C., and Schiffman, P., 1983, Submarine hydrothermal metamorphism of the Del Puerto Ophiolite, California: American Journal of Science, v. 283, p. 289–340.

Ewart, A., and Bryan, W. B., 1972, Petrography and geochemistry of the igneous rocks from Eua, Tongan Islands: Geological Society of America Bulletin, v. 83, p. 3381–3298.

Fisher, R. L., and Engel, C. J., 1969, Ultramafic and basaltic rocks dredged from the nearshore flank of the Tonga Trench: Geological Society of America Bulletin, v. 80, p. 1373–1378.

Fox, P. J. Schreiber, E., Rowlett, H., and McCamy, K., 1976, The geology of the Oceanographer Fracture Zone: A model for fracture zones: Journal of Geophysical Research, v. 81, p. 4117–4128.

Fritz, D. M., 1975, Ophiolite belt west of Paskenta, northern California Coast Range [M.S. thesis]: University of Texas at Austin, 63 p.

Geary, E. E., and Kay, R. W., 1983, Petrological and geochemical documentation of ocean floor metamorphism in the Zambales Ophiolite, Philippines, *in* Hayes, D., ed., The tectonic and geologic evolution of southeast Asian seas and islands, Part 2: American Geophysical Union Geophysical Monograph 27, p. 139–156.

Hawkins, J., and others, 1979, Mariana arc-trench system: Petrology of the inner trench wall: EOS, v. 60, p. 968.

Hekinian, R., 1982, Petrology of the ocean floor: Amsterdam, Elsevier, 393 p.

Hekinian, R., and Aumento, F., 1973, Rocks from the Gibbs fracture zone and the Minia seamount near 53°N in the Atlantic Ocean: Marine Geology, v. 14, p. 47–72.

Helmstaedt, H., and Allen, J. M., 1977, Metagabbronorite from DSDP hole 334: An example of high-temperature deformation and recrystallization near the Mid-Atlantic Ridge: Canadian Journal of Earth Sciences, v. 14, p. 886–898.

Hopson, C. A., and Frano, C. J., 1977, Igneous history of the Point Sal Ophiolite, southern California, *in* Coleman, R. G., and Irwin, W. P., eds., North American Ophiolites: Oregon Department of Geology and Mineral Industries Bulletin 95, p. 161–183.

Hopson, C. A., Mattinson, J. M., and Pessagno, E. A. Jr., 1981, Coast Range Ophiolite, Western California, *in* Ernst, W. G., ed., Geotectonic Development of California (The Rubey Volume, Vol. I): Englewood Cliffs, N.J. Prentice-Hall, Inc., p. 418–510.

Hussong, D., Uyeda, S., and others, 1978, Leg 60 ends in Guam: Geotimes, October 1978, p. 19–22.

Ingersoll, R. V., 1976, Evolution of the Late Cretaceous fore-arc basin of northern and central California [Ph.D. Thesis]: Palo Alto, Calif., Stanford University, 200 p.

Ingersoll, R. V., 1978a, Submarine fan facies of the Upper Cretaceous Great Valley Sequence, northern and central California: Sedimentary Geology, v. 21, p. 205–230.

Ingersoll, R. V., 1978b, Paleogeography and paleotectonics of the Late Mesozoic forearc basin of northern and central California, *in* Howell, D. G., and McDougall, K., eds., Mesozoic paleogeography of the western United States: Pacific Section, Society of Economic Paleontologists and Mineralogists, Pacific Coast Paleogeography Symposium 2, p. 471–482.

Ingersoll, R. V., 1979, Petrofacies and provenance of the lower part of the Great Valley Sequence, Sacramento Valley, California: Geological Society of America Abstracts with Programs, v. 11, p. 85–86.

Ingersoll, R. V., and Dickinson, W. R., 1981, Great Valley Group (Sequence), Sacramento Valley, California, *in* Frizzell, Virgil, ed., Upper Mesozoic Franciscan Rocks and Great Valley Sequence, Central California Coast Ranges, California: Pacific Section SEPM Field Trip Guide, p. 1–33.

Karig, D. E., 1970, Ridges and basins of the Tonga-Kermadec island arc system: Journal of Geophysical Research, v. 75, p. 239–254.

Karig, D. E., and Sharman, G. F., III, 1975, Subduction and accretion in trenches: Geological Society of America Bulletin, v. 86, p. 377–389.

Karig, D. E., and Wageman, J. H., 1975, Structure and sediment distribution in the northwestern corner of the west Philippine Basin: *in* Ingle, J. C., and others, Initial Reports of the Deep Sea Drilling Project, v. 31, p. 615–620.

Karig, D. E., Caldwell, J. G., and Parmentier, E. M., 1976, Effects of accretion on the geometry of the descending lithosphere: Journal of Geophysical Research, v. 81, p. 6281–6291.

Karig, D. E., Anderson, R. N., and Bibee, L. D., 1978, Characteristics of back arc spreading in the Mariana Trough: Journal of Geophysical Research, v. 83, p. 1213–1226.

Karig, D. E., Lawrence, M. B., Moore, G. F., and Curray, J. R., 1980, Structural framework of the fore-arc basin, NW Sumatra: Journal of the Geological Society of London, v. 137, p. 77–91.

Karson, J., and Dewey, J. F., 1978, Coastal Complex, western Newfoundland: An Early Ordovician oceanic fracture zone: Geological Society of America Bulletin, v. 89, p. 1037–1049.

Lanphere, M. A., 1971, Age of the Mesozoic oceanic crust in the California Coast Ranges: Geological Society of America Bulletin, v. 82, p. 3209–3212.

Lawton, J. E., 1956, Geology of the north half of the Morgan Valley quadrangle and the south half of the Wilbur Springs quadrangle (California) [Ph.D. thesis]: Palo Alto, California, Stanford University, 259 p.

Lawver, L. A., and Hawkins, J. W., 1978, Diffuse magnetic anomalies in marginal basins: Their possible tectonic and petrologic significance: Tectonophysics, v. 45, p. 323–339.

Le Pichon, X., and Angelier, J., 1981, The Aegean Sea: Philosophical Transactions of the Royal Society of London, Series A, v. 300, p. 357–372.

Liou, J. G., and Ernst, W. G., 1979, Oceanic ridge metamorphism of the East Taiwan Ophiolite: Contributions to Mineralogy and Petrology, v. 68, p. 335–348.

Lockwood, J. P., 1971, Sedimentary and gravity-slide emplacement of serpentinite: Geological Society of America Bulletin, v. 82, p. 919–936.

Lockwood, J. P., 1972, Possible mechanisms for the emplacement of Alpine-type serpentinite: Geological Society of America Memoir 132 (The Hess Volume), p. 273–287.

Loney, R. A., Himmelberg, G. R., and Coleman, R. G., 1971, Structure and petrology of the alpine-type peridotite at Burro Mountain, California, U.S.A.: Journal of Petrology, v. 12, p. 245–309.

Lonsdale, P., 1978, Near-bottom reconnaissance of a fast-slipping transform fault zone at the Pacific-Nazca plate boundary: Journal of Geology, v. 86, p. 451–472.

Macdonald, K. C., and others, 1979, Deep-tow studies of the Tamayo transform fault: Marine Geophysical Research, v. 4, p. 37–70.

Malcolm, F. L., 1981, Microstructures of the Cayman Trough gabbros: Journal of Geology, v. 89, p. 675–688.

Maxwell, J. C., 1974, Anatomy of an orogen: Geological Society of America Bulletin, v. 85, p. 1195–1204.

McLaughlin, R. J., 1977, The Franciscan assemblage and Great Valley Sequence: *in* The Geysers-Clear Lake region of northern California: Geological Society of America Cordilleran Section Field Trip Guide, p. 3–24.

McLaughlin, R. J., 1981, Tectonic setting of pre-Tertiary rocks and its relation to geothermal resources in the Geysers-Clear Lake area, *in* McLaughlin, R. J., and others, eds., Research in the Geysers-Clear Lake Geothermal Area, Northern California: U.S. Geological Survey Professional Paper 1141, p. 3–23.

Melson, W. G., and van Andel, Tj.H., 1966, Metamorphism in the Mid-Atlantic Ridge, 22°N latitude: Marine Geology, v. 4, p. 165–186.

Melson, W. G., and Thompson, G., 1971, Petrology of a transform fault and adjacent ridge segments: Philosophical Transactions of the Royal Society of London, Series A, v. 268, p. 423–441.

Miyashiro, A., Shido, F., and Ewing, M., 1971, Metamorphism in the Mid-Atlantic Ridge near 24° and 30° N: Philosophical Transactions of the Royal Society of London, Series A, v. 268, p. 589–603.

Moiseyev, A., (Moiseef), 1966, Geology and geochemistry of the Wilbur Springs quicksilver deposits, Lake and Colusa Counties, California [Ph.D. thesis]: Palo Alto, California, Stanford University.

Moiseyev, A. N., 1970, Late serpentinite movements in the California Coast Ranges: New evidence and its implications: Geological Society of America Bulletin, v. 81, p. 1721–1732.

Mrozowski, C. L., and Hayes, D. E., 1979, The evolution of the Parece Vela Basin, Eastern Philippine Sea: Earth and Planetary Science Letters, v. 46, p. 49–67.

Mrozowski, C. L., and Hayes, D. E., 1980, A seismic reflection study of faulting in the Mariana fore arc: *in* Hayes, D. E., ed., The Tectonic and Geologic Evolution of Southeast Asian Seas and Islands: American Geophysical Union Geophysical Monograph 23, p. 223–234.

Page, B. M., 1972, Oceanic crust and mantle fragment in subduction complex near San Luis Obispo, California: Geological Society of America Bulletin,

v. 83, p. 957–972.

Phipps, S. P., 1981, Confirmation of imbricate thrusting in California Coast Ranges: Geology of Mysterious Valley area, Napa County: American Association of Petroleum Geologists Bulletin, v. 65, p. 970–971.

Phipps, S. P., 1983, Structural analysis of imbricate thrusting in Napa County, Northern California Coast Ranges [abs.]: Geological Society of America Abstracts with Programs, v. 15, p. 313.

Phipps, S. P., Kleinspehn, K. L., and Suppe, J., 1979, Ophiolitic olistostromes in the basal Great Valley Sequence, Napa County, California Coast Ranges [abs]: Geological Society of America Abstracts with Programs, v. 9, p. 122.

Ploshko, V. V., Bogdanov, Yu.A., and Knyazeva, D. N., 1970, Gabbro-amphibolite from the abyssal Romanche Trench, Atlantic region: Doklady Akademii Nauk SSSR, v. 192, p. 40–43 (translation).

Raleigh, C. B., and Paterson, M. S., 1965, Experimental deformation of serpentinite and its tectonic implications: Journal of Geophysical Research, v. 70, p. 3965–3985.

Raney, J. A., 1976, Geology of the Elk Creek-Stonyford area, northern California [Ph.D. thesis]: The University of Texas at Austin, 166 p.

Rea, D. K., 1975, Model for the formation of topographic features of the East Pacific Rise: Geology, v. 3, p. 77–80.

Rich, E. I., 1968, Geology of the Wilbur Springs Quadrangle, Colusa and Lake Counties, California [Ph.D. thesis]: Palo Alto, California, Stanford University, 101 p.

Rich, E. I., 1971, Geologic map of the Wilbur Springs Quadrangle, Colusa and Lake Counties, California: U.S. Geological Survey Map I-538.

Saleeby, J., 1978, Kings River ophiolite, southwest Sierra Nevada foothills, California: Geological Society of America Bulletin, v. 89, p. 617–636.

Saleeby, J., 1979, Kaweah serpentinite melange, southwest Sierra Nevada foothills, California: Geological Society of America Bulletin, Part I, v. 90, p. 29–46.

Saleeby, J. B., 1982, Polygenetic ophiolite belt of the California Sierra Nevada: geochronological and tectonostratigraphic development: Journal of Geophysical Research, v. 87, p. 1803–1824.

Saleeby, J. B., 1984, On the tectonic significance of serpentinite mobility and ophiolitic melange (this volume).

Scholl, D. W., von Huene, R., Vallier, T. L., and Howell, D. G., 1980, Sedimentary masses and concepts about tectonic processes at underthrust ocean margins: Geology, v. 8, p. 564–568.

Schweickert, R. A., and Cowan, D. S., 1975, Early Mesozoic tectonic evolution of the western Sierra Nevada, California: Geological Society of America Bulletin, v. 86, p. 1329–1336.

Scott, R., and Kroenke, L., 1980, Evolution of back arc spreading and arc volcanism in the Philippine Sea: Interpretation of Leg 59 DSDP results: *in* Hayes, D. E., ed., The Tectonic and Geologic Evolution of Southeast Asian Seas and Islands: American Geophysical Union Geophysical Monograph 23, p. 283–291.

Shouldice, J. R., 1947, Geology of a part of the St. Helena quadrangle, California [M.S. thesis]: University of California at Berkeley.

Silling, R. M., and Cowan, D. S., 1977, A dynamic, scaled model of accretion at trenches and its implications for Franciscan tectonics [abs.]: Geological Society of America Abstracts with Programs, v. 9, p. 499–500.

Snavely, P. D., Jr., Wagner, H. C., and Lander, D. L., 1980, Geologic cross section of the central Oregon continental margin: Geological Society of America Map and Chart Series MC-28J.

Speed, R. C., and LaRue, D. K., 1982, Barbados: Architecture and implications for accretion: Journal of Geophysical Research, v. 87, p. 3633–3643.

Spray, J. G., and Roddick, J. C., 1981, Evidence for Upper Cretaceous transform fault metamorphism in West Cyprus: Earth and Planetary Science Letters, v. 55, p. 273–291.

Stockmal, G. S., 1983, Modeling of large-scale accretionary wedge deformation: Journal of Geophysical Research, v. 88, p. 8271–8287.

Suppe, J., 1979a, Structural interpretation of the southern part of the Northern Coast Ranges and Sacramento Valley, California: Summary: Geological Society of America Bulletin, Part I, v. 90, p. 327–330.

Suppe, J., 1979b, Cross section of the southern part of the Northern Coast Ranges and Sacramento Valley, California: Geological Society of America Map and Chart Series, MC-28B.

Suppe, J., and Foland, K. A., 1978, The Goat Mountain Schists and Pacific Ridge Complex: A redeformed but still-intact Late Mesozoic Franciscan schuppen complex, *in* Howell, D. G., and McDougall, K. A., eds., Mesozoic Paleogeography of the Western United States: Society of Economic Paleontologists and Mineralogists, Pacific Section, p. 431–451.

Swe, W., and Dickinson, W. R., 1970, Sedimentation and thrusting of late Mesozoic rocks in the Coast Ranges near Clear Lake, California: Geological Society of America Bulletin, v. 81, p. 165–188.

Taliaferro, N. L., 1943, Franciscan-Knoxville problem: American Association of Petroleum Geologists Bulletin, v. 27, p. 109–219.

Taylor, B., 1979, Bismarck Sea: Evolution of a back-arc basin: Geology, v. 7, p. 171–174.

Wagner, D. L., 1975, Mesozoic geology of the Walter Springs area, Napa County, California [M.S. thesis]: San Jose, California, San Jose State University, 68 p.

Manuscript Accepted by the Society July 12, 1984

Printed in U.S.A.

Geological Society of America
Special Paper 198
1984

Composition and origin of the Kanar Melange, southern Pakistan

*Ghulam Sarwar**
Kees A. DeJong
Department of Geology
University of Cincinnati
Cincinnati, Ohio 45221

ABSTRACT

The Kanar Melange is an ophiolitic melange underlain by Cretaceous shelf-slope argillites and overlain by a thrust sheet of Cretaceous ophiolites. The melange consists of a chaotic mixture of millimeter to kilometer-size clasts embedded in a dominantly argillaceous matrix. The affinities of the clasts are continental (e.g., Mesozoic shelf carbonate rocks, clastic and sedimentary rocks, mafic-ultramafic and alkali-volcanic rock bearing conglomerates) and oceanic (serpentinite, gabbro, and basalt). The matrix is mildly deformed to strongly folded and foliated. The melange was emplaced during the Paleocene in a trough located on the western margin of the Indo-Pakistan subcontinent.

A two stage model for the origin of melange is proposed. It involves Paleocene oblique convergence between the western margin of the Indian Plate and the adjoining Neo-Tethys along a Cretaceous transform boundary. This resulted in the formation of a zone of tectonically mixed continental and oceanic rocks (proto-Kanar Melange) bounded by a trough (tectonic foredeep) on the continental margin. Debris from the mixed zone was shed into the trough, leading to the formation of the Kanar Melange. Tectonic mixing of contrasting rock types is considered an important step in the genesis of the Kanar Melange. However, the presence of sedimentary features in the matrix of the melange and its depositional relationship with the underlying Sembar Formation suggest that the emplacement of the melange may have been controlled by sedimentary processes. Oblique convergence culminated in the obduction of ophiolite nappes above the melange and regional-scale folding of the whole area. These events produced a tectonic overprint that obscured primary stratigraphic and structural relationships within the melange belt.

INTRODUCTION

The Kanar Melange occurs in the Bela Ophiolite Belt located in the southwestern part of the Kirthar-Sulaiman mountain belt (Fig. 1). The Bela Ophiolites (DeJong and Subhani, 1979; Gansser, 1979; and Sarwar, 1981, 1982) form a narrow belt approximately 380 kilometers long and are bounded to the west by the Ornach-Nal-Chaman fault system, which is at present the active western tectonic boundary of the Indo-Pakistan subcontinent (Figs. 1 and 2; Gansser, 1966; Quittmeyer et al., 1979; Lawrence and Yeats, 1979; and Kazmi, 1979). The name Kanar Melange was introduced by DeJong and Subhani (1979) for an ophiolitic melange exposed a few kilometers east of the village of Kanar in the central part of the ophiolite belt (Fig. 2). Sarwar (1981) applied the term Kanar Melange to the whole ophiolitic melange, which is tectonically overlain by the Bela Ophiolites, and which is located west of the Mor Range (Fig. 2). The melange is underlain by the Sembar Formation, which consists of Cretaceous shelf-slope type argillites (Fig. 3). The Sembar Formation is underlain, in turn, by the Shirinab Formation, a unit of

*Present Address: Conoco Inc., P.O. Box 2197, Houston, Texas 77252.

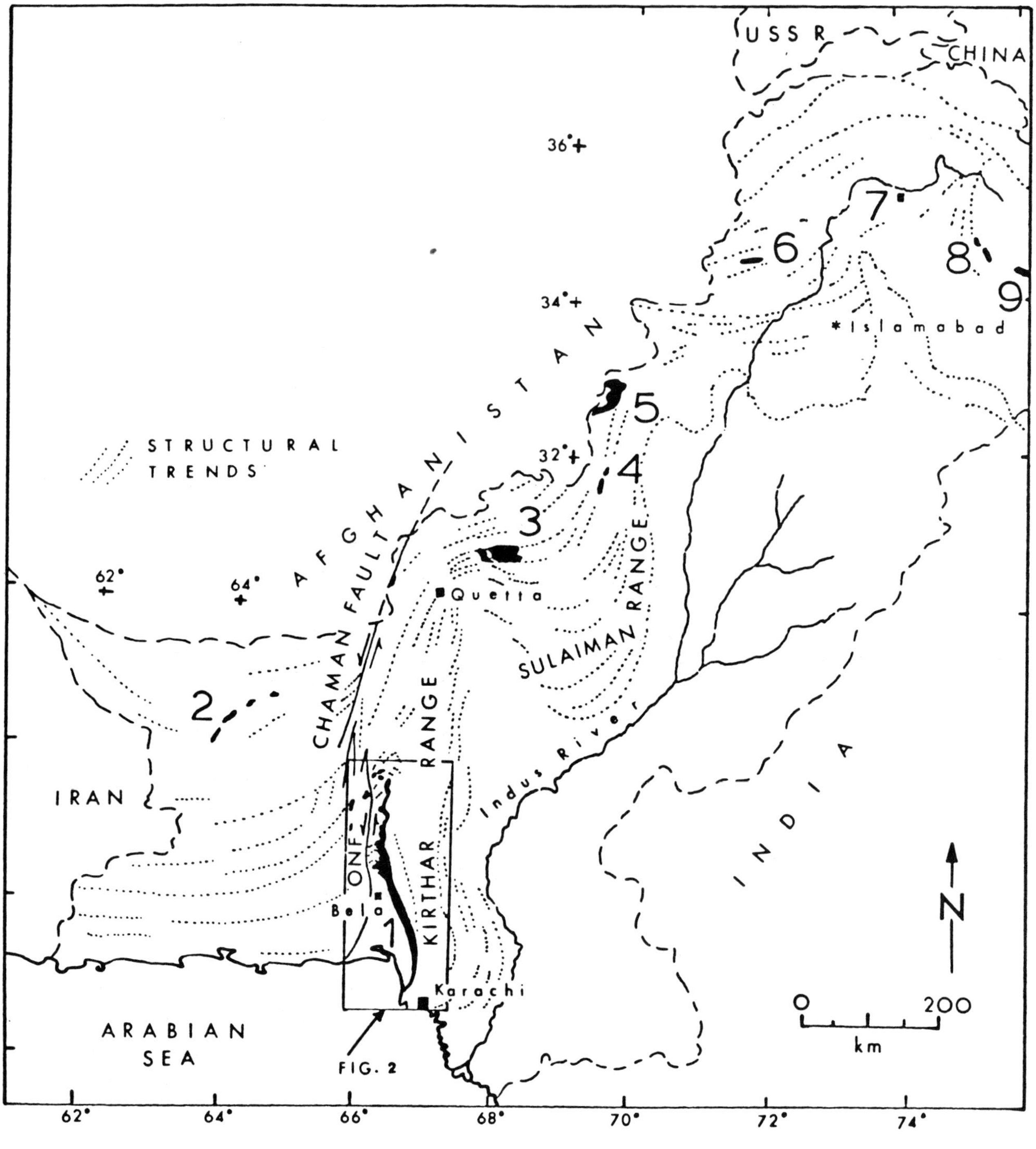

Figure 1. Map showing the ophiolite occurrences (black areas) in Pakistan; the Kirthar-Sulaiman Mountain belt, and the Chaman-Ornach-Nal (ONF) fault system. 1: Bela ophiolite belt: 2: Ras Koh ophiolite: 3: Muslimbagh ophiolite: 4: Zhob ophiolite: 5: Waziristan ophiolite: 6: Dargai ophiolite: 7-8-9: Main Mantle Thrust (MMT) - Indus Suture ophiolites (7: Chilas; 8: Buril; 9: Dras).

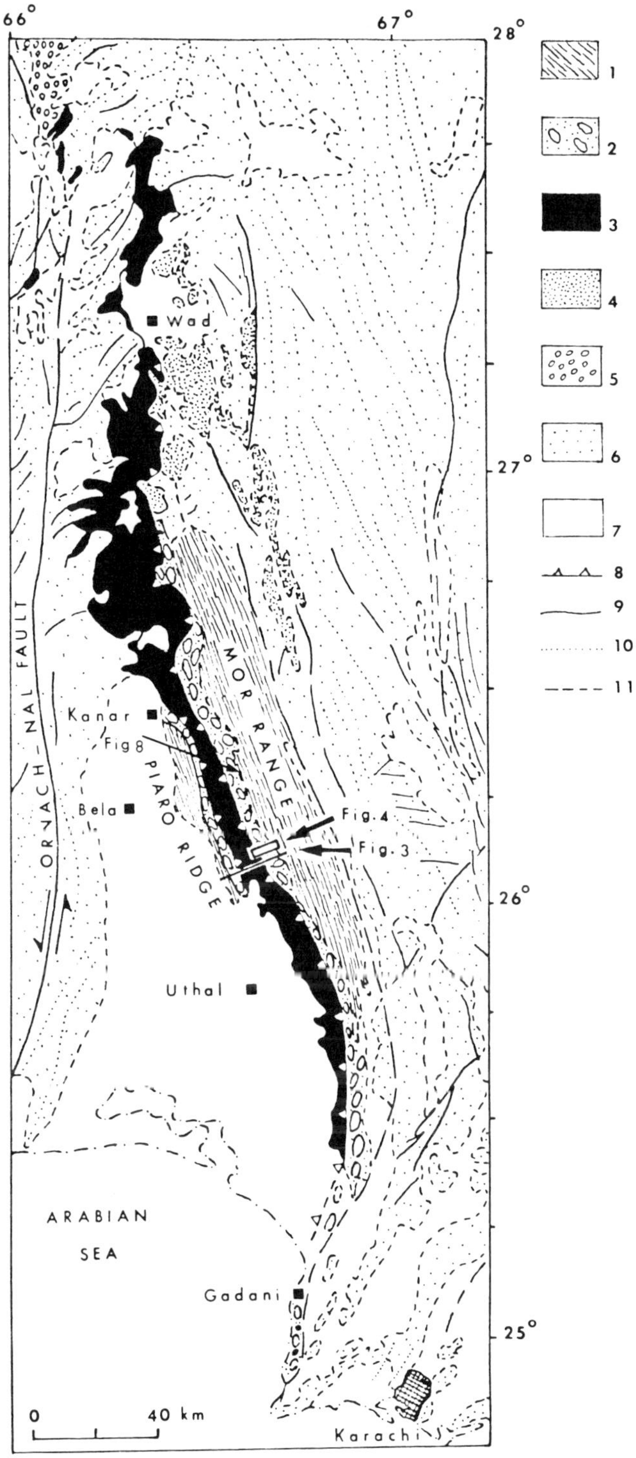

Figure 2. Schematic geologic map of the Bela Ophiolite Belt (modified after DeJong and Subhani, 1979). 1: Mesozoic sedimentary rocks (Shirinab and Sembar Formations) of the Mor Range and Piaro Ridge. 2: Kanar Melange (Paleocene). 3: Bela ophiolites (Cretaceous). 4: Thar and Bad Kachu Formations (Late Cretaceous-Paleocene). 5: Gidar Dhor Group (Late Cretaceous-Early Eocene). 6: Undifferentiated Mesozoic and Tertiary sedimentary rocks. 7: Quaternary. 8: Basal ophiolite thrust. 9: Fault. 10: Axial trace of fold. 11: Stratigraphic boundary.

Triassic and Jurassic platform-type carbonate and clastic rocks.

The Shirinab Formation, Sembar Formation, and the Kanar Melange are intruded by ultramafic and mafic sills and dikes (Mor intrusions). The Bela Ophiolites consist of Aptian-Maestrichtian mafic and ultramafic igneous rocks, and associated sedimentary rocks (Sarwar, 1981), which were obducted during the Paleocene-Early Eocene on the western margin of the Indo-Pakistan subcontinent (Allemann, 1979).

Between the latitudes 26°N and 26°30′N, the major structure of the ophiolite belt is an open NNW-SSE trending syncline. The Kanar Melange is exposed in two linear belts along the limbs (Figs. 2 and 3). The eastern melange belt extends along the western edge of the Mor Range and can be traced southwards to the village of Gadani on the Arabian Sea (Fig. 2). The western belt is exposed only along the eastern edge of Piaro Ridge, which is an upthrust mass of the Jurassic Shirinab rocks (Fig. 3).

The present paper is intended to draw the attention of melange enthusiasts to a newly discovered melange developed along a major plate boundary. The ideas expressed here are tentative in nature and are open to modification in accordance with future work.

COMPOSITION

The Kanar Melange consists of a chaotic mixture of clasts—ranging from a millimeter to several kilometers in length—of a variety of rocks, in a dominantly argillaceous matrix. Figure 4 presents a detailed map view of a part of the melange.

Clasts

Most clasts are derived from the Shirinab Formation, the Sembar Formation, and the ophiolites, but some are exotic. The Sembar clasts are difficult to identify, because they blend in with the melange matrix. Local melanocratic sills are present within the Shirinab clasts. The sills are clearly of pre-melange age and should not be confused with those occurring in the matrix of the melange itself.

Clasts of Shirinab affinity consist of grayish, thin- to thick-bedded micritic limestone (mudstone, wackestone) with or without quartzitic sandstone and interbedded splintery shale (Fig. 5). Lithologically, they resemble the rocks of the Mor Range and Piaro Range, and some large blocks contain barite-sulphide mineralization, which is characteristic of the Shirinab Formation. The Shirinab clasts occur as boulders, thin slivers, or as mountainous blocks of several cubic kilometers (e.g., the Bujji and Garo-Tutaka Mountains, Figs. 4-5).

Certain clasts consist of interlayered, finely laminated to thinly bedded limestone and calcareous shale. They occur mainly in the eastern melange belt and form folded or foliated slivers of up to a kilometer in length: these are tentatively regarded as Shirinab rocks.

Large Shirinab blocks are generally strongly deformed (Figs. 4-5). Locally, these rocks show tight, disharmonic folds with amplitudes of up to a few tens of meters. The limestone and

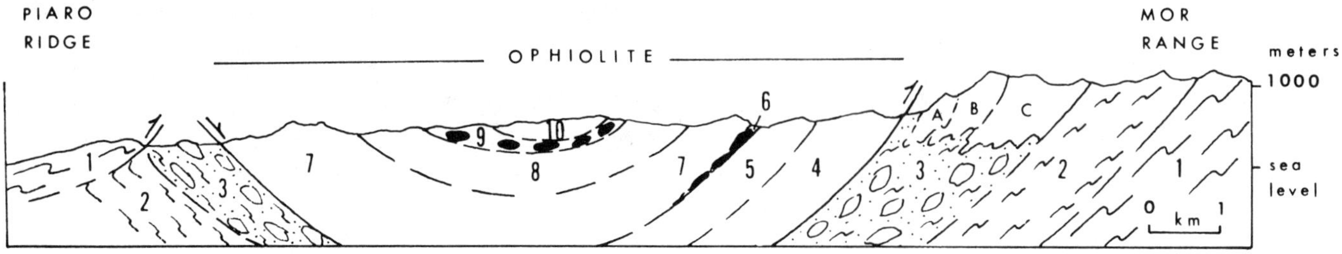

Figure 3. Schematic section across the Bela Ophiolite Belt along the location shown in Figure 2. *Allochthonous:* 5, 7, 10: Diabase - gabbro sills and host sedimentary rocks: 6, 9: inter-ophiolite melange (millimeter-kilometer size debris, mainly of deformed serpentinite, serpentine-carbonate breccia, with variable amounts of such rock types as massive/pillowed lava, diabase, gabbro, chert, marble, amphibolite, etc. set in a pelitic matrix): 4, 8: pillowed basalt and associated argillite, limestone, and chert. *Autochthonous:* 3: Kanar Melange. A = Porali Volcanic Conglomerate, B = pelagic limestone, C = Shirinab limestone, 2 = Sembar Formation-Lower-Upper Cretaceous, 1 = Shirinab Formation (Triassic-Jurassic).

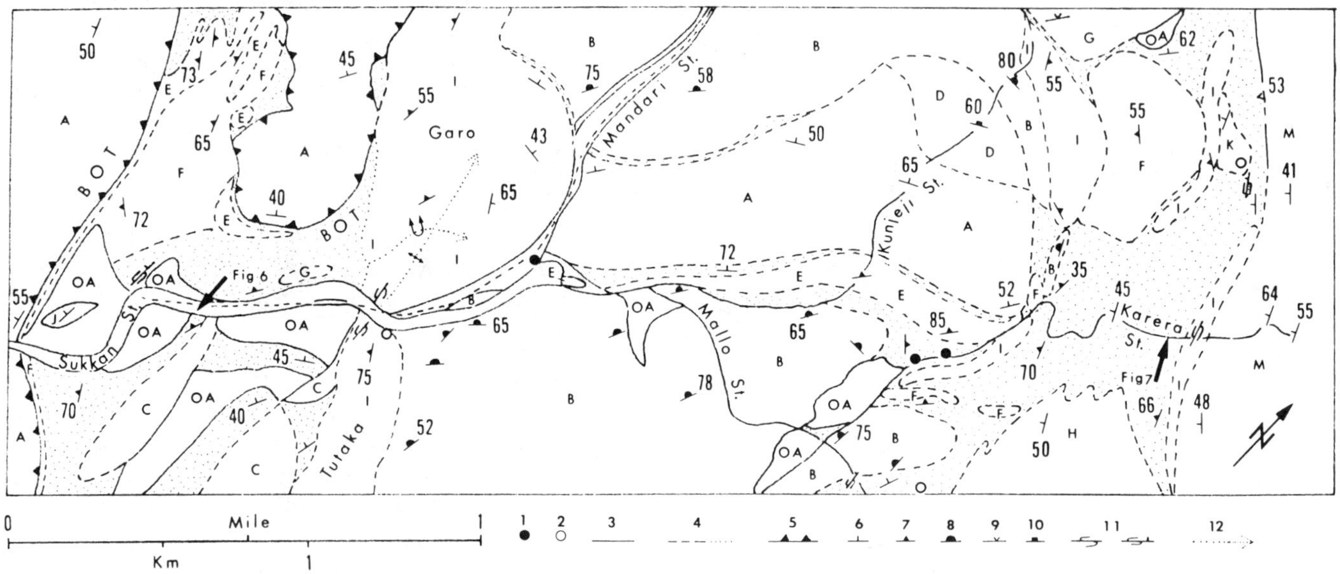

Figure 4. Geological map of a part of the Kanar Melange (location shown on Figure 2). OA: older alluvium; forms high terraces. Stipple: matrix of melange (dominantly pelitic, with a few basaltic sills). A: Diabase-gabbro sills and host chert, argillite, and limestone. B: basaltic pillow lava, chert, argillite, and limestone. C: Massive basalt. D: Basalt with columnar joints. E: Serpentinite-carbonate breccia (ophicalcite) and minor detrital serpentinite. F: Serpentinite. G: Pegmatitic gabbro with multiple dikes of basalt. H: Volcanic conglomerate (Porali Conglomerate). I: Shirinab and Shirinab (?) rocks: Limestone, sandstone, and shale. J: Chert. K: Limestone, consists of oolites and biogenic clasts. M: Sembar Formation: mudstone and shale. 1: small exposure of serpentinite. 2: fossil locality. 3: normal contact. 4: block contact, dotted where covered. 5: basal ophiolite thrust (upper contact of melange). 6: bedding. 7: foliation. 8: pillow top. 9: dike. 10. sill. 11: tight complex folds (with and without dip of axial plane). 12: plunging anticline.

sandstone beds are commonly fractured and boudinaged along the limbs of folds, whereas the shale is pervasively sheared. Bed attitudes are irregular because of refolding (e.g., Garo block, Fig. 4). The deformation within the Shirinab blocks is more intense than that of the pelitic matrix that surrounds them. The same is true for some serpentinite blocks described below.

The ophiolitic clasts found in the Kanar Melange consist of serpentinized ultramafic rocks, serpentinite-carbonate breccia (ophicalcite), detrital serpentinite, diabase, gabbro, and massive or pillowed basalt. Chert and other associated sedimentary rocks are also present. The ultramafic rocks (serpentinites) occur as small to large (± 1 km long) lenticular masses, thin slivers, and, less commonly, as blocks. They are generally foliated and occur mostly in the eastern melange belt. The rocks are mottled green in color and commonly contain 0.5 mm to 1 cm pseudomorphs of serpentine after pyroxene that have a silky luster and a well-

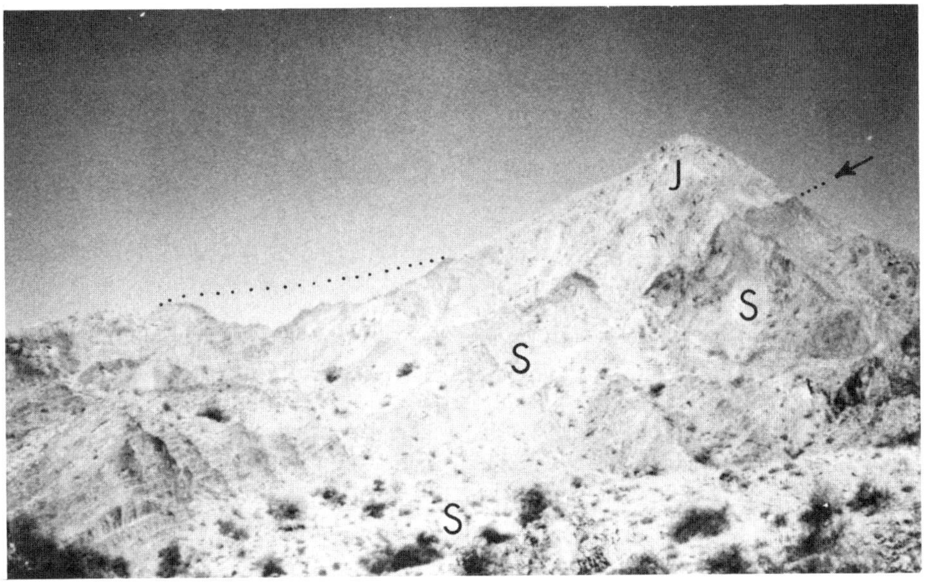

Figure 5. View of Bujji Mountain, a 3-kilometer-long block of irregularly folded Shirinab limestone (J), which occurs along the contact of the Sembar Formation (S) and the Kanar Melange. Arrow marks the lower contact of the block. The Sembar Formation here consists of westward dipping, thin bedded mudstone and shale and is only mildly deformed except along the contact with the block, where it is chaotically deformed and contains a few small blocks of limestone. (Location: 6 kilometers south of Fig. 3, looking north).

preserved cleavage. The foliation is irregularly folded and is discordant with the foliation of the surrounding matrix. Small deformed masses of pegmatitic amphibole, rodingite, and pyroxenite dikes are found within a serpentinite block. At places, serpentinite may be cut by a network of deformed carbonate and asbestos veinlets.

Serpentinite also occurs as foliated slivers interfingered with a dirty red or gray rock composed of serpentinite-carbonate breccia (ophicalcite). The breccia also occurs as separate bodies. It consists of carbonate or marly matrix cementing angular to subrounded, pebble- to jeep-sized clasts of serpentinite with subordinate chert, marl, altered massive or pillowed basalt, limestone, and lithic sandstone. The serpentinite clasts in the breccia are generally foliated, and the breccia may give the impression of being of tectonic origin. However, we believe some of it has a sedimentary origin, as suggested by crudely developed sedimentary features such as bedding, lamination, and size-sorting. Calcite crystals in the matrix are commonly undeformed as observed in thin section. Blocks of detrital serpentinite resemble the ophicalcite, although their sedimentary features, such as graded bedding and cross bedding, are more distinct than those of the ophicalcite.

The basalts, diabases, gabbros, and their associated sedimentary rocks (e.g., chert, limestone, and argillite) occur in the melange both as composite and separate blocks of up to several hundred meters in dimension. The basalts are generally pillowed and are petrographically similar to the spilite-basaltic andesite-keratophyre suite of low-K tholeiites of the Bela Ophiolites from which they were probably derived (Sarwar, 1981).

The diabases and gabbros typically show a subophitic texture and are very similar in bulk chemistry to the lava flows. The clasts of these rocks were derived from the sills that occur in the volcanic-sedimentary sequence of the ophiolites. The blocks of basalt, diabase-gabbros, and their associated sedimentary rocks are not internally deformed, in striking contrast to the Shirinab and serpentinite blocks.

Commonly, the melange itself has been reworked. The reworked material occurs as thin lithic beds or as pebble- to large house-sized blocks in the melange.

Exotic Clasts

Clasts are considered exotic if they consist of rock types which occur exclusively in the melange, and not in the overlying rocks (Bela Ophiolite) or underlying rocks (Sembar and Shirinab Formations, Mor intrusions). The comparison with the overlying and underlying rock types is restricted to the southern part of the Bela Ophiolite belt. Exotic clasts include, in order of decreasing abundance: the Porali Volcanic Conglomerate, certain gabbroic and basaltic rocks, certain biogenic limestone and lithic sandstones, two boulders of schistose metamorphic rocks, a small block of brachiopod-bearing limestone, and a block of grayish white granite. Blocks of pink granite, similar to granites occurring in the Indian Shield, have also been reported from the melange near Gadani (Fig. 2; Gansser, 1979, Fig. 36).

The Porali Conglomerate consists of clasts of massive, aphanitic or porphyritic alkali basalt with subordinate alkaline ultramafic clasts such as limburgite, and clasts of sedimentary rocks (red marl, pelagic limestone, chert, and reworked conglomerate).

The clasts are angular to well-rounded and vary in size from a small pebble to a large house, the average size being about 0.3 m. All clasts are embedded in a sandy calcareous/volcaniclastic matrix.

Clasts of the Porali Conglomerate range from small boulders up to blocks several kilometers long, all enveloped in the matrix of the Kanar Melange. The largest block lies south of the Karera stream in the eastern melange belt (Fig. 4). It is 6 km long and reveals a 1.2 km stratigraphic thickness of the conglomerate sequence. In the western belt, isolated small hills and ridges of conglomerate form most of the melange outcrop along the eastern edge of Piaro Ridge.

The petrology and chemistry (major and trace elements, including REE) of the volcanic clasts of the Porali Conglomerate are similar to that of the sills and dikes (Mor intrusions) which cut the Shirinab and Sembar Formations of the western margin of the Indo-Pakistan subcontinent (Sarwar, 1981). Thus, it is likely that the Porali volcanism represents continental volcanism that is not related to the Bela Ophiolites. The monotonous lithology, unsorted nature, and tremendous thickness of the Porali Conglomerate suggest that the conglomerate was deposited in thick piles around volcanic centers.

Exotic blocks of gabbro and basalt are found locally in the Kanar Melange. Two examples from the Kunjeji stream (Fig. 4) are described here. One is a large (1.5 km) block of a medium-grained to pegmatitic gabbro, cut by numerous basaltic dikes and probably derived from the transitional zone between the gabbro layer and the sheeted dike complex in an ophiolite sequence. The basalt varies in color (light greenish gray to gray) and texture (aphanitic to prophyritic) from dike to dike. From their mutual cross-cutting relations, three generations of dikes were identified. The dikes are up to two meters thick and arranged broadly subparallel to one another in strike. The proportion of dike to host rock is highly variable, and both extremes are seen within the block. The second example is a large block of a unique type of massive gray basalt. The basalt shows excellent hexagonal and rectangular columnar joints. Pegmatitic gabbro and massive basalt with columnar joints were not observed in the Bela Ophiolites.

Matrix

The matrix of the melange commonly consists of pelitic rocks such as shale, mudstone, and marl. Coarse material such as lithic sandstone is less common. All matrix-forming rocks are of variegated colors, with red, maroon, gray, green, purple, and brown dominant. Melanocratic sills and a few dikes (Mor intrusions) are present locally (e.g., Karera stream gorge—Fig. 4) and indicate that at least some intrusive activity either postdates melange formation or was coeval with it.

Rock fragments of different kinds and sizes are irregularly distributed through the matrix. The proportion of matrix to clasts is variable but generally low. In places, blocks are chaotically jumbled against one another with thin screens of matrix filling the inter-block spaces. Elsewhere, the melange consists of only a few blocks set in a relatively 'clean' matrix.

The deformation within the matrix is irregular and varies considerably: the rocks may be mildly deformed to strongly folded and foliated. The mildly deformed rocks show only poorly developed bedding parallel cleavage. In strongly deformed rocks, this is replaced by well-developed foliation characterized by wavy anastomosing surfaces that obliterate bedding (Fig. 6). However, distinct beds are present elsewhere in the melange (see Fig. 7). Tight or isoclinal folds are not a dominant structural feature of the matrix, but are present. Fold amplitude and axial length are small (up to a few meters and few tens of meters, respectively), and fold axes commonly plunge. Folds are asymmetrical with generally west-dipping axial planes. Locally, small, tight, and disharmonic folds, confined only to a few thin beds, are present. They do not affect the overlying or underlying horizons (Fig. 7). Such folds are probably of slump origin.

The internal folds, foliation, and the orientation of the long axis of large slab-shaped blocks are frequently discordant with that of the surrounding matrix (Fig. 4). Smaller clasts are tightly wrapped by the foliation and tend to be parallel with the fabric of the enclosing matrix (Fig. 6).

AGE

The Kanar Melange contains ophiolitic debris and is overlain by an ophiolite nappe. The ophiolites are of Aptian-Early Maestrichtian age (Sarwar, 1981). The nappe is locally weathered at the top and, near latitude 26°30′, overlain by Middle Eocene or uppermost Lower Eocene rocks (Allemann, 1979). The youngest microfauna in limestone clasts in the melange are of Maestrichtian age (Sarwar, 1981). Thus, the time span in which the ophiolite nappe had been emplaced ranges from the Early Paleocene through the early part of the Early Eocene. Because the melange is older than the final emplacement of the ophiolite nappe, it was probably formed in the Paleocene.

ORIGIN

The Kanar Melange was formed in a trough on the western margin of the Indo-Pakistan subcontinent, a part of the Indian Plate. The western boundary of the Indian Plate is presently marked by a transform zone (Fig. 8) that consists of the oceanic Owen Fracture Zone-Murray Ridge and the intracontinental Ornach-Nal-Chaman fault zone (Hunting Survey, 1960; Matthews, 1966; Barker, 1966; Gansser, 1966; Lawrence and Yeats, 1979; Jacob and Quittmeyer, 1979; and Powell, 1979). Several workers believe that the western edge of the Indian Plate has been a transform boundary since the Late Cretaceous (McKenzie and Sclater, 1971; Powell, 1979; and Norton and Sclater, 1979). It is noteworthy that the Bela Ophiolites lie adjacent to the Ornach-Nal fault zone (Figs. 2 and 8) and may have been formed near this transform boundary. There is evidence (e.g., abundant debris of serpentinite and other oceanic rocks in numerous inter-

Figure 6: View of Kanar Melange showing subangular to subrounded clasts of limestone, chert, and basalt embedded in a foliated pelitic matrix. Note hammer for scale. (Location: see Fig. 4).

Figure 7. Slump fold in a lenticular bed of thinly laminated calcareous mudstone. Note lack of bedding in the surrounding pebbly material. (Location: see Fig. 4).

ophiolite melange horizons, Fig. 9) that the Bela Ophiolites probably originated in a large oceanic fracture zone during the Cretaceous (Sarwar, 1981, 1982). Such an origin is also compatible with the rather anomalous chemistry of the lava flows (i.e., low K-Tholeiites, enriched in Fe, TiO_2, LREE, etc.) in the Bela Ophiolites.

There is thus some evidence that, since the Early Cretaceous, the Indo-Pakistan subcontinent has been bounded to the west by a transform fault. However, to allow for the ophiolite obduction and melange formation, some Paleocene-Early Eocene convergence must have taken place along this plate boundary. Yet no major convergence can be advocated, because no such features as blueschist-eclogite-bearing melange, calc-alkaline volcanics, and volcanogenic flysch have been found in the Kirthar mountain belt (Fig. 1). The magmatism that took place on the continental margin, represented by the Mor intrusions and the Porali Volcanic Conglomerate, was alkaline in nature and probably of rift type (Sarwar, 1981). We believe that oblique convergence along a

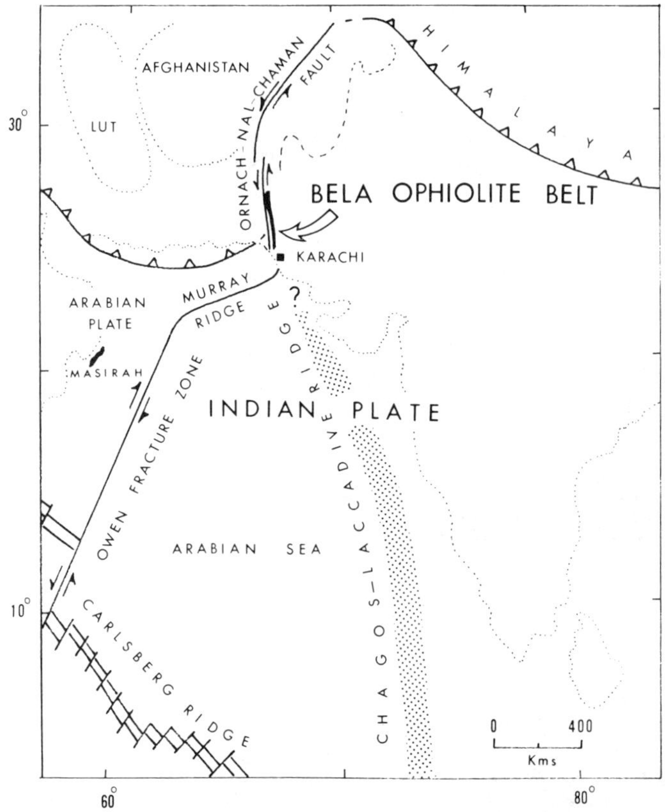

Figure 8: Present-day plate tectonic sketch map of the Indo-Arabian region. Toothed lines in the northern Arabian Sea and northern India represent the Makran subduction zone and the Himalayan collisional belt, respectively. The western tectonic boundary of the Indian Plate is marked by the Owen Fracture Zone-Murray Ridge-Ornach-Nal-Chaman fault. Note the position of the Bela Ophiolite Belt with respect to the above plate boundary. (Modified after Jacob and Quittmeyer, 1979).

transform boundary may provide the answer to the question of melange formation and ophiolite obduction on the western margin of the Indo-Pakistan subcontinent.

The Kanar Melange offers several clues to its origin.

1) The contact of the Kanar Melange and the underlying Sembar Formation is stratigraphic (Fig. 10; cf. Gucwa, 1975). The contact is irregular and well-exposed along the Mor Range. However, its irregularity appears to be due to the formation of channels in the Sembar Formation at the time of melange formation (cf. Shearman, 1976). The contact is locally disturbed because of thrusting.

2) Locally, bedding, graded bedding, and slump features occur in the matrix of the melange (Fig. 7).

3) The shapes of pebbles and boulders disposed along certain horizons are well-rounded.

4) Reworked melange material occurs in the melange. This material occurs as lithic horizons and pebble- to house-size debris (cf. Cowan and Page, 1975; Smith et al., 1976).

5) Highly deformed blocks are present in the relatively undeformed or mildly deformed melange matrix. Examples of this are the highly brecciated and tightly folded blocks of the Shirinab limestone/sandstone and the foliated serpentinite and serpentinite-carbonate breccia blocks.

There is evidence that the deformed blocks were deformed largely prior to their emplacement in the melange. The Shirinab limestone/sandstone is often thick bedded and appears to be far more resistant and mechanically competent than the pelitic matrix of the melange. Yet it was observed that large hill-size blocks of brecciated and tightly folded Shirinab limestone were embedded in the matrix, showing only open folds with incipient cleavage. The serpentinite bearing clasts in the melange were undoubtedly derived from the Bela Ophiolites, in which they occur either as large (up to a km long) slabs or as smaller clasts in the numerous debris horizons interbedded with the volcanic-sedimentary sequence. The serpentinite in the Bela Ophiolites is generally highly brecciated and/or foliated (Fig. 9) and was definitely deformed in an oceanic regime (i.e., probably a fracture zone, Sarwar, 1981, 1982). Obviously, the serpentinite was deformed long before the Ophiolite obduction and melange formation.

We believe that three major stages in the evolution of the Kanar Melange can be distinguished:

1) Deformation (folds, foliation, and cleavage) and tectonic mixing of the continental and oceanic rocks (formation of proto-Kanar Melange);

2) Emplacement of clasts in the melange matrix (formation of Kanar Melange); and

3) Deformation (folds, foliation, faults, and cleavage) associated with the emplacement of the Bela Ophiolites.

It appears that the deformation of stage (3) did not cause the fragmentation and mixing of clasts, which rules out melange formation as a direct result of the emplacement of ophiolites in the Bela area. Instead, it is more likely that the melange fabric was merely tectonically overprinted by the ophiolite obduction. We suggest that the Kanar Melange was emplaced, largely by gravity, in a trough located on the western margin of the Indo-Pakistan subcontinent. The blocks in the melange, even those measuring several kilometers in length (e.g., Fig. 5), are thus slide blocks or olistoliths. Such blocks are known from other melange belts as well (Elter and Trevisan, 1973; Robertson, 1977; and Leitch and Cawood, 1980). Also, large olistoliths and sediment slides do occur along present day active and passive margins (Moore *et al.,* 1976; Embley and Jacobi, 1977; and Jacobi, this volume).

The occurrence of the Shirinab-Porali clasts (both of continental origin) and the ophiolitic clasts in the melange implies that the continental and oceanic rocks had already been juxtaposed and mixed before they slid into the melange trough. We propose that the Kanar Melange was derived from a tectonic mixture of oceanic and continental rocks (proto-Kanar Melange) that was formed along a transform boundary between the Indian and Neo-Tethys Plates.

A possible model for the proto-Kanar Melange may be the Masirah-Oman Melange (Moseley and Abbotts, 1979; see Fig. 8

Figure 9: Inter-ophiolite melange (M), underlain by pillow lava (P) and overlain by shale-mudstone beds (Sh) with a gabbro sill (G). The melange consists of clasts of foliated serpentinite (S), serpentinite-carbonate-breccia (B), and chert, gabbro, and basalt (not shown) embedded in an argillaceous matrix. The contact of the melange and overlying shale-mudstone is depositional, as indicated by the absence of deformation in the shale, which fills in the upper irregular surface of both the foliated serpentinite and breccia. (Location: Fig. 3, Unit 9).

Figure 10: Lower contact of Kanar Melange (K) with Sembar Formation (S) in upper Khinar stream valley. The contact is marked by the sudden appearance of debris of Shirinab (?) limestone (J) in an otherwise continuous section of argillites. (Location: see Fig. 2).

for location). It forms a 5 km wide melange zone perpendicular to the trend of a sheeted dike complex which it truncates. The melange is parallel to the Owen fracture zone in the Indian Ocean and is apparently the result of brecciation and tectonic mixing of oceanic crust along a fracture zone. Subsequently, the melange has been thrust upon the southeast coast of Arabia due to oblique transform movements between the Arabian and Indian Plates.

CONCLUSIONS

A model for the obduction of the Bela Ophiolites and the

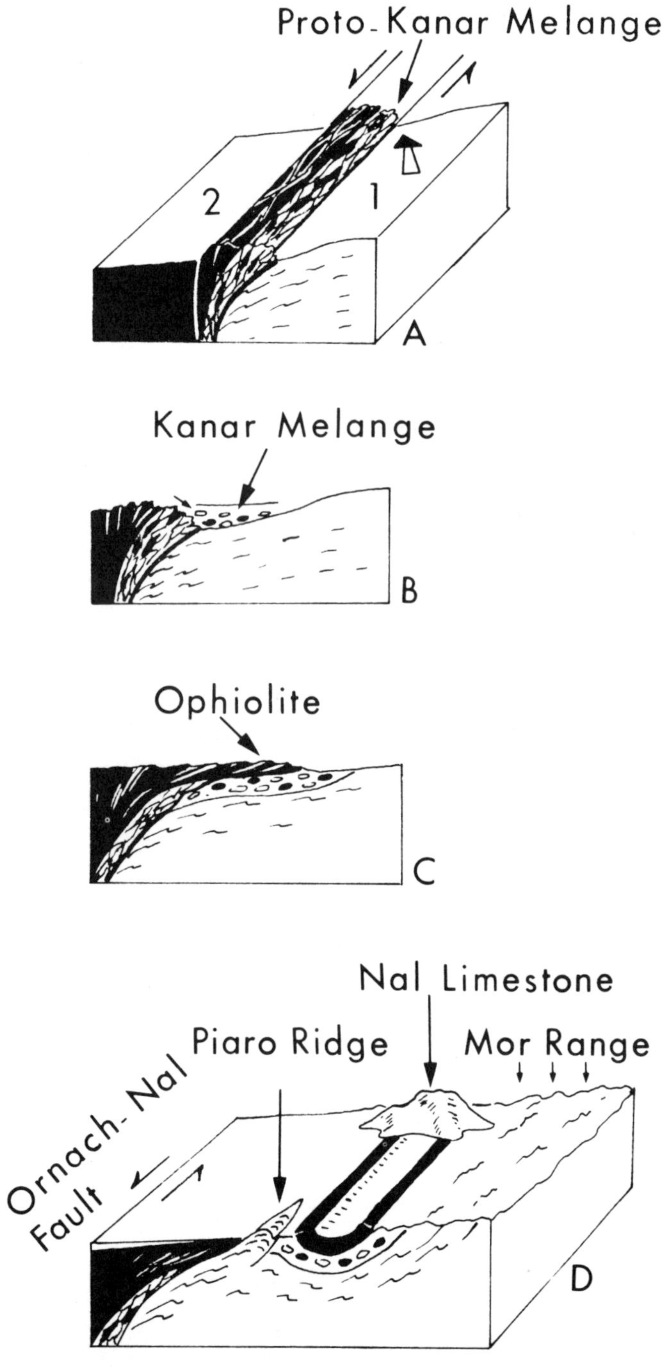

Figure 11: Model for the formation of the Kanar Melange: (A), Paleocene oblique convergence between the Indian Plate (1) and the Neo-Tethys (2) along a Cretaceous transform boundary, creates an emergent belt of tectonic melange (Proto-Kanar Melange). (B), debris derived from the tectonic melange belt is deposited as Kanar Melange in a tectonic foredeep on the continental margin. (C), progressive oblique convergence leads to obduction of the Bela Ophiolites above the Kanar Melange, inducing a tectonic fabric. Ophiolite obduction is complete by the Early Eocene (D); further convergence results in folding of the melange/ophiolite belt and upthrusting of the Piaro Ridge. Nal limestone is neo-autochthonous, deposited on the folded and eroded surface of the ophiolites in the Middle Eocene-Oligocene.

formation of the Kanar Melange is presented in Fig. 11. According to this model, the Kanar Melange formed on the continental margin adjacent to a transform boundary, in response to ophiolite obduction. The obduction resulted from an oblique convergence between the Indo-Pakistan subcontinent and the adjacent Neo-Tethys along the transform boundary. A possible cause of this oblique convergence could have been a change in the direction of motion of the Indian Plate, induced by the Early Tertiary collision of its northern continental edge with the Eurasian blocks after the closure of the Neo-Tethys (Powell, 1979). This interpretation of the obduction of the Bela Ophiolites is in accordance with Brookfield's theory (1977) that many ophiolites are emplaced along transform zones bordering continental margins during changes in the direction of plate motion, and with Fitch's suggestion (1972) that convergence with slip that is oblique to the plate margin is resolved into parallel zones of transcurrent faulting and thrusting.

According to the above model, the Kanar Melange developed in an elongated trough on the continental margin in front of the eastward moving ophiolite nappe. The mixing of continental and oceanic rock types was achieved in the transform zone between the Indian Plate and the Neo-Tethys (Fig. 11A). The tectonic melange along this transform was subjected to erosion, emplaced by sedimentary processes (Fig. 11B), and subsequently overridden by the ophiolite nappe (Fig. 11C). During Early Middle Eocene-Oligocene, neoautochthonous marine carbonates were deposited on top of the ophiolites (Fig. 11D; Allemann, 1979). The synclinal folding of the ophiolite nappe, eastward thrusting of the Piaro Ridge, and uplift of the whole area probably all resulted from the Neogene deformation which is thought to be responsible for shaping the major tectonic features of the Kirthar-Sulaiman Belt (Hunting Survey, 1960; Powell, 1979; and Sarwar and DeJong, 1979).

ACKNOWLEDGMENT

We thank Waheed-ud-Din Ahmad, Akhtar Subhani, Ishaq Durrazai, and Jafar Qureshi, all of the Geological Survey of Pakistan, for their help with the fieldwork. We also thank Karen Christensen, Carol Kazmer, Robin Lease, Loren Raymond, Jeff Rosenfeld, Pinar Yilmaz, and two anonymous GSA reviewers for their criticism on an early draft of the manuscript. Finally, we thank Ms. Joan Harman, who typed the manuscript and Anglicized our Urdu-Dutch expressions. The study was funded by grants NSF-INT 7814553 and 8013158 of the National Science Foundation, Washington, D.C.

REFERENCES CITED

Allemann, F., 1979, Time of emplacement of the Zhob Valley ophiolites and Bela ophiolites, *in* Farah, A., and DeJong, K. A., eds., Geodynamics of Pakistan: Geological Survey of Pakistan. Quetta, 1979, p. 215–242.

Barker, P. F., 1966, A reconnaissance survey of the Murray Ridge: Philosophical Transactions of the Royal Society of London, v. 259, p. 187–197.

Brookfield, M. E., 1977, The emplacement of giant ophiolite nappes, 1. Mesozoic-Cenozoic examples: Tectonophysics, v. 37, p. 247–303.

Cowan, D. S., and Page, B. M., 1975, Recycled Franciscan material in Franciscan melange west of Paso Robles, California: Geological Society of America Bulletin, v. 86, p. 1089–1095.

DeJong, K. A., and Subhani, A. M., 1979, Note on the Bela Ophiolites with special reference to the Kanar area, *in* Farah, A., and DeJong, K. A., eds., Geodynamics of Pakistan: Geological Survey of Pakistan, Quetta, 1979, p. 263–270.

Elter, P., and Trevisan, L., 1973, Olistostromes in the tectonic evolution of the northern Apennines, *in* DeJong, K. A., and Scholton, R., eds., Gravity and Tectonics: John Wiley and Sons, New York, 502 p.

Embley, R. W., and Jacobi, R. D., 1977, Distribution and morphology of large submarine sediments slides and slumps on Atlantic continental margins: Marine Geotechnology, v. 2, Marine slope stability, p. 205–228.

Fitch, T. J., 1972, Plate convergence, transcurrent faults and internal deformation adjacent to southeast Asia and the western Pacific: Journal of Geophysical Research, v. 77, p. 4432–4460.

Gansser, A., 1966, The Indian Ocean and the Himalayas—a geologic interpretation: Eclogae Geological Helvetiae, v. 59, p. 831–848.

——1979, Reconnaissance visit to the ophiolites in Baluchistan and the Himalaya, *in* Farah, A., and DeJong, K. A., eds., Geodynamics of Pakistan: Geological Survey of Pakistan, Quetta, p. 193–214.

Gucwa, P. R., 1975, Middle to Late Cretaceous sedimentary melange, Franciscan Complex, northern California: Geology, v. 3, p. 105–108.

Hunting Survey, 1960, Reconnaissance geology of West Pakistan: A Colombo Plan Cooperative Project: Government of Canada, Toronto, 550 p.

Jacob, K. H., and Quittmeyer, R. C., 1979, The Makran region of Pakistan and Iran: Trench-arc system with active plate subduction, *in* Farah, A., and DeJong, K. A., eds., Geodynamics of Pakistan: Geological Survey of Pakistan, Quetta, p. 305–318.

Kazmi, A. H., 1979, Active fault systems in Pakistan, *in* Farah, A., and DeJong, K. A., eds., Geodynamics of Pakistan: Geological Survey of Pakistan, Quetta, p. 285–294.

Lawrence, R. D., and Yeats, R. S., 1979, Geological reconnaissance of the Chaman fault in Pakistan, *in* Farah, A., and DeJong, K. A., eds., Geodynamics of Pakistan: Geological Survey of Pakistan, Quetta, p. 351–358.

Leitch, E. C., and Cawood, P. A., 1980, Olistoliths and debris flow deposits at ancient consuming plate margins: An Eastern Australian example: Sedimentary Geology, v. 25, p. 5–22.

Matthews, D. H., 1966, The Owen fracture zone and the northern end of the Carlsberg Ridge: Philosophical Transactions of the Royal Society of London, v. 259, p. 172–186.

McKenzie, D., and Sclater, J. G., 1971, The evolution of the Indian Ocean since the Late Cretaceous: Royal Astronomical Society, Geophysical Journal, v. 24, p. 437–528.

Moore, D. G., Curray, J. R., and Emmel, F. J., 1976, Large submarine slide (olistostrome) associated with Sunda Arc subduction zone, Northeast Indian Ocean: Marine Geology., v. 21, p. 211–226.

Moseley, R., and Abbotts, I. L., 1979, The ophiolite melange of Masirah, Oman: Journal of the Geological Society of London, v. 136, p. 713–724.

Norton, I. O., and Sclater, J. G., 1979, A model for the evolution of the Indian Ocean and the breakup of Gondwanaland: Journal of Geophysical Research, v. 84, p. 6803–6830.

Powell, C. M., 1979, A speculative tectonic history of Pakistan and surroundings: Some constraints from the Indian Ocean, *in* Farah, A., and DeJong, K. A., eds., Geodynamics of Pakistan: Geological Survey of Pakistan, Quetta, p. 5–24.

Quittmeyer, R. C., Farah, A., and Jacob, K. H., 1979, The seismicity of Pakistan and its relation to surface faults, *in* Farah, A., and DeJong, K. A., eds., Geodynamics of Pakistan: Geological Society of Pakistan, Quetta, p. 271–284.

Robertson, A.H.F., 1977, The Moni Melange, Cypress: An olistostrome formed at a destructive plate margin: Journal of the Geological Society of London, v. 133, p. 447–466.

Sarwar, G., 1981, Geology of the Bela Ophiolites in the Wayaro Area, Las Bela District, South Central Pakistan: (Ph.D. Dissertation): University of Cincinnati, Cincinnati, Ohio, 375 p.

—1982, Bela Ophiolites, south central Pakistan—A Cretaceous plate boundary transform complex: Geological Society of America Abstracts with Programs, v. 14, no. 7, p. 608.

——and DeJong, K. A., 1979, Arcs, oroclines, syntaxes: The curvatures of mountain belts in Pakistan, *in* Farah, A., and DeJong, K. A., eds., Geodynamics of Pakistan: Geological Survey of Pakistan, Quetta, p. 341–350.

Shearman, D. J., 1976, The geological evolution of southern Iran: The report of the Iranian Makran Expedition: The Geographical Journal, v. 42, Part 3, p. 393–413.

Smith, G. W., Howell, D. G., and Ingersoll, R. V., 1979, Late Cretaceous trench-slope basins of central California: Geology, v. 7, p. 303–306.

Manuscript Accepted by the Society July 12, 1984

Printed in U.S.A.

Geological Society of America
Special Paper 198
1984

An example of an obduction melange: The Alakır Çay unit, Antalya Complex, southwest Turkey

P. O. Yılmaz*
Mobil Oil Corporation
P.O. Box 5444
Denver, Colorado 80217

J. C. Maxwell
Department of Geological Sciences
The University of Texas
Austin, Texas 78712

ABSTRACT

The Antalya Complex in southwestern Turkey lies along the Cyprus-Hatay ophiolitic belt of the Eastern Mediterranean. The Antalya Complex contains, from west to east, the autochthonous Mesozoic-Cenozoic platform carbonates of the Bey Dağları massif and three allochthonous units: the Kumluca Complex of imbricate Mesozoic continental-margin sedimentary rocks; the Alakır Çay unit, a mélange of oceanic and carbonate margin rocks; and the Late Cretaceous Tekirova (partial) ophiolite.

One of the allochthonous units, the Alakır Çay, is interpreted here as a mélange. The Alakır Çay unit contains a great variety of clasts, derived from continental margin to pelagic environments. Ophiolitic rocks are the most conspicuous component. The matrix varies from pelagic marls to pillow-basalt flows and serpentinite and is essentially absent locally. The Alakır Çay mélange contains ophiolitic debris of probable Late Cretaceous age and is overlain by a conglomerate sequence with clasts containing a Maestrichtian fauna.

The Alakır Çay mélange formed during obduction in front of advancing ophiolite sheets, thrust over a relatively flat foreland of the Kumluca continental margin. The initiation of deformation is assumed to be a result of thrusting related to compression. The mélange movement took place as submarine slides that successively occupied flysch basins developed in front of the advancing thrust mass during Late Cretaceous time. The Alakır Çay mélange was emplaced on the Bey Dağları autochthon during the late Miocene.

INTRODUCTION

This study presents details of the stratigraphy and structure of the extensive Alakır Çay mélange, composed largely of ophiolitic rocks, which was obducted with other allochthonous units onto continental crust of the Bey Dağları region of southwest Turkey (Fig. 1). An evolutionary model is proposed for the formation and emplacement of the Alakır Çay mélange, developed on the basis of nine months of detailed mapping, compilations based on published material, and laboratory work.

As defined in this paper, the Antalya Complex in southwest Turkey (Fig. 2) consists of the following, beginning with the structurally lowest unit: the autochthonous Mesozoic-Cenozoic continental platform rocks of Bey Dağları; the Kumluca unit of allochthonous, imbricated, Mesozoic continental margin rocks; the Alakır Çay unit of oceanic and carbonate margin rocks, assembled during the latest Cretaceous by tectonic and gravitational mechanisms into multiple slices interlayered with mélange, which

*Present Address: Exxon Production Research Co., P.O. Box 2189, Houston, Texas 77001.

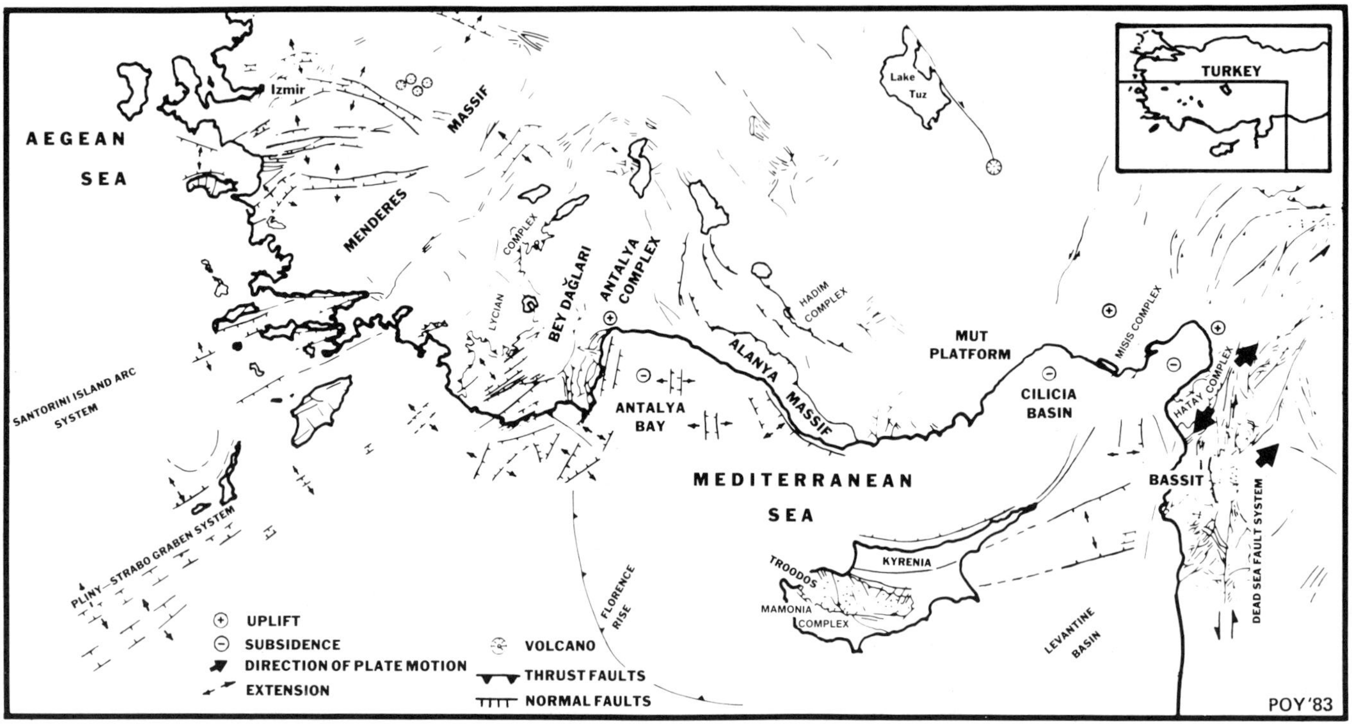

Figure 1. Index map showing the Antalya Complex in relation to the present tectonic elements of the Eastern Mediterranean.

includes as a subdivision the Teke-Tahtalı Dağ unit of exotic Paleozoic-Mesozoic platform rocks; and the Tekirova unit, a Late Cretaceous partial ophiolite sequence (Table 1). The area discussed covers approximately 2600 km^2 lying along the western shore of the Gulf of Antalya in southwest Turkey (Fig. 2).

The Antalya Complex is part of the southwest Taurid Mountains of the Alpine system, which continues westward to the Hellenides and eastward to the Zagros. Much of the Alpine-Himalayan system is characterized by the presence of slivers of ophiolites, believed to be remnants of the Mesozoic-Cenozoic Tethys oceanic crust. The Antalya Complex (after Woodcock and Robertson, 1977) was originally termed the Antalya Nappes by Lefévre (1967) and "Antalya Birliği" by Özgül (1976). The dominant north-trending tectonic grain of the Complex recognized by previous workers (Colin, 1962; Juteau, 1968, 1970, 1977, 1979; Juteau and others, 1973, 1977; Brunn and others, 1970, 1971; Baykal and Kalafatçıoğlu, 1973; Dumont and others, 1972; Poisson, 1977, 1978; Marcoux, 1976, 1977; Woodcock and Robertson, 1977; Yılmaz, 1978, 1981; Yılmaz and others, 1981) is due to major faults that juxtapose components of the Complex having contrasting lithology, age, and structure.

All of the above workers considered the Antalya Complex to be an allochthonous assemblage dominated by Mesozoic ophiolitic and sedimentary rocks. There is, however, no general agreement about the places of origin, nor the direction of transport, of the allochthons. Lefévre (1967) first proposed that the alochthons were nappes, transported eastward over the Bey Dağları autochthonous carbonate platform. Brunn and others (1977) and Ricou and others (1974, 1975) proposed that the Antalya Nappes originated in a Mesozoic ocean basin located to the north. However, because sedimentation was continuous on the autochthon from Jurassic to Miocene time (Poisson, 1978; Table 1), an internal origin from the north requires either great distances of transport (250 km, Ricou and others, 1974) or that part of the Bey Dağları platform be allochthonous (Ricou and others, 1979). An external origin for the Complex in a Mesozoic ocean basin located to the south was proposed by Dumont and others (1972), Brunn (1974), Monod and others (1974), and Monod (1976). Recent works by Delaune-Mayere and others (1977) and Ricou and others (1979) favor an internal origin from the north, whereas Robertson and Woodcock (1980a) propose emplacement from an external, southern basin. Paleogeographic reconstructions by Şengör and Yılmaz (1981) propose emplacement of the Antalya Nappes from an external southern basin, a remnant of the Neo-Tethys ocean.

Robertson and Woodcock (1980a, 1980b) and Woodcock and Robertson (1981a, 1981b) have postulated that north-trending strike-slip faults dominate the structure. They suggest that the Antalya Complex was an originally intact part of the Mesozoic continent-ocean boundary tectonically disrupted by wrench faulting during the Late Cretaceous and Tertiary. Elongate masses of sheared rocks separated by anastomosing shears, said to be of strike-slip origin, are considered by them to be field evidence supporting their interpretation. Mylonitic shear zones,

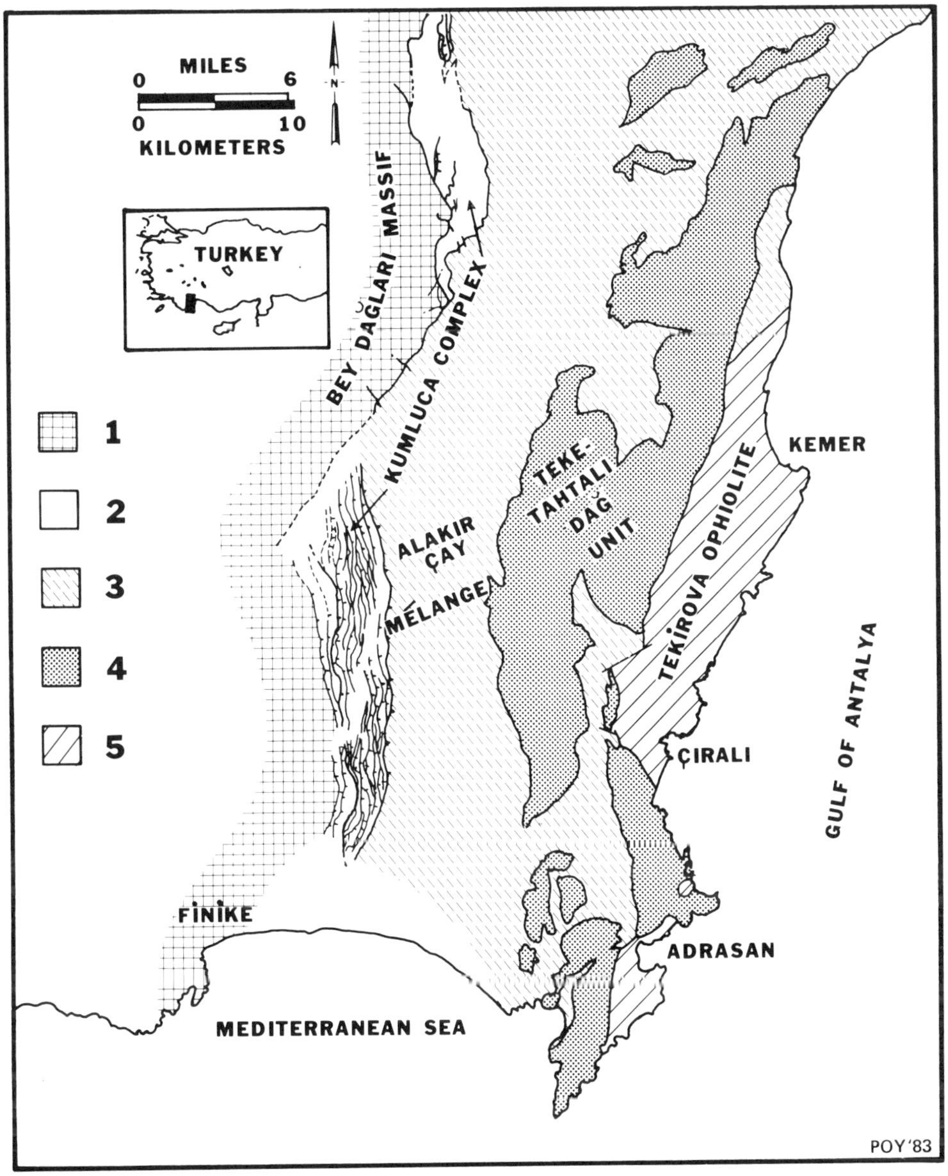

Figure 2. Generalized map of the Antalya Complex showing distribution of major units: 1 - Bey Dağları autochthon; 2 - Kumluca Complex; 3 - Alakır Çay mélange; 4 - Teke-Tahtalı Dağ unit; 5 - Tekirova ophiolite.

on millimeter to kilometer scales, within the tectonized rocks in the southeastern part of the Antalya Complex, have been related to a postulated transform fault by Reuber (1982).

Subdivision of the Antalya Complex into north-trending units, separated by major faults, has been based, by all authors, on differences in constituent rocks, tectonic style, and age (Fig. 3). Brunn and others (1970, 1971) and Dumont and others (1972) defined three nappe units, transported over large distances. Robertson and Woodcock (1981a, 1981b, 1981c) recognize four units, described as rocks of a deformed Mesozoic carbonate continental margin which also lapped over adjacent mafic crust. Yılmaz (1981) and Yılmaz and others (1981) followed Robertson and Woodcock's (1981a) division for the Antalya Complex but interpreted one of the allochthonous units as a mélange. This unit, the Alakır Çay mélange, is the topic of this paper.

The Alakır Çay mélange is equivalent to the Middle Nappe of Brunn and others (1970, 1971), Dumont and others (1972), Ricou and others (1974, 1975, 1979), Marcoux (1976, 1977), and Juteau (1975) and is described by them as consisting of Late Triassic Ophiolite and Late Triassic to Upper Cretaceous pelagic sedimentary rocks. Robertson and Woodcock (1980a, 1981a, 1981b, 1981c) interpreted the Alakır Çay unit as tectonized Late Cretaceous ophiolite with Upper Triassic alkaline basalts and pelagic sediments, originating on mafic crust. In this paper and

TABLE 1: GENERALIZED STRATIGRAPHIC COLUMN

ANTALYA COMPLEX, SW TURKEY

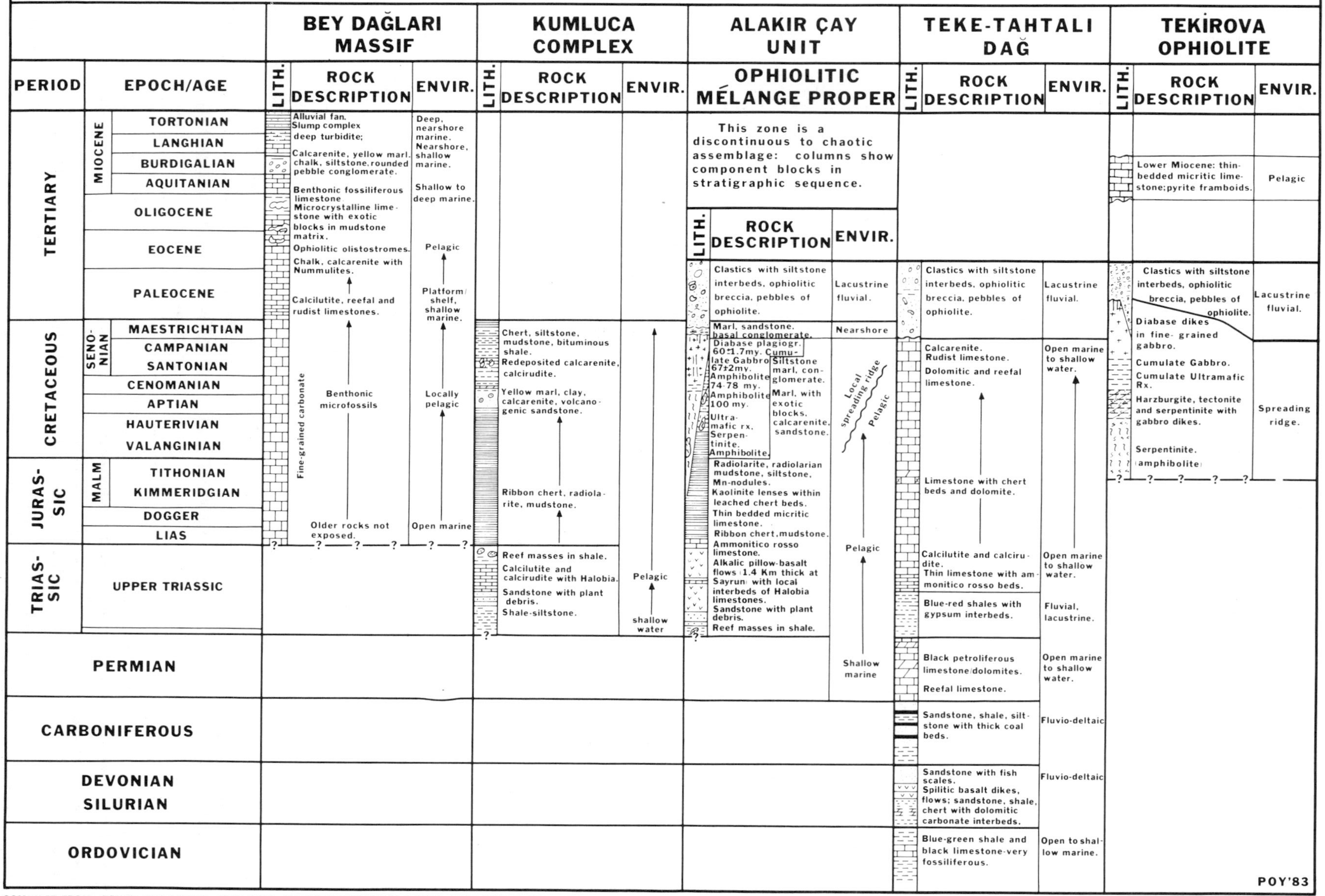

COMPILED FROM: BRUNN AND OTHERS, (1970; 1971); DELAUNE-MAYERE AND OTHERS, (1977); JUTEAU (1975; 1977); MARCOUX (1976; 1977); POISSON (1977); ROBERTSON AND WOODCOCK (1977; 1980; 1981); YILMAZ (1978; 1981).

GENERAL COMPARISON OF DIFFERENT TECTONIC SUBDIVISIONS OF THE ANTALYA COMPLEX

Juteau, 1975 Juteau & Whitechurch, 1979 Delaune-Mayere et al, 1977	Robertson & Woodcock, 1980a; 1981a,b,c	Yılmaz, 1981 and this study
Bey Dağları platform:	Bey Dağları zone:	Beg Dağları Massif
Upper Mesozoic-Tertiary carbonate platform.	Upper Mesozoic-Tertiary carbonate platform.	Upper Mesozoic-Tertiary carbonate platform.
Lower Antalya nappe:	Kumluca zone:	Kumluca unit:
Çatal Tepe fm.: Mesozoic shelf and pelagic sedimentary rocks.	1) Northern part: Bilelyeri/Dereköy Group sediments (Late Triassic-Upper Cretaceous). Southern part: Alakır Çay Group sediments (Late Triassic-Lower Cretaceous).	Imbricated Mesozoic pelagic sediments.
Middle Antalya nappe:	Gödene zone:	Alakır Çay unit:
Alakır Çay fm.: Late Triassic ophiolite and Late Triassic-Upper Cretaceous pelagic sedimentary rocks. Late Cretaceous ophiolite at Adrasan-Tekirova with no genetic relationship to the Alakır Çay ophiolite.	Tectonized Late Cretaceous ophiolite and Alakır Çay Group sediments. Upper Triassic alkaline basalts and pelagic sediments origined on a mafic crust.	Late Cretaceous tectonically imbricated ophiolitic mélange. Upper Triassic-Jurassic alkaline basalts with pelagic sediments and Upper Jurassic-Early Cretacous alkaline basalts with pelagic sediments origined on oceanic-type crust. Late Cretaceous ophiolite which consists of serpentinite, ultramafic cumulates, gabbro & diabase.
Upper Antalya nappe:	Kemer zone:	Teke-Tahtalı Dağ trend rocks:
Kemer fm.: Paleozoic-Mesozoic shelf/platform sedimentary rocks	Paleozoic-Mesozoic sediments and carbonate build-ups on continental basement.	Paleozoic-Mesozoic sedimentary sequences origined on continental crust. (a subdivision of the Alakır Çay unit)
	Tekirova zone:	Tekirova ophiolite:
	Late Cretaceous partial ophiolite.	Late Cretaceous partial ophiolite with basal tectonites and ultramafic/mafic cumulate rocks overlain unconformably by conglomerates with Maastrichtian-age fauna bearing clasts.

Figure 3. General comparison of different tectonic subdivisions of the Antalya Complex.

elsewhere (Yılmaz, 1978, 1981, 1983, Yılmaz and Maxwell, 1983), the Alakır Çay unit is postulated to be an ophiolitic mélange of Late Cretaceous age.

COMPOSITION OF THE ALAKIR ÇAY UNIT

The Alakır Çay unit includes a great variety of clasts, of unknown provenance, derived from environments ranging from shallow water continental margin to pelagic. Ophiolitic rocks are the most conspicuous component. Other clasts include Paleozoic clastic and carbonate rocks, Mesozoic limestone and dolomite, Triassic evaporite-bearing to deep marine sedimentary rocks with basalt, Jurassic and Lower Cretaceous mudstones and radiolarites, and Cretaceous volcanogenic sedimentary rocks. Clast size ranges from millimetric to kilometric. The mélange matrix varies from pelagic marls to pillow-basalt flows and serpentinite. Locally, matrix is essentially absent.

Clasts

Ophiolite clasts consist of serpentinite, partly serpentinized ultramafic rocks, layered and massive gabbro, diabase, plagiogranite, massive and layered pillow-basalt flows, amphibolite, and pelagic sedimentary rocks. The serpentinite clasts crop out in linear belts with tectonized contacts. Serpentinites tend to weather reddish, but on fresh surfaces are light to dark green. Others, where intensively sheared, are green to steel-blue. No large bodies of unserpentinized ultramafic rocks were observed. However, small clasts of fresh orthopyroxene-clinopyroxene pyroxenite occur locally within serpentinite. Petrologically, serpentinites consist of chrysotile serpentine with mesh, vein, and hour-glass textures. Relict orthopyroxene grains persist in the chrysotile serpentine.

Amphibolites are found enclosed in highly deformed serpentinites, most commonly associated with probable thrusts. They are medium grained (1 to 3 mm) with fresh blue-green hornblende and enstatite, zoned and twinned plagioclase, sphene, and opaques, and an interstitial mosaic of grains of altered feldspar, abundant small anhedral apatite, and opaques. Secondary feldspar grains indicate recrystallization. The amphibolites apparently formed by thermal metamorphism of mafic igneous rocks at the base of thrust sheets. K-Ar ages for the amphibolites average 73.7 $\pm$ 1.1 and 78.5 $\pm$ 1.7 Ma (hornblende dates, Table 2; Yılmaz and Maxwell, 1982; Yılmaz, 1984), suggesting a Late Cretaceous age for the metamorphism.

Gabbros with prominent alternating layers rich in pyroxene and plagioclase are exposed along the Kumluca-Gödene road (Fig. 4). The largest mass, 1 km by 2 km, at Karkadın in the south (Fig. 4), is cut by numerous diabase dikes. The gabbros contain pods (0.5 m across) and dikes (5 cm wide and 2 to 50 m long) of gabbroic pegmatite with pyroxene and plagioclase crystals up to 10 cm across. All of the gabbroic samples show cumulate texture, with the plagioclase crystals as the cumulus, and pyroxene crystals with apatite, olivine, and magnetite as the ground mass. Most samples show alteration, including apparently late deuteric conversion of pyroxene to actinolitic amphibole, and conversion of olivine to a variety of alteration products. Some gabbros have been extensively sheared and mylonitized. A plagioclase separated from a quartz-bearing hornblende gabbro gave a K-Ar age of 60 Ma (Table 2; Yılmaz and Maxwell, 1982; Yılmaz, 1984). A plagiogranite whole-rock K-Ar date of 65 Ma (Table 2) may represent the youngest igneous activity in the Alakır Çay unit, although plagioclase and whole-rock ages are suspect (Yılmaz, 1984).

Diabase dikes with chill margins are found within serpentinite, gabbro, basalt flows, and pelitic rocks. The dikes range from 2 to 15 m in width. East of Karkadın, in the southeast (Fig. 4), up to 40 dikes, counted within an exposure 15 m across, form an irregular sheeted dike complex overlying gabbro. Some dikes extend from the gabbro into the sheeted complex, and dike clusters occur in the upper gabbroic layers. A K-Ar date of 81 $\pm$ 3 Ma for Feldspar (T7) separated from a diabase sill within pelitic rocks was reported by Thuizat and others (1978).

Most basalts are pillowed flows a few tens of meters thick. The maximum total thickness measured is 1.4 km at Sayrun (Fig. 4), where basaltic feeder dikes can be traced into and through the flows. In some flows, pillows are set in a deuterically altered matrix, which is either red and iron-rich or green and chlorite-rich. Basalts, in the north along the Gönen Çay (Fig. 4), occur as tabular flows. All the basalts consist of albite, augite, chlorite, and opaques. The augite is titanium-rich, which is unusual for basalts in ophiolites. The chemical and mineralogical characteristics of basalts, in particular the high content of sodium, magnesium, and calcium, and low potassium and silica, reflect strong similarities to spilites (Fig. 5).

The basalt flows are locally interbedded with and directly overlain by sediments, in part pelagic, including limestone, turbiditic sandstone, ribbon chert with limestone interbeds, and manganiferous chert. At Karabük (Fig. 4), sedimentary rocks

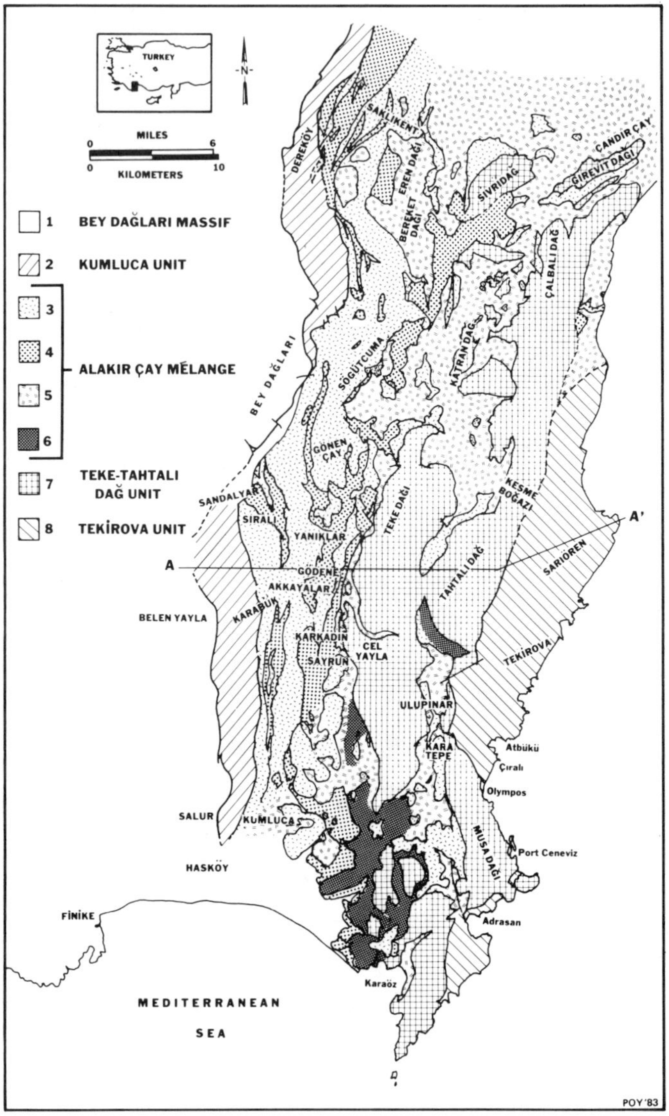

Figure 4. General geology of the Antalya Complex. 1 - Upper Cretaceous-Eocene platform carbonate rocks; 2 - Mesozoic sandstone-pelagic limestone-radiolarite-bituminous shale-blue/green shale; 3 - Mélange: pillow-basalt flows dominant - Halobia limestone, pelagic limestone-ammonitico rosso-radiolarite-siliceous mudstone; 4 - Mélange: serpentinite dominant - amphibolite - ultramafic rocks -gabbro-diabase-plagiogranite; 5 - Mélange: reefal carbonate rocks dominant - pelagic sediments - shale with gypsum - shale with graptolite - limestone and dolomite with macrofossils - clastics with ophiolitic debris; 6 - Alluvial clastics rudites - carbonate breccia - ophiolite debris; 7 - Paleozoic - Mesozoic pelagic and neritic sediments; 8 - Mesozoic ophiolite-harzburgite-dunite-chromite-pyroxenite-gabbro-diabase.

interbedded with pillow-basalt flows include Upper Carnian to Upper middle Norian pelagic limestones with calcareous nannofossils, foraminifera and radiolaria, dated by Pessagno (1980, pers. comm.). Late Triassic volcanic rocks are widespread. The Sayrun volcanic massif (1.4 km thick) is interbedded with Upper Triassic Halobia-bearing and ammonitico rosso facies limestones (Juteau, 1975; Marcoux, 1976; Robertson and Woodcock,

TABLE 2: K-Ar ANALYTICAL DATA FOR TEKİROVA AND ALAKIR ÇAY UNITS.

Sample #	Locality	Rock Type	Mineral	%K	$\%Ar^{40}_{rad}$	v/m	Calculated Age* (my)
POY-2A	Kozaklı Tepe-Coastal Tekirova	amphibolite	hornblende	0.048	8.7 7.6	$(2.50 \pm 0.22)10^{-4}$ $(2.21 \pm 0.21)10^{-4}$	130.4 ± 11.5 114.4 ± 11.0
POY-3	Arap Yalısı - Coastal Tekirova	diabase	hornblende	0.050	8.9 7.1	$(1.95 \pm 0.17)10^{-4}$ $(1.83 \pm 0.18)10^{-4}$	98.0 ± 8.7 92.1 ± 9.2
POY-51	Tekerlek Tepe-Coastal Tekirova	quartz diabase	whole rock	0.158	15.9 15.1	$(4.40 \pm 0.16)10^{-4}$ $(4.17 \pm 0.16)10^{-4}$	70.4 ± 3.9 66.8 ± 3.8
TT-1	Tekerlek Tepe-Coastal Tekirova	quartz hornblende diabase	whole rock	0.116	77.3 77.8 77.5	$(3.49 \pm 0.08)10^{-4}$ $(3.59 \pm 0.08)10^{-4}$ $(3.55 \pm 0.08)10^{-4}$	91.0 ± 1.8 92.3 ± 1.8 92.4 ± 1.8
GB-1	Göynük Çeşme - Tekirova	quartz hornblende gabbro	whole rock	0.183	58.6 52.8 55.4	$(4.90 \pm 0.07)10^{-4}$ $(4.86 \pm 0.06)10^{-4}$ $(4.84 \pm 0.06)10^{-4}$	67.6 ± 1.9 67.2 ± 1.8 66.8 ± 1.8
GB-2	Göynük Çeşme-Tekirova	quartz hornblende gabbro	plagioclase	0.297	69.9 69.2 58.4	$(7.88 \pm 0.10)10^{-4}$ $(7.64 \pm 0.11)10^{-4}$ $(8.04 \pm 0.12)10^{-4}$	68.1 ± 0.9 65.2 ± 1.0 68.5 ± 1.1
GB-3	Göynük Çeşme-Tekirova	quartz hornblende diabase	whole rock	0.126	80.8 56.8 78.0	$(4.03 \pm 0.08)10^{-4}$ $(4.01 \pm 0.08)10^{-4}$ $(4.08 \pm 0.08)10^{-4}$	84.4 ± 2.2 84.0 ± 2.2 85.5 ± 2.2
POY-117B	Dedeler mah.-Kemer	gabbro	plagioclase	0.070	8.8 9.0	$(1.56 \pm 0.10)10^{-4}$ $(1.58 \pm 0.13)10^{-4}$	67.8 ± 4.8 68.6 ± 6.0
POY-117C	Dedeler mah.-Kemer	gabbro	plagioclase	0.026	4.0 3.6	$(5.13 \pm 0.77)10^{-4}$ $(5.05 \pm 0.76)10^{-4}$	49.5 ± 9.9 48.7 ± 9.7
POY-118C	Aladağlı Boğazı-Kemer	diabase	hornblende	0.386	26.0 27.3	$(8.24 \pm 0.35)10^{-4}$ $(8.68 \pm 0.33)10^{-4}$	54.1 ± 3.2 56.9 ± 3.4
POY-64	Karkadındömek Tepe-Alakır Çay	gabbro	whole rock	0.032	5.3 5.1	$(7.09 \pm 1.35)10^{-5}$ $(7.93 \pm 1.19)10^{-5}$	55.5 ± 10.9 61.8 ± 9.8
K-4	Karkadındömek Tepe-Alakır Çay	quartz-bearing hornblende gabbro	plagioclase	0.212	50.5 37.8 49.2	$(5.01 \pm 0.08)10^{-4}$ $(4.90 \pm 0.08)10^{-4}$ $(5.05 \pm 0.08)10^{-4}$	59.8 ± 1.7 58.5 ± 1.7 60.3 ± 1.7
8-D	Sarsaldömek Tepe-Alakır Çay	plagiogranite	whole rock	0.214	61.5 60.8 60.2	$(5.63 \pm 0.08)10^{-4}$ $(5.36 \pm 0.08)10^{-4}$ $(5.45 \pm 0.08)10^{-4}$	66.5 ± 2.1 63.5 ± 2.0 64.5 ± 2.2
POY-14	Akkayalar-Alakır Çay	amphibolite	hornblende	0.793	70.0 69.1	$(1.89 \pm 0.03)10^{-3}$ $(1.96 \pm 0.03)10^{-3}$	72.4 ± 1.1 75.0 ± 1.1
POY-59	Zincirkıracak Dere-Alakır Çay	amphibolite	hornblende	0.529	68.3 69.0	$(1.34 \pm 0.03)10^{-3}$ $(1.40 \pm 0.03)10^{-3}$	76.8 ± 1.7 80.2 ± 1.7

* $\lambda_{\epsilon} = 0.581 \times 10^{-10} yr^{-1}$, $\lambda_{\beta} = 4.962 \times 10^{-10} yr^{-1}$, K^{40}/K total $= 1.167 \times 10^{-4}$ atom/atom. Estimated error of all %K Values is 1%.

v/m = Volume of radiogenic Ar^{40} in mm^3/gram. $\sigma = 2$

Argon analyses by FM Consultants LTD., Cambridge, England.

Potassium analyses by Mobil Research and Development Corp.

1980b, 1981b; Yılmaz, 1981, 1984). Pelagic sediments interbedded with alkaline pillow-basalt flows locally yield late Jurassic to Early Cretaceous ages. These have been reported in the Ulupınar/Kara Tepe area by Yilmaz (1981, 1984, Fig. 4).

Sedimentary rock clasts consist of sandstone with plant debris, calcilutite, pelagic limestone, ammonitico rosso limestone, radiolarite, siliceous mudstone, and reefal carbonate. Exposures of sandstones with plant debris vary from 20 to 200 m in thickness, with graded beds 25 to 30 cm thick. The sandstone is made up of medium-size, sub-angular to rounded grains of quartz with numerous lithic fragments in a muddy matrix. Calcareous mudstones very rich in plant debris occur as lenticular beds in the sandstones. Beds of green shale and limestones 30 to 50 cm thick occur locally toward the top of this unit. All exposures show sole markings, including flute and groove casts, as well as ripple marks, worm tracks, and plant impressions. The sandstone locally grades upward into Halobia- and radiolarian-bearing hemipelagic limestone. Sandstones from Karabük (Fig. 4) are of Late Triassic

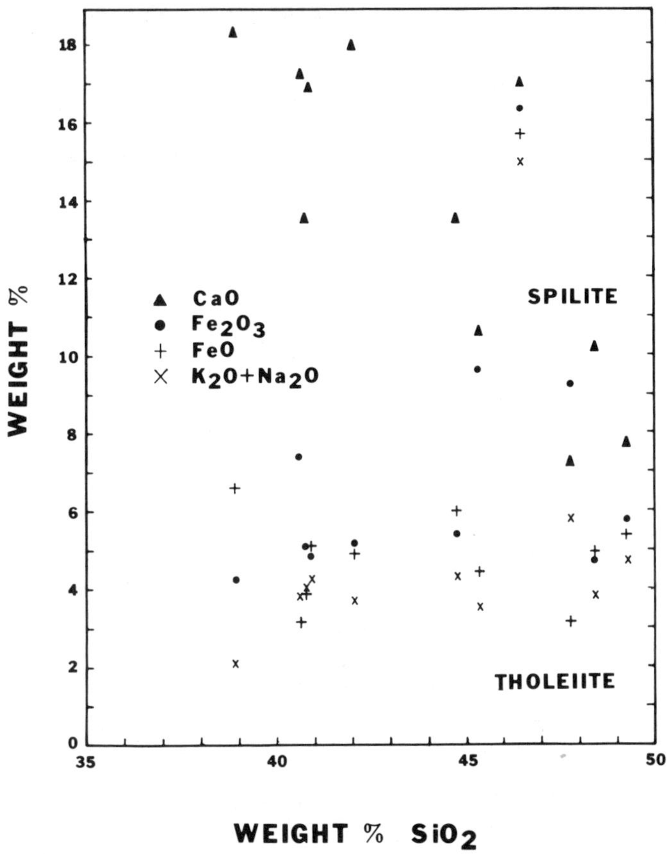

Figure 5. Chemical analyses of alkaline lavas from the Alakır Çay unit showing spilite trend.

age (Mobil Exploration and Producing Services Inc.-MEPSI Stratigraphy Laboratory, Appendix I: Fossil Identification Lists, Yılmaz, 1981).

The limestone clasts are calcilutite, locally very silty and argillaceous, thin to medium bedded, and light gray to yellow weathering, with a local thickness up to 90 m. The beds show convolute laminations and local slump structures. The matrix is partially recrystallized micrite. At higher levels, calcilutites which formed as deep water oozes alternate with calcirudites, deposited on a periplatform, in an open-marine environment (Robertson and Woodcock, 1981a). The limestones are locally interbedded with graded conglomerates containing fossiliferous limestone clasts up to 40 cm across. The clasts contain algal debris, echinoderm plates, ooids, and benthonic foraminifera, suggesting derivation from a carbonate platform sequence, possibly the Bey Dağları platform to the west.

Pelagic limestones locally are interlayered with and overlie basalt flows. This is best seen west of Sayrun (Fig. 4), where limestone was deposited on the rounded pillow surfaces. These limestones contain abundant Halobia fossils of late Triassic age (Brunn and others, 1970, 1971; Marcoux, 1976, 1977). In Yanıklar, the pelagic limestones locally grade upward into radiolarite sections several tens of meters thick that persist laterally for several kilometers (Fig. 4). The radiolarite beds show micro-cross lamination, and some are lenticular. Manganese is a common constituent, occurring as pyrolusite in the form of disseminated grains, nodules, and layers.

The siliceous mudstones are calcareous, have abundant rounded manganese nodules, and weather orange-red. Bedding varies from cross-bedded to graded to finely laminated. Locally, the mudstone contains kaolinite-rich layers and kaolinitic siltstones up to 25 cm thick. The total thickness of the largest block is more than 100 m. Toward the top, that unit becomes chalky and marly; the bottom contact is covered. The marls and chalks are intercalated with thin, graded calcarenites containing planktonic foraminifera. Radiolaria, foraminifera, nannofossils, and polynomorphs, identified by Pessagno and workers of the MEPSI Stratigraphy Laboratory, give ages for these pelagic units ranging from Late Triassic to Early Cretaceous (Kimmeridgian-Tithonian through lower Hauterivian) (Appendix I: Fossil Identification Lists; Yılmaz, 1981; Robertson and Woodcock, 1981a, 1981b, 1981c).

The large blocks of the Teke-Tahtalı Dağ unit include clasts of graptolitic shale, sandstone with fish scales, coal sequences, evaporites, platform and reefal limestone, and dolomites ranging in age from Early Paleozoic to Late Cretaceous. These largest blocks of the mélange are sub-divided from the rest of the mélange on the basis of their size and distinctive topographic expression. Many of the same rock types are present, however, as smaller clasts within the mélange.

The oldest Paleozoic rocks, Ordovician graptolitic blue-green shales (Şenel, 1980), are exposed in a faulted Paleozoic sequence at Kesme Boğazı (Fig. 4). The thickness of the Ordovician shale unit has not been measured, but several meters are present in the one known locality. A richly fossiliferous black, thick-bedded (2 to 3 m) limestone with Ordovician gastropods, brachiopods, and pelecypods (K. Young, 1980, pers. comm.) is in fault contact on the shale. Faulted on to the limestone are brown, coarse-grained sandstones containing fish scales of Silurian-Devonian age (M. Şenel, 1979, pers. comm.). The sandstones are cut locally by sills and dikes of altered basalt (Özgül, 1979, pers. comm.). The sandstones are overlain, in fault contact, by thick-bedded (1 to 2 m) brownish gray sandstones with interbedded blue-green shales, small pebble conglomerates, siltstones, and coal seams. The coals are 10 cm to 2 m thick at Kesme Boğazı. This unit is of Carboniferous age based on brackish water brachiopods (Özgül, 1979, pers. comm.). Similar coal beds from Pamucak Yayla, to the north of the study area, have yielded a Tethyan fauna of Permian age (E. Akyol, 1981, pers. comm.).

Black to gray, medium to thick-bedded (15 cm to 1 m) limestone and dolomite are seen along a north-trending linear belt on the east margin of the Teke-Tahtalı Dağ (Fig. 4). The rock has a sulfurous and petroliferous odor when broken and is the most likely source for gas seeps near Çıralı (Fig. 4). The limestone is locally interbedded with marl and siltstone rich in fossils (bryzoans, brachiopods, crinoids, etc.) and is of Permian age (K. Young, 1979, pers. comm.).

Blue and red fissile shales with 10 to 50 cm thick gypsum

interbeds crop out north of Ulupınar (Fig. 4). Based on palynology, this unit is of Triassic age (MEPSI Stratigraphy Laboratory, Appendix I: Fossil Identification Lists; Yılmaz, 1981).

The ammonitico rosso limestone is pink, very fossiliferous, and 7 to 10 m thick, overlying the Halobia-bearing limestones at Saklıkent (Fig. 4). This unit is of Carnian-Norian age (E. T. Tozer, 1980, pers. comm.). The ammonitico rosso limestone grades upward into masive reefal limestone, rich in Jurassic and Cretaceous microfossils (M. Serdaroğlu, 1977, pers. comm.).

Thick-bedded reefal limestones and dolomites form the largest discrete blocks in the mélange. Minimum total stratigraphic thickness is 1500 m. These rocks form the Teke Dağı, Tahtalı Dağ, and other mountain peaks of the Antalya allochthon that reach heights of more than 2000 m above sea level. They are generally north-elongated giant blocks with steep internal dips and near vertical contacts. The carbonate rocks range in age from Late Triassic to Late Cretaceous (M. Serdaroğlu, 1980, pers. comm.) and show a transition from Late Triassic pelagic to Jurassic-Cretaceous neritic facies. These rocks may have formed as carbonate buildups on pre-Triassic basement slivers (Robertson and Woodcock, 1980a, 1980b, 1981c).

Matrix of Mélange

The matrix of the mélange consists most commonly of serpentinite, but locally is basalt or pelitic rocks such as marl, claystone, and shale. Coarse clastic rocks such as lithic sandstone are less common. Bedding may be obscured, but distinct beds are present and interlayered with chaotic zones. The serpentinite matrix is best observed north of Yanıklar (Fig. 4), where it includes clasts of gabbro, Permian to Cretaceous platform carbonates, Triassic Halobia limestones, pillow-basalt flows, and radiolarites. The serpentinite is highly sheared and brecciated. Included clasts locally show low-grade metamorphism, as in the vicinity of Kumluca (Fig. 4), where radiolarite and limestone were metamorphosed under greenschist facies conditions (A. H. F. Robertson, 1979, pers. comm.). A yellow-gray weathering clay-marl with interbedded platy, greenish limestone, exposed at Karabük along the Gödene road, is the matrix for blocks of serpentinite. At Cel Yayla (Fig. 4), these claystones tectonically overlie the massive Cretaceous carbonate rocks of the Teke Dağı.

The matrix/clast ratio is quite uneven. Matrix and clasts are estimated to be equal in volume in the western exposures, whereas matrix forms only about 25 per cent of the mélange in the central and eastern parts. Locally, the matrix is essentially absent, the mélange being made almost entirely of clasts.

Rocks of the matrix are mildly to strongly folded, sheared, and foliated. Millimeter to kilometer scale zones of strongly sheared serpentinites, locally with amphibolite lenses, are widespread. The serpentinites show well-developed foliation characterized by wavy anastomosing surfaces.

Tight and isoclinal folds are not a dominant structural feature of the matrix but are present in and near radiolarite and limestone blocks. Locally, small, tight, and disharmonic folds, confined only to a few thin beds, are present. They are apparently the product of intrastratal sliding and slumping. The attitude of bedding within large blocks is generally discordant to that of recognizable beds in the surrounding matrix. Bedding in smaller clasts tends to be oriented parallel with the fabric of the enclosing matrix.

STRUCTURE OF MÉLANGE

The mélange shows a general north-trending structural grain. The larger blocks, mostly limestones and dolomites, tend to be elongated north-south with steeply dipping to vertical internal bedding and block boundaries. These blocks are faulted against each other and against mélange matrix, but, on a large scale, are "floating" in the mélange, as are most smaller blocks (Fig. 6). Mud-flows and olistostromes, subunits within the mélange, are common, especially in the Akkayalar region (Fig. 4).

The northerly grain is intersected by major east-trending and minor northwest-trending faults. Normal faults just east of Karabük (Fig. 4) are downthrown 10 to 15 m to the west. These faults cut a thrust surface between Mesozoic sedimentary rocks and basalt flows and serpentinite. The sedimentary rocks are

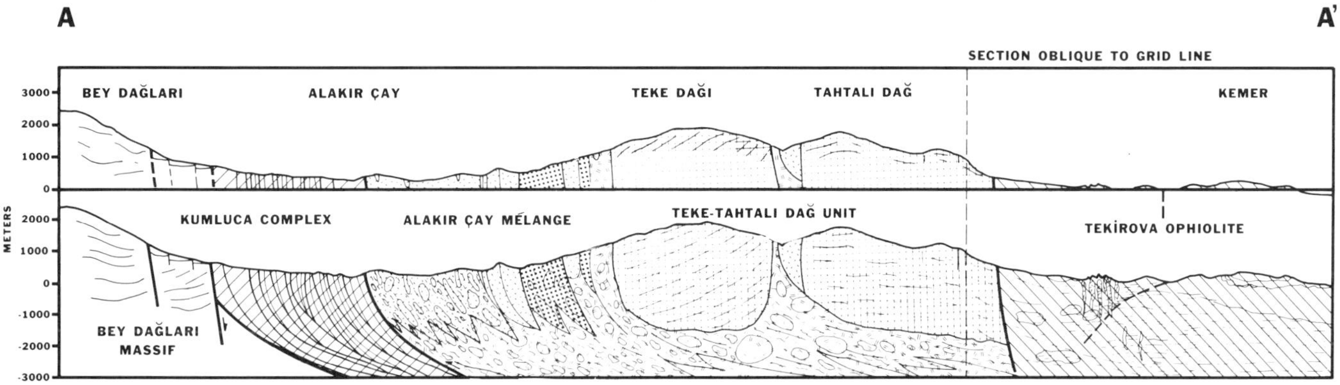

Figure 6. Schematic cross section A-A′ across the Antalya Complex. Location indicated on Figure 4. The upper section indicates field data, while the lower is the structural interpretation. The vertical and horizontal scales are the same. Symbols as on figure 4.

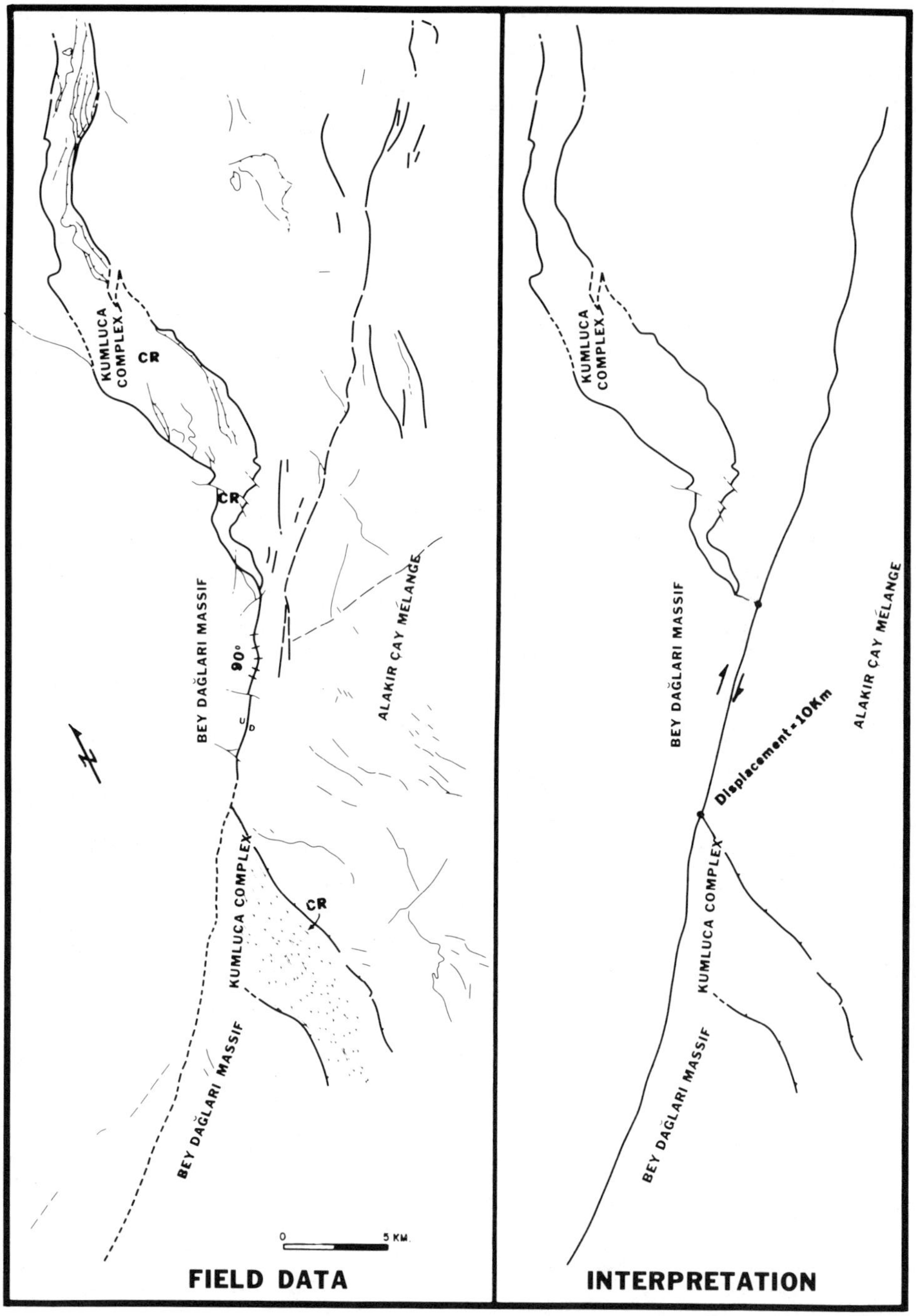

Figure 7. The fault contact between the Alakır Çay mélange - Kumluca Complex and the Bey Dağları. The left figure shows field data while the right figure shows interpretation as an oblique strike-slip fault.

tightly folded, the folds being overturned to the west. Faulting is common, parallel to the east-dipping limbs of the overturned folds.

The Alakır Çay mélange is in fault contact with rocks of the Bey Dağları massif in the center of the study area (Fig. 4). West of Söğütcuma, highly deformed Jurassic cherts of the mélange are faulted against relatively undeformed Eocene carbonates of the Bey Dağları massif. The contact, originally a low angle thrust, is now a high angle reverse fault, east side up, with the fault plane at an angle of 30-45° to bedding planes in adjacent Eocene limestone. The reverse fault has been cut by a younger normal fault with the east side down. The normal fault is interpreted to have a right lateral slip component based on field relationships (Yılmaz, 1981, Fig. 7). The contact between the Kumluca unit and the mélange is also a fault. This contact is best exposed in the south, where the fault plane is near-vertical or dips steeply to the east. The eastern contact of the mélange is faulted against the partial ophiolite sequence of the Tekirova unit and the sedimentary

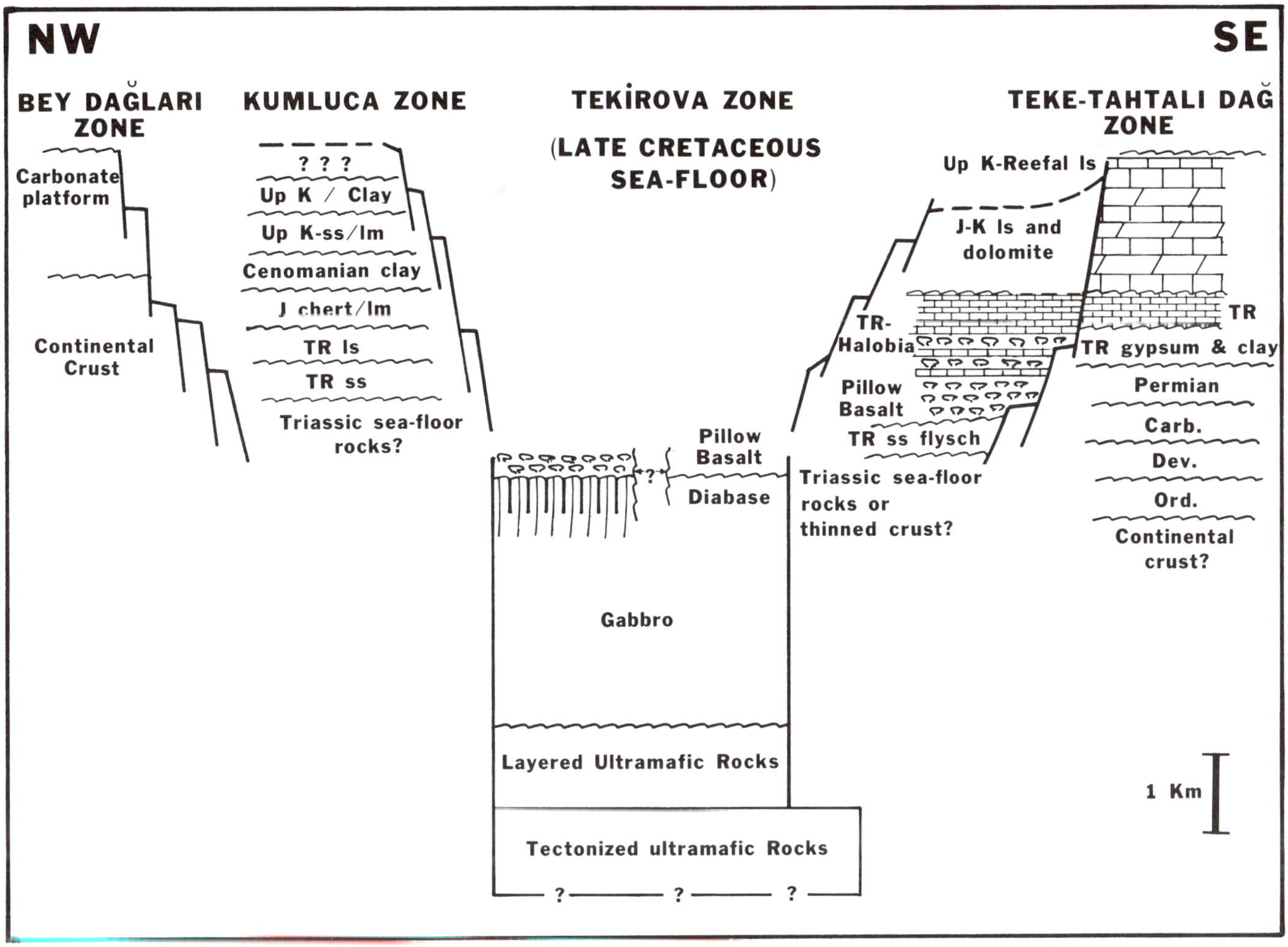

Figure 8. Schematic representation of the Antalya Complex during the latest Cretaceous. Boundaries between different zones, and between most rock units, are tectonic; the rocks of any zone may have moved by strike-slip displacement with respect to any other zone.

rocks of the Teke-Tahtalı Dağ unit. These faults are near vertical in most areas (Fig. 6).

AGE OF MÉLANGE

As detailed in the description of clasts, the Alakır Çay mélange contains materials ranging in age from Early Paleozoic to Late Cretaceous. Formation of the mélange may not have started before Late Cretaceous, although in many obduction mélanges it can be demonstrated that beds deposited on mélange are later incorporated into it. The mélange is unconformably overlain by a conglomeratic sequence with clasts containing a Maestrichtian fauna (Appendix I, Fossil Identification Lists, Yılmaz, 1981). This is the only known direct evidence for the age of completion of mélange formation. It is deduced that the mélange probably began to form during the Late Cretaceous, based in part on the suggested age of metamorphism of the amphibolites (Table 2). The process probably continued well into the Tertiary.

The time of emplacement of the Alakır Çay mélange on the Bey Dağları massif is constrained stratigraphically to the Miocene. The Aquitanian shallow water carbonates of the Bey Dağları massif are considered to pre-date the emplacement. The overlying Burdigalian-Langhian clastic section includes olistostromes of ophiolitic mélange, implying that the Alakır Çay mélange was close enough to have been a sediment source (Hayward, 1982; Hayward and Robertson, 1982).

FORMATION AND EMPLACEMENT OF MELANGE

The geologic setting of the Alakır Çay mélange and associated allochthonous units closely resembles that of other obducted sequences in orogenic belts surrounding the Mediterranean. Elements of similarity exist with the Betic-Rif-Tell trend, the North Corsica-central Apennine region, the Sicilian-Calabrian trend, and the Kyrenia range of northern Cyprus. In each area, deformed allochthonous sequences of differing provenance have apparently moved outward from areas now occupied by youthful marine basins or troughs (see cross sections in Biju-

Duval and others, 1979). Two very different mechanisms have been favored: the first involves thrusting related to collision, followed by sinking of the hinterland caused by spreading; in the second, doming related to thermal expansion is accompanied by centrifugal gravitational denudation of the domed region, which subsequently cooled and subsided (see, for example, Kligfield, 1979; Loomis, 1975). Some combination of the two mechanisms is probable for most areas cited above.

In the Antalya Complex, the distribution throughout the mélange of very large blocks characteristic of the Teke-Tahtalı Dağ trend clearly indicates that this mélange has traveled a considerable distance, carrying along the broken up (or breaking up) carbonate platform sequence. In many respects, the geometry of the Alakır Çay mélange and its incomplete cover of large slabs and blocks of more rigid carbonate rocks resemble the western part of the northern and southern Apennines of Italy (Merla, 1951; D'Argenio and others, 1973), where thick, ophiolite-bearing mélanges have moved from the area of an existing marine basin onto a continental platform. These mélanges continued to move, carrying blocks of cover rocks many tens of kilometers (Abbate and others, 1981; Migliorini, 1952). It appears that the later stages of mélange movement took place essentially as submarine slides or olistostromes that encroached on successively younger flysch basins in front of the advancing mass (Maxwell, 1959, 1963, Naylor 1981). Apparently, mélange formation in the Apennines also predated or accompanied the emplacement of the largely carbonate cover rocks. Opinions are divided, however, as to whether the initiation of deformation resulted from thrusting related to crustal compression (or collision), or whether the process from the beginning was largely a result of gravitational spreading and sliding. Collision is currently the favored explanation, though little hard evidence supports it in the areas cited above.

The presence of an ocean basin south-southeast of the present Antalya Complex area during the Late Cretaceous has been postulated by various workers (Robertson and Woodcock, 1980a, 1980b, 1981a, 1981b; Hayward and Robertson, 1982; Şengör and Yılmaz, 1981; Yılmaz, 1981; and Yılmaz and others, 1981). The idea is supported by sedimentologic, paleontologic, K-Ar radiometric dating, structural fabric, and field data. Cyprus and adjacent areas were also on oceanic crust (Rocci and Lapierre, 1969; Desmet and Lapierre, 1977; Moores and Vine, 1971; and Miyashiro, 1973), as was Hatay (Delaloye and others, 1977; Tinkler and others, 1981; and Delaune-Mayere and others, 1977), during the Late Cretaceous. In Figure 8, we suggest a schematic representation of the distribution of Antalya area allochthonous units during Late Cretaceous time, just prior to mélange formation and inclusion of Teke-Tahtalı Dağ rocks. In Figure 8, boundaries between different zones and between most rock units are tectonic. The rocks of any zone may have moved by strike-slip displacement with respect to any other zone.

The basin closure and obduction of ophiolites in the Antalya area began during Late Cretaceous time. The Alakır Çay mélange formed during northwestward obduction in front of the emplacing ophiolite thrust sheets over a relatively flat foreland of the Kumluca continental margin. The mélange movement took place as submarine slides that successively occupied flysch basins developed in front of the advancing thrust mass. This interpretation is supported by the presence of Cenomanian marls that occur along or near the contact with the Alakır Çay mélange. These marls are the youngest sedimentary rocks in the stacked sequence of the Kumluca unit underlying the mélange.

The carbonate deposition on the Bey Dağları was terminated during Burdigalian time. The clastic rocks with ophiolitic debris, deposited on the Bey Dağları during Burdigalian-Langhian, are the first indication that the Antalya allochthons were near the Bey Dağları platform.

The uplift of the Bey Dağları during Tortonian time (Table 1; Poisson, 1978) folded and faulted the Antalya allochthons. The Alakır Çay mélange was deformed also by later Neogene normal faulting with oblique-slip components.

CONCLUSIONS

The Alakır Çay unit is a mélange similar to obduction sequences in orogenic belts surrounding the Mediterranean. Mélange formation started during Late Cretaceous time, and the mélange traveled a considerable distance carrying along a broken up carbonate platform sequence. We favor a mechanism that involves thrusting related to collision, followed by sinking of the hinterland area caused by spreading. The later stages of mélange movement took place as olistostromes or submarine slides. The Alakır Çay mélange formed an elevated land mass to the east of Bey Dağları by Early Miocene. The final phase of westward thrusting took place during the pre-Tortonian-Late Miocene.

ACKNOWLEDGMENTS

Funding was provided by W. A. Belden, The University of Texas at Austin, and the Mobil Exploration and Producing Services Inc. (MEPSI). Logistic support in the field was through the Mineral Research and Exploration Institute of Turkey (MTA). Earlier drafts of this paper were reviewed by A. W. Bally, Teoman Norman, C. Gans, and R. E. Cobb.

REFERENCES CITED

Abbate, E., Bortolotti, V., and Sagri, M., 1981, An approach to olistrome interpretation, in Ricci Lucchi, A. F., ed., Excursion Guidebook, International Association of Sedimentologists: 2nd European Regional Meeting, Bologna, Italy.

Baykal, F., and Kalafatçıoğlu, A., 1973, Antalya Körfezi batısında yeni jeolojik müşahedeler: Maden Tetkik ve Arama Enstitüsü Bülteni, sayı 80, p. 33–42.

Biju-Duval, B., Létouzèy, J., and Montadert, L., 1979, Variety of margins and deep basins in the Mediteranean, *in* Watkins, J. S., Montadert, L., and Dickerson, P. W., eds., Geological and Geophysical Investigations of Continental Margins: American Association of Petroleum Geologists, Memoir 29, p. 293–317.

Brunn, J. H., 1974, Le probleme de l'origine des nappes et de leurs translations dans les Taurides occidentales: Bulletin de la Societe Geologique de France, v. XVI (7), p. 101–106.

Brunn, J. H., De Graciansky, P. C., Gutnic, M., Juteau, T., Lefévre, R., Marcoux, J., Monod, O., and Poisson, A., 1970, Structures majeures et correlations stratigraphiques dans les Taurides occidentales: Bulletin de la Societe Geologique de France, v. 12, p. 515–556.

Brunn, J. H., Dumont, J. F., De Graciansky, P. C., Gutnic, M., Juteau, T., Marcoux, J., Monod, O., and Poisson, A., 1971, Outline of the geology of the western Taurides, *in* Campbell, A. S., ed., Geology and history of Turkey: Petroleum Exploration Society of Libya, Tripoli, p. 225–255.

Colin, H. J., 1962, Geologische Untersuchungen in Roume Fethiye-Antalya Kas-Finike (SW Anatolien): Maden Tetkik ve Arama Enstitüsü Bülteni, v. 59, p. 19–61.

D'Argenio, B., Pescatore, T., and Scandone, P., 1973, Schema geologico dell' Appennino Meridionale (Campania e Lucania): Estratto dal Quaderno no. 183, Atti del Convegno sul Tema: Moderne vedute sulla geologia dell'Appennino, University of Naples, Italy.

Delaloye, M., Vuagnat, M., and Wagner, J. J., 1977, K-Ar dates from the Kızıl Dağ ophiolitic complex (Hatay, Turkey) and their interpretation, *in* Biju-Duval, B., and Montadert L., eds., Structural History of the Mediterranean basins: Editions Technip, Paris, p. 73–78.

Delaune-Mayere, M., Marcoux, J., Parrot, J. F., and Poisson, A., 1977, Modele d'Evolution Mesozoique de la Paleo-Marge Tethysienne au niveau des nappes radiolaritiques et ophiolitiques du Taurus Lycien, d'Antalya et du Baer-Bassit, *in* Biju-Duval, B. and Montadert, L., eds., Structural History of the Mediterranean basins: Editions Technip, Paris, p. 79–94.

Desmet, A., and Lapierre, H., 1977, Les Manifestations Magmatiques du Troodos et des nappes de Mamonia (Chypre) comparaison et discussion geotectonique: Sciences de la Terre, v. XXI, no. 1, p. 3–16.

Dumont, J. F., Gutnic, M., Marcoux, J., Monod, O., and Poisson, A., 1972, Le Trias des Taurides occidentales (Turquie). Definition du bassin pamphylien: Un nouveau domaine a ophiolites a la marge externe de la chaine taurique: Zeitschrift des Deutschen Vereins fuer die Geschichte Maehrens und Schlesiens, v. 123, p. 385–409.

Hayward, A. B., 1982, Türkiye 'nin güneybatısındaki Bey Dağlari ve Susuz Dağ masiflerinde Miyosen yaşlı kırıntılı tortulların stratigrafisi: Türkiye Jeoloji Kurumu Bülteni, v. 25, no. 2, p. 81–89.

Hayward, A. B., and Robertson, A.H.F., 1982, Direction of ophiolite emplacement inferred from Cretaceous and Tertiary sediments of an adjacent autochthon, the Bey Dağları, southwest Turkey: Geological Society of America Bulletin, v. 93, p. 68–75.

Juteau, T., 1968, Kumluca (Güney Türkiye, Likya Torosları) Bölgesinin ofiolitlerine ait jeolojik haritanın açıklanması: Strüktürel kadro ile yatak şekilleri ve ofiolitli korteje ait belli başli fasiyeslerin tasviri: Maden Tetkik ve Arama Enstitüsü Bülteni, sayı 70, p. 83–103.

Juteau, T., 1970, Petrogenese des ophiolites des Nappes d'Antalya (Taurus Lycien Oriental). Leur liaison avec une phase d'expansion ocenique active au Trias Superieur: Sciences de la Terre, v. 15, p. 265–288.

Juteau, T., 1975, Les ophiolites des nappes d'Antalya (Tauride occidentales, Turquie): Sciences de la Terre, v. 32, 692 p.

Juteau, T., 1977, The ophiolitic complex of Antalya, *in* Güvenç, T., and others, eds., Western Taurus Excursion Geological Guidebook: VI. Colloquium on the Geology of the Aegean Region, Sept. 19-25, 1977, Izmir, 22 p.

Juteau, T., 1979, Ophiolites des Taurides: Essai sur leur historie oceanique: Revue de Geographie Physique et de Geologie Dynamique, v. 21, no. 3, p. 191–214.

Juteau, T., Lapierre, H., Nicolas, A., Parrot, J. F., Ricou, L. E., Rocci, G., and Rollet, M., 1973, Idees actuelles sur la constitution, l'origine et l'evolution des assemblages ophiolitiques mesogeens: Bulletin de la Societe Geologique de France, v. XV, no. 5-6, p. 478–493.

Juteau, T., Nicolas, A., Debussy, J., Fruchard, J. C., and Bouchez, J. L., 1977; Structural relationships in the Antalya ophiolite complex, Turkey. Possible model for an oceanic ridge: Geological Society of America Bulletin, v. 88, p. 1740–1748.

Juteau, T., and Whitechurch, H., 1979, The magmatic cumulates of Antalya (Turkey): Evidence of multiple intrusions in an ophiolitic magma chamber, *in* Panayiotou, A., ed., Ophiolites, Proceedings, International ophiolite Symposium Cyprus, 1979, p. 377–391.

Kligfield, R., 1979, The Northern Apennines as a collisional orogen: American Journal of Science, v. 279, p. 676–691.

Lefévre, R., 1967, Un nouvel element de la geologie du Taurus Lycien: Les nappes d'Antalya (Turquuie): Comptes Rendus de l'Academie des Sciences, (7), v. Ser. D265, p. 1365–1368.

Loomis, T. P., 1975, Tertiary mantle diapirism, orogeny, and plate tectonics east of the Strait of Gibraltar: American Journal of Science, v. 275, p. 1–30.

Marcoux, J., 1976, Les series des nappes a radiolarites et ophiolites d'Antalya (Turquie): Homologies et signification probable: Bulletin de la Societe Geologique de France, v. XVIII (7), no. 2, p. 511–512.

Marcoux, J., 1977, Geological Sections of the Antalya region, *in* Güvenç, T., and others, eds., Western Taurus Excursion Geological Guidebook: VI, Colloquium on the Geology of Aegean Region, Sept. 19-25, 1977, Izmir, 22 p.

Maxwell, J. C., 1959, Turbidite, tectonic, and gravity transport, northern Apennine mountains, Italy: American Association of Petroleum Geologists Bulletin, v. 43, p. 2701–2719.

Maxwell, J. C., 1962, Structural geology of the Ottone area, Piacenza and Genova: Memoir Societe Geologica Italiana, v. IV, p. 69–92.

Merla, G., 1951, Geologia dell'Appennino Seltentrionale: Societe Geologia Italiana Bolletino, v. 20, no. 3, p. 95–382.

Migliorini, C. I., 1952, Composite wedges and orogenic landslips in the Apennines: International Geological Congress Report of Sessions, Great Britain, 1948, Part XIII, p. 186–198.

Miyashiro, A., 1973, The Troodos ophiolitic complex was probably formed in an island arc: Earth and Planetary Science Letters, v. 19, p. 218–224.

Monod, O., Marcoux, J., Poisson, A., and Dumont, J. F., 1974, Le domaine d'Antalya, temoin de la fracturation de la plateforme africaine au cours du Trias: Bulletin de la Societe Geologique de France, v. 16, p. 116–127.

Monod, O., 1976, La "courbure d'Isparta": Une mosaique de blocs autochthones surmontes de nappes composites a la jonction de l'arc hellenique et de l'arc taurique: Bulletin de la Societe Geologique de France, v. 18, p. 521–532.

Moores, E. M., and Vine, F. J., 1971, The Troodos Massif, Cyprus and other ophiolites as oceanic crust: Evaluation and implications: Royal Society of London Philosophical Transactions, v. 268, p. 443–466.

Naylor, M. A., 1981, Debris flow (Olistostromes) and slumping on a distal passive continental margin: The Palombini limestone-shale sequence of the Northern Apennines: Sedimentology, v. 28, p. 837–852.

Özgül, N., 1976, Torosların bazı jeolojik özellikleri: Türkiye Jeoloji Kurumu Bülteni, sayı 19, p. 65–78.

Poisson, A., 1977, The massif of Bey Dağlari, *in* Güvenç, T., and others, eds., Western Taurus Excursion Geological Guidebook: VI. Colloquium on the Geology of the Aegean Region, Sept. 19-25, 1977, Izmir, 22 p.

Poisson, A., 1978, Recherches geologiques dans les Taurides occidentales (Turquie). (These de Docteur des Sciences.) Universite de Paris-Sud.

Reuber, I., 1982, Mylonitic shear zones of millimetric to kilometric scale related to a transform fault in an ophiolitic complex (Antalya, Turkey): International Conference on Planar and Linear fabrics of deformed rocks, in Mitt. EH Zurich, Neue Folge, v. 239a, p. 233–236.

Ricou, L. E., Marcoux, J., and Poisson, A., 1979, L'allochtonie des Bey Dağları orientaux: Reconstruction palinspastique des Taurides occidentales: Bulletin de la Societe Geologique de France, v. XXI (7), no. 2, p. 125–133.

Ricou, L. E., Argyriadis, I., and Marcoux, J., 1975, L'Axe calcaire du Taurus, un alignement de fenetres arabo-africaines sous des nappes radiolaritiques, ophiolitiques et metamorphiques: Bulletin de la Societe Geologique de France, v. XVII (7), p. 1024–1044.

Ricou, L. E., Argyriadis, I., and Lefévre, R., 1974, Proposition d'une origine interne pour les nappes d'Antalya et le massif d'Alanya (Taurides Occidentales, Turquie): Bulletin de la Societe Geologique de France, v. XVI, (7), p. 107–111.

Robertson, A.H.F., and Woodcock, N. H., 1980a, Strike-slip related sedimentation in the Antalya Complex, SW Turkey, *in* Ballance, P. F., and Reading, H. G., eds., Sedimentation in Oblique-Slip Mobile Zones: International Association of Sedimentologists, Special Publication 4, p. 127–145.

Robertson, A.H.F., and Woodcock, N. H., 1980b, Tectonic setting of the Troodos massif in the east Mediterranean, *in* Panayiotou, A., ed., Ophiolites: Proceedings, International Ophiolite Symposium, Cyprus, 1979, p. 36–49.

Robertson, A.H.F., and Woodcock, N. H., 1981a, Bilelyeri Group, Antalya Complex: Deposition on a Mesozoic passive continental margin, southwest Turkey: Sedimentology, v. 28, p. 381–399.

Robertson, A.H.F., and Woodcock, N. H., 1981b, Alakır Çay Group, Antalya Complex, SW Turkey: A deformed Mesozoic carbonate margin: Sedimentary Geology, v. 30, p. 95–131.

Robertson, A.H.F., and Woodcock, N.H., 1981c, Gödene Zone, Antalya Complex, SW Turkey: Volcanism and sedimentation on Mesozoic marginal ocean crust: Geologische Rundschau, v. 70, no. 3, p. 1177–1214.

Rocci, G., and Lapierre, H., 1969, Etude comparative des diverses manifestations du volcanisme preorogenique au Sud de Chypre: Schweizeriche Mineralogische Petrographische Mittelilunger, v. 45, p. 31–46.

Şenel, M., 1980, Teke Torosları Güneydoğusunun Jeolojisi, Finike-Kumluca Kemer (Antalya), Unpublished Report, M.T.A. Ankara, Turkey, 106 p.

Şengör, A.M.C., and Yılmaz, Y., 1981, Tethyan Evolution of Turkey: A plate tectonic approach: Tectonophysics, v. 75, p. 181–241.

Thuizat, R., Montigny, R., Çakir, U., and Juteau, T., 1978, K-Ar Investigations on two Turkish ophiolites, *in* Zartman, R. E., ed., Short papers of the 4th International Conference, Geochronology, Cosmochronology, Isotope geology: Geological Survey Open-File Report 78-701, p. 430–432.

Tinkler, C., Wagner, J. J., Delaloye, M., and Selçuk, H., 1981, Tectonic history of the Hatay ophiolites (South Turkey) and their relation with the Dead Sea Rift: Tectonophysics, v. 72, p. 23–41.

Woodcock, N. H., and Robertson, A.H.F., 1977, Imbricate thrust belt tectonics and sedimentation as a guide to emplacement of part of the Antalya Complex, SW Turkey: International Colloquium on Aegean Geology, 6th, Izmir, Proceedings.

Woodcock, N. H., and Robertson, A.H.F., 1981a, Wrench-related thrusting along a Mesozoic-Cenozoic continental margin: Antalya Complex, SW Turkey, *in* McClay, K. R., and Price, N. J., eds., Thrust and Nappe Tectonics: Geological Society of London, Special Publication v. 9, p. 359–362.

Woodcock, N. H., and Robertson, A.H.F., 1982, Wrench and thrust tectonics along a Mesozoic-Cenozoic continental margin: Antalya Complex, SW Turkey: Journal of Geological Society of London, v. 193, p. 147–163.

Yılmaz, P. O., 1978, Alakır Çay unit of the Antalya Complex (SW Turkey): An example of ocean floor obduction [M. A. thesis]: Bryn Mawr College, 91 p.

Yılmaz, P. O., 1981, Geology of the Antalya Complex, SW Turkey, [Ph.D. Dissertation]: Austin, The University of Texas, University Microfilms International, 8208281, 268 p.

Yılmaz, P. O., 1984, Fossil and K-Ar data for the age of the Antalya Complex, SW Turkey, *in* Dixon, J., and Robertson, A.H.F., eds., Evolution of the Eastern Mediterranean: London, Blackwell Publications, p. 97–109.

Yılmaz, P. O., Maxwell, J. C., and Meuhlberger, W. R., 1981, Antalya Kompleksinin yapısal evrimi ve Doğu Akdeniz'deki yeri: Yerbilimleri Dergisi: Ankara, Hacettepe Universitesi, v. 7, p. 119–127.

Yılmaz, P. O., and Maxwell, J. C., 1982, K-Ar investigations from the Antalya Complex ophiolites, SW Turkey: Ofioliti, v. 7, no. 2/3, p. 527–538.

MANUSCRIPT ACCEPTED BY THE SOCIETY JULY 12, 1984

Printed in U.S.A.

Geological Society of America
Special Paper 198
1984

Tectonic significance of serpentinite mobility and ophiolitic melange

Jason B. Saleeby
Division of Geological and Planetary Sciences
California Institute of Technology
Pasadena, California 91125

ABSTRACT

Serpentinite is known to be an important constituent of oceanic crust and is particularly abundant along zones of primary structural weakness. Such primary weaknesses occur along fracture zones where large ultramafic massifs are commonly exposed, and along ridge-parallel serpentinite belts that are emplaced near slow-spreading ridges. Tectonic disruption of newly formed crust along transform segments of fracture zones in conjunction with, and followed by, the solid state injection of serpentinitic rocks leads to the development of serpentinite-matrix melange as a primary constituent of oceanic crust. Slumping of rocks along transform axial deep walls and ponding of pelagic and, locally, turbiditic sediments produce an array of additional chaotic rocks, which also belong to the fracture zone regime. Inasmuch as the total length of fracture zone crust on the ocean floors far exceeds the length of subduction and collision zones, and whereas fracture zones are commonly on the order of 50 km wide, geologists working in melange belts of orogenic zones should be aware of fracture zone processes and products.

Geological relations in a number of on-land ophiolitic melange and nappe belts, and deformation patterns in modern oceanic environments, where changes in rotational poles have occurred over the recent geologic past, suggest that ophiolite obduction and perhaps intra-oceanic subduction zones develop along primary structural weaknesses, particularly fracture zones. The initial imbricate structural sequences in such environments consist of ophiolitic melange, large ultramafic slices and ophiolite sheets along fracture zones, and ophiolite sheets with basal dynamothermal aureoles where thrusting initiates along or near spreading ridge segments. Such imbricate sequences constitute oceanic orogenic welts that may experience a considerable oceanic residence time prior to accretion within a collisional suture belt or along a subducting trench or that may slide by gravitational forces onto deeply submerged continental shelves or oceanic plateaus. Oceanic orogenic welts, along with large unmodified fracture zone massifs, should be considered similar to oceanic plateaus in regard to their susceptibility for tectonic accretion. Tectonic accretion of the welts often disrupts the initial imbricate sequence, leaving high temperature metamorphic blocks and slices of ophiolitic melange scattered along the base of ultramafic or ophiolitic nappes. In many instances, it is clear that dynamothermal metamorphism and ophiolitic melange mixing pre-date continental margin obduction and the related suturing event.

Ophiolitic melange may constitute an important element in uplifted subduction complexes, but it is not clear that "normal" ridge-crest ocean crust is tectonically mixed into melange within inner trench wall environments. Pre-mixed fracture zone assemblages are highly susceptible to accretion in such environments, because of their pre-weakened state, bathymetric expression, and their gravitational instability with respect to "normal" oceanic crust. In addition to *en masse* accretion of fracture zone

crust, ophiolitic melange components in subduction complexes are probably also derived by the diapiric movement of oceanic serpentinites away from the downgoing slab. Such rising serpentinites may carry with them deep-seated metamorphic blocks and are susceptible to intermixing with higher-level components of the accretionary prism by solid state injection, surface breachment, and olistostromal recycling. The point stressed in this paper is that many serpentinitic assemblages of orogenic zones are the result of polystage developmental histories with heritages that commonly stem from primary serpentinitic assemblages of the ocean floor.

INTRODUCTION

Serpentinite mobility is considered a fundamental process in the tectonics of oceanic environments and in the behavior of ocean floor fragments in orogenic zones. The importance of ocean floor serpentinization was first pointed out by Hess (1962) when he hypothesized that serpentinized peridotite was produced as crustal layer 3 by hydration of the upper mantle during ocean-floor spreading. In subsequent years, a number of geophysical and geological arguments had mounted which, in effect, refuted the Hess model and replaced it with a mafic layer model (Cann, 1968; Oxburgh and Turcotte, 1968; Coleman, 1971a; Christensen, 1972). With the wide acceptance of plate tectonics and renewed interest in ophiolite sequences, the layered mafic model of oceanic crust was widely accepted throughout the earth science community.

Key papers of the early 1970's (Coleman, 1971b; Dewey and Bird, 1971; Davies, 1971; Moores and Vine, 1971; Church, 1972) established the conceptual framework for the production of ophiolites by ocean-floor spreading and for the detachment and emplacement of ophiolites into continental margins by plate convergence and ocean-basin closure. Even though the ophiolitic model of the ocean floor satisfies a wide range of observational and theoretical constraints, the ubiquitous occurrence of serpentinite belts in orogenic zones, with or without ophiolite sequences, leads us to reconsider the importance and distribution of serpentinite in oceanic crust. Since the impact of the early 1970 papers, there has been a significant amount of additional field mapping done in serpentinite and ophiolite belts, and also a significant amount of marine geological data gathered on serpentinites of the ocean floor. Such new data should at least refine the ophiolite concept as a model for oceanic crust and shed light on oceanic serpentinites and their expression in orogenic zones.

This paper considers the concepts of ophiolitic ocean floor and its tectonic emplacement in light of what is known about serpentinite mobility, both in orogenic zones and ocean floor environments. With these considerations, speculations are made about the importance of serpentinite in the tectonic emplacement of ophiolites and the production of chaotic mixtures of ophiolitic rocks known as ophiolitic melange (Gansser, 1974). It is suggested that the worldwide occurrences of ophiolitic melange, as well as serpentinite belts, may in many cases stem from a common heritage. Primary ocean floor serpentinization and associated tectonic disruption of young mafic crust are considered the first phases in a complex structural chronology that can lead to the emplacement of serpentinite belts, ophiolitic melange, and perhaps large ophiolite nappes.

SERPENTINITE MOBILITY

Serpentinite is potentially mobile under conditions that span its entire temperature and pressure stability field. A review of serpentinite mobility and its implications as a mechanism for the vertical migration of serpentinite in orogenic zones was presented in Lockwood (1972). He argued for the use of the term protrusion, which was coined by Lyell for masses of crystalline rock tectonically injected into stratified rock. Lockwood also reviewed the relations between serpentinite protrusions and sedimentary accumulations of serpentinite and discussed possible multi-stage protrusive activity with the remobilization of sedimentary serpentinite upon burial and tectonism. Unfortunately, the views set forth by Lockwood have not been widely adopted by geologists working in melange terranes, which commonly contain serpentinites of probable protrusive and sedimentary origin. One of the aims of this paper is to briefly reiterate the ideas set forth by Lockwood and other workers in the field and to apply them to ocean floor processes and the development of ophiolitic melange.

Upward protrusion of light serpentinite is driven by buoyant forces imparted by density contrasts with heavier wallrock materials. Externally induced tectonic forces contribute to serpentinite mobility by progressive development of shear fabrics as well as direct transport. At intermediate to great depths, serpentinite mobility is greatly increased by dehydration reactions that lead to significant embrittlement and reduction in strength (Raleigh and Paterson, 1965). It was pointed out by Coleman (1971a) that such dehydration weakening could commence at temperatures as low as 200°C depending on protolith and serpentine mineral species.

The significance of a strong shear fabric in the mechanical behavior of serpentinite was demonstrated by Cowan and Mansfield (1970) in their calculations and discussions on surficial serpentinite flows derived from the New Idria ultramafic massif of the California Coast Ranges. Using derivations and analysis procedures developed by Johnson (1970), they calculated serpentinite shear strengths between 1 and 5 bars from morphological parameters measured on surface flows and their source exposures. Their results appeared to be in conflict with experimental data on non-sheared and mesh-textured serpentinite, particularly antigoritic rocks, which show considerable strength below 300°C by yielding at 8 to 12 Kb (Raleigh and Paterson, 1965). These experimental data suggest that serpentinite possesses high shear strength under low-temperature conditions; thus, diapiric and tectonic injection (or protrusion) under such conditions is unlikely. However, Cowan and Mansfield (1970) argued cogently that a shear fabric reduces the low temperature strength of serpentinite significantly, particularly in chrysotile-lizardite species, and that strength can decrease drastically as deformation proceeds.

Viewed together, in a context that considers fabric evolution and mineralogy, the analyses of Raleigh and Paterson (1965) and Cowan and Mansfield (1970) demonstrate the extreme range of behaviors that serpentinite possesses. Such extremes are often exhibited at outcrop-scale, where conspicuously hard, mesh-textured serpentinite blocks are encased in flaky-textured, sheared serpentinite that has obvious been quite mobile. The competent non-sheared blocks are transported within such serpentinite masses without significant deformation. On a larger scale, massive serpentinite and serpentinized peridotite massifs are often mantled by strongly-sheared serpentinite. High-temperature peridotite fabrics are commonly preserved in olivine-pyroxene remnants and chromite aggregates of the massive cores, whereas fabrics reflecting high level crustal emplacement are developed primarily in the sheared rinds. In the words of Cowan and Mansfield (1970), such rinds act as "lubricants" during low-temperature transport of the significantly stronger core assemblage. One could envision such lubricating action occurring along the margins of a completely serpentinized and sheared body during diapiric rise, or along the margins of a partially to completely serpentinized body during tectonic transport along a major fault zone.

The concept of serpentinite protrusion was broadened by Casey and Dickinson (1976) to include, as well as solid state injections, bodies that breach the earth's surface and flow downslope. This is considered an important concept because of the observation that, in some well-preserved protrusions, the shape fabric and foliation surfaces acquired at depth are structurally continuous with the downslope flow structure (Dickinson, 1966; Cowan and Mansfield, 1970; Casey and Dickinson, 1976). Such inherited fabrics demonstrate continuity in the dynamics of diapiric rise, surface breaching, and downslope flow. Fabric continuity may persist downslope for on the order of 10 km before progressive disruption and clastic sedimentological processes begin to significantly alter the serpentinite fabric (Casey and Dickinson, 1976; Carlson, 1981). Subsequent burial, compaction, folding, recrystallization, and/or possible remobilization make serpentinite bodies that lie concordantly within layered sedimentary and metasedimentary sequences difficult to interpret.

Complexities that may be typical of protrusive serpentinites in orogenic zones are displayed in the Kaweah ophiolitic melange in the California Sierra Nevada foothills (Saleeby, 1979, 1982). Here at least three episodes of serpentinite protrusion are distinguished. The first episode occurred within the ocean floor regime, whereupon serpentinitic olistostromes and detrital rocks were interstratified and chaotically mixed with ophiolitic pillow flows, chert, and ophicalcite (carbonate-cemented ultramafic ± mafic breccia). Subsequently, in Middle to Late Jurassic time, quartz-rich turbidites were shed across this ophiolitic melange basement. Basement-derived serpentinite protrusions and olistostromes within the turbidite sequence represent second-stage mobilization which occurred in the continental margin regime. Subsequent tight folding and slaty cleavage development during the Late Jurassic Nevadan orogeny were accompanied by local remobilization of earlier protruded serpentinite along axial surfaces and oppressed limbs of folds within the turbidite sequence. This suggests that some or many concordant serpentinite bodies mapped within slate belts as fault slices may actually be deformed or remobilized sedimentary serpentinite beds, or perhaps remnants of syn-depositional protrusion dikes. Additional pertinent features of the Kaweah ophiolitic melange will be reviewed below. The point emphasized here is the multi-stage history of serpentinite mobilization, starting with serpentinization processes on the ocean floor.

SERPENTINITES OF THE OCEAN FLOOR

Serpentinite was recovered from dredge hauls over the Mid-Atlantic Ridge as early as 1947 (Shand, 1949). These early findings helped influence Hess when he developed his model for ocean floor serpentinization, although the structural settings of ocean floor serpentinites were unknown at the time. At about the same time that his model was being refuted, significant amounts of serpentinite were being recovered from a wide range of locations in ocean floor environments (Van Andel and others, 1969; Aumento and Loubat, 1971; Bonatti and others, 1971; Melson and Thompson, 1971; Melson and others, 1972; Thompson and Melson, 1972; Engel and Fisher, 1975). The majority of these ocean floor serpentinites were recovered from fracture zones. Consequently, an argument developed over whether exposures of "normal" or idealized oceanic crust would be exposed in such environments. However, if the width of fracture zones and the density of their distribution in the ocean floors are considered in such a context, then the basic concept of "normal" oceanic crust should be restructured. It is suggested below that serpentinization processes not entirely different from those envisaged by Hess contribute significantly to the production of oceanic crustal rocks and that ophiolitic melange develops in conjunction with ocean floor serpentinization along fracture zones. Oceanic crust is considered in its primary state to be a mosaic of ophiolitic slabs bounded by belts of serpentinite and ophiolitic melange.

A series of bathymetric profiles is shown in Figure 1 for some of the better studied fracture zones. Also shown are generalized symbols for rocks recovered from dredge-hauls and some structural relations deduced from seismic profiles. A striking feature of these sections is the bathymetry across the fracture zones accentuated by steep-sided ridges and trenches that define the fracture trend. Relief variations in the range of 2 to 4 km are common for lightly sedimented fracture zones. At St. Pauls frac-

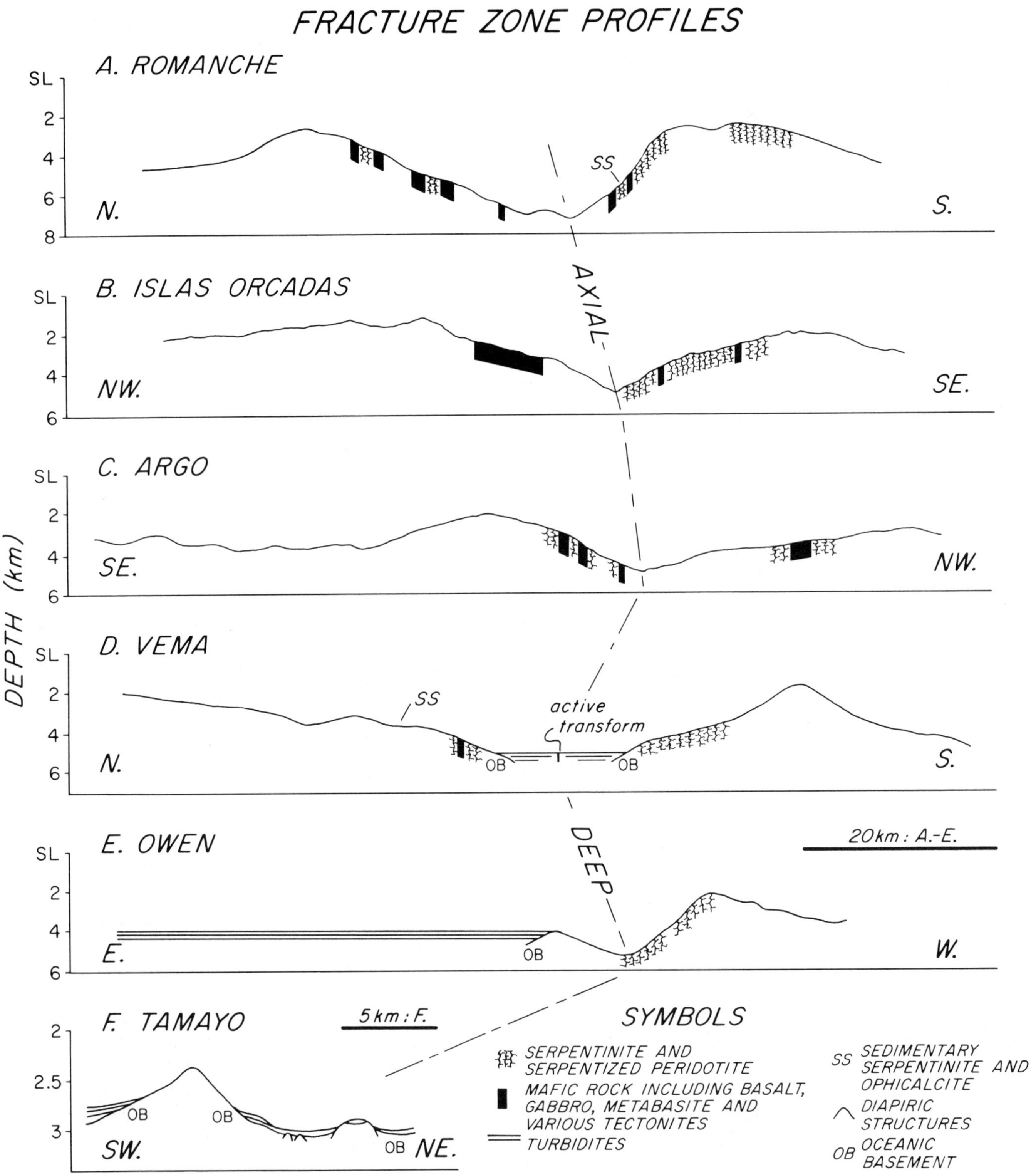

Figure 1. Bathymetric profiles and diagrammatic summary of dredge data for some better-studied oceanic fracture zones. A. Romanche (equatorial Atlantic) after Bonatti and others (1973, 1974), Bonatti and Honnorez (1976). B. Islas Orcadas (Antarctic plate boundary) after Sclater and others (1978). C. Argo (Mid-Indian Ridge) after Engel and Fisher (1975). D. Vema (equatorial Atlantic) after Bonatti and others (1973, 1974), Bonatti and Honnorez (1976), Eittreim and Ewing (1975). E. Owen (Mid-Indian Ridge) after Bonatti (1978). F. Tamayo (Gulf of California) after Macdonald and others (1979).

ture zone in the equatorial Atlantic (not shown in Figure 1), mylonitic and serpentinized peridotite is uplifted above sea-level (Melson and others, 1972). Other examples of emergent fracture zones will be discussed below.

Uplift and exposure of large ultramafic massifs along fracture zones are shown clearly in sections a through e of Figure 1. The Romanche fracture zone of the equatorial Atlantic (Figure 1a) shows overall relief of 4500 m. Substantial amounts of serpentinized peridotite and minor metagabbro constitute the southern wall and ridge top. Greenschist to amphibolitic and mylonitic mafic rock with lesser ultramafic rock constitute most of the northern wall. In a summary of the petrology of rocks dredged from the Romanche fracture zone, Bonatti and Honnorez (1976) point out the predominance of strong tectonitic fabrics in all rock types dredged. They also make special note that serpentinized peridotite was present in 34 out of 59 dredge hauls across various points in the fracture. Of special interest with respect to the problem of ophiolitic melange is the abundant sedimentary serpentinite and ophicalcite dredged from the lower slopes of the south wall (Bonatti and others, 1973, 1974). The thickness of slumped and tectonically brecciated material present along the axial deep of the fracture zone is estimated as hundreds of meters by analogy with *in situ* observations made from submersibles along smaller fracture zones of the FAMOUS area (Choukroune and others, 1978). The Romanche fracture zone may be characterized as a vast terrane of exposed serpentinized ultramafic rocks, with slices of largely tectonitic and metamorphosed mafic rock and sedimentary derivatives of ultramafic rock. Such an assemblage, with little or no tectonic modification, may be mapped on land as ophiolitic melange.

Additional ultramafic massifs of the ocean floor are shown in sections b and c (Figure 1), which cross the Islas Orcadas fracture zone of the Antarctic plate boundary (Sclater and others, 1978) and the Argo fracture zone of the western Indian Ocean (Engel and Fisher, 1975). The density of dredge hauls across these fractures is significantly lower than across the Romanche fracture, but nevertheless the volumetric importance of serpentinized peridotite is clear. The Islas Orcadas fracture yields primarily ultramafic rock in extensive dredging across its southeast wall and primarily basalt along limited dredges from its northwest wall. Both the southeast and northwest walls of the Argo fracture yield predominately serpentinized ultramafic rocks with lesser mafic rocks, some of which are highly tectonitic.

From the analysis of marine magnetic anomaly patterns and fracture zone morphology in the northeast Pacific, Menard and Atwater (1968, 1969) suggested that fracture zones may undergo significant extensional or compressional deformations as a function of minor changes in the pole of relative motion. Similar considerations led Van Andel and others (1969) and Thompson and Melson (1972) to suggest rifting mechanisms for the emplacement of ultramafic massifs along equatorial Atlantic fracture zones. A similar mechanism involving rifting and perhaps compressive closure in conjunction with diapirism was suggested for the dismemberment of mafic crustal rocks and the emplacement of ultramafic rocks along the Oceanographer fracture zone of the North Atlantic (Fox and others, 1976). Extensive upper mantle-lower crustal serpentinization and related protrusion within discrete domains of the Atlantic oceanic crust were suggested by Bonatti (1976) and Bonatti and Honnorez (1976). They emphasized the importance of voluminous protrusions along fracture zones in conjunction with smaller protrusions along deep faults that parallel the spreading axis.

The orthogonal pattern of serpentinitic rocks as fracture zone massifs and smaller ridge-parallel belts is considered to be a product of slow-spreading ridges (Bonatti, 1976; Bonatti and Honnorez, 1976; Francis, 1981). Francis (1981) named these structures serpentinization faults. Such structures are thought to greatly enhance the deep circulation of ocean water by providing large horizontal temperature gradients by either juxtaposing different ages of young warm crust (transform faults) or disrupting low-dipping isothermal surfaces along steep, ridge-parallel normal faults (Francis, 1981). Pore water pressures will vary significantly across steep horizontal temperature gradients driving a circulation pattern in which water sinks through sheared rocks along the cooler side of the fault and ascends through sheared rocks along the warmer side. In a detailed study of stable isotopes in the Oman ophiolite, Gregory and Taylor (1981) showed that ocean water was hydrothermally circulated to upper mantle depths near the spreading axis, although temperatures remained above the serpentine stability field for the duration of hydrothermal activity. They concluded that ocean floor serpentinization is not important within intact oceanic crust and upper mantle. In the serpentinization fault model, the deep structural breaks provide a permeable pathway resulting in high water-rock ratios at upper mantle depths. Convective heat loss through water-rock interactions accelerates cooling in proximity to the fault zone, thereby facilitating serpentinization. If serpentinization proceeds far enough to yield a significant density differential, protrusive activity may commence along the fault zone. The serpentinization fault model explains the observation that fracture zone massifs often begin their major uplift on the cooler side of transform segments as they migrate past their adjacent ridge segment (Van Andel and others, 1971; Phillips and Fleming, 1973). It also predicts that massif uplifts will be greatest for large transform offsets of slower spreading ridges, which would provide the largest horizontal temperature gradients. In a general survey of marine geological literature, it appears that fracture zone ultramafic massifs are more common in the slower spreading Atlantic and Indian Oceans than along the rapidly spreading east Pacific rise. Finally, the serpentinization fault model predicts that development of the ridge parallel species will be inhibited along fast-spreading ridges due to the presence of a continuous large magma chamber (Cann, 1974; Orcutt and others, 1976; Herron and others, 1978), which restricts the depth of normal faulting in proximity to the spreading axis.

Consideration of the serpentinization fault model in conjunction with the structural complexities that are thought to develop along fracture zones to accommodate both minor changes in the pole of relative motion (Menard and Atwater, 1968, 1969; Dewey, 1975) and the simplest of ridge-ridge transform fault displacement patterns (DeLong and others, 1977; Karson and Dewey, 1978) leads to the conclusion that vast strips of tectonically mixed mafic and serpentinized ultramafic rocks line the ocean floors. For instance, in the north Atlantic Ocean south of the Azores transform offsets greater than 20 km occur for about every 55 km of ridge segment (Fox and others, 1969; Vogt and others, 1971; Johnson and Vogt, 1973; Vogt and Avery, 1974). The width of crustal disruption across the very large fracture zones of the equatorial Atlantic, Indian Ocean, and Antarctic plate boundary is typically about 50 km (examples are given in Figure 1). It seems likely that ophiolite melange belts totaling many thousands of square kilometers are packaged into the earth's oceanic crust prior to arrival in continental margin orogenic zones. Any realistic view of the ophiolitic melange problem in orogenic zones must consider the probable role of primary ocean floor melange.

Geologists working in orogenic zones may wish for some diagnostic features to distinguish primary ocean floor melange from ophiolitic melanges mixed solely by continental margin tectonics. One may think that the intermixing of terrigenous and ophiolitic material would be such a criterion. Examination of sections d through f of Figure 1 casts some degree of uncertainty on such a view. The profile across the Vema fracture zone of the equatorial Atlantic (Figure 1d) is similar in complexity to profiles across the Romanche, Islas Orcadas, and Argo fracture zones. However, at least 500 m of turbidites derived from the Amazon fan system lie ponded within the axial deep of the Vema fracture zone (Eittreim and Ewing, 1975). The surface of the turbidite sequence is broken by steep faults apparently related to active transform motions. The width of tectonic disruption within the turbidite sequence widens at depth to the point that structural details are acoustically irresolvable. It is not unlikely that turbiditic and ophiolitic materials are progressively intermixed at depth along this axial deep environment. The Vema fracture zone assemblage thus consists of disrupted ocean floor rocks, continent-derived turbidites, and large ultramafic slices, which together may be destined for accretion onto the inner wall of the Puerto Rico trench. The question of how a geologist could distinguish a Vema-like assemblage from an assemblage that was disrupted and mixed solely along an inner trench wall environment is not trivial.

In Figure 1e, a profile is shown for the Owen fracture zone of the western Indian Ocean (after Bonatti, 1978). The Owen fracture shows about 3000 m of relief. Dredge hauls along the steep inner west slope yield ultramafic rocks exclusively. A more subdued basement ridge along the west wall of the fracture is damming turbidites of the western Indian abyssal plain. Structural modification of the rock assemblages shown in this profile by convergence-related accretion would also lead to a structural package with an ambiguous petrotectonic heritage.

A profile across the Tamayo fracture zone at the mouth of the Gulf of California is shown in Figure 1f (after Macdonald and others, 1979). This relatively small fracture is also heavily sedimented with basement exposures occurring primarily along uplifted areas. Of interest here are diapiric structures that pierce the sediments along the axis of the fracture. Two of these structures lie buried and show evidence of deforming the lower levels of the turbidite sequence during their rise. To their northeast, another diapir has protruded through the surface of the turbidite sequence. Vestiges of turbidite beds lie perched in the summit area of the protrusion. Magnetic studies across the diapiric structures limit their composition to serpentinite or, possibly, basalt (Macdonald and others, 1979). Dredge data are not available for the Tamayo profile.

The profiles of Figure 1 are intended to demonstrate the sizes of fracture zones, some of their structural complexities, and the importance of uplifted and protruded serpentinized ultramafic rocks. The view expressed here is that complex rock assemblages, which on land would appear as ophiolitic melange and ultramafic massifs, could be a common product of fracture zone tectonics. It also seems clear that such assemblages may be deformed and intermixed with turbidite sequences within the fracture zone environment. The widespread distribution of fracture zones in modern oceanic crust demands cognizance of fracture zone processes and products for those working in melange terranes. Attention will now focus on ophiolitic melange belts exposed in orogenic zones.

OPHIOLITIC MELANGE IN OROGENIC ZONES

The worldwide importance of ophiolitic melange within orogenic zones was reviewed by Gansser (1974). He discussed the tectonic settings of Tethyan belts and stressed a number of points in his definition of ophiolitic melange. Such points included both tectonic and olistostromal mixing of ophiolitic blocks into serpentinitic or pelagic and flyschoidal matrix materials. He also pointed out that ophiolitic melanges are often overthrust by ultramafic or ophiolite nappes, and that the melanges are prone to remobilization as olistostromes and protrusions. Gansser stressed that ophiolitic melange, with or without ophiolite nappes, often marks major sutures within continental crust, which are the remnants of closed ocean basins. Another point that he stressed, but did not explore the significance of, is that ophiolitic melanges are often transgressed by pelagic and/or flysch sedimentary sequences. The implications of such stratigraphic relations are that the transgressed melanges formed and resided for some time in deep marine settings, and thus they are not directly related to ocean floor obduction during collisional suturing events. Such a relation is clear in the Kaweah ophiolitic melange, where serpentinite-matrix melange was transgressed by a turbidite sequence (flysch) after substantial ocean floor residence (Saleeby, 1979, 1982).

The Kaweah ophiolitic melange lies within the Late Jurassic Nevadan suture belt, which is considered to have formed by ocean basin closure and related collisional orogeny (Moores, 1970; Schweickert and Cowan, 1975). However, the Kaweah melange clearly underwent its major mixing phases long before the Nevadan orogeny. Remnants of primary relations within the melange record depositional sequences consisting of pillowed lava and breccia, sedimentary serpentinite, ophicalcite, chert, and scattered hypabyssal intrusives. Such sequences were constructed on or immediately adjacent to previously disrupted mafic and ultramafic ophiolitic assemblages. Further north along the Kaweah belt, in the Kings River area, the ophiolitic melange contains strongly mylonitized peridotite and flaser gabbro massifs, and blocks and slabs with the remnants of sheeted dike-pillow lava sequences (Saleeby, 1978, 1982). Mylonitization and protoclastic deformation of the peridotite-gabbro massifs and emplacement of sheeted dikes and commensurate pillow eruptions are pre-Jurassic in age, as are the hypabyssal rocks associated with the remnants of pillow lava, sedimentary serpentinite, and chert sequences of the Kaweah area. Protoclastic deformation and tectonic brecciation of oceanic crustal rocks in conjunction with protrusion of mylonitic and serpentinized mantle materials in marine environments are processes best documented along oceanic fracture zones. It thus appears that ophiolitic melange from the southern segment of the Nevadan suture belt is derived from a pre-Nevadan fracture zone assemblage. Similar pre-Nevadan ophiolitic melange occurs beneath Upper Jurassic slate in several other locations along the Nevadan suture belt (Saleeby, 1982).

Ophiolitic melange and flaser gabbro and peridotite massifs have been interpreted as the remnants of accreted fracture zones in a number of other locations. As with the Kaweah melange, structural modification and olistostromal reworking related to orogenic emplacement are superimposed over the effects of ocean floor disruption processes. Ophiolitic melange assemblages of the northern Apennines in Italy, for the most part, originated as oceanic crustal sequences consisting of brecciated and disrupted mafic assemblages and protruded serpentinite (Gianelli and Principi, 1979; Abbate and others, 1980). Flaser gabbro which formed under protoclastic or hot-subsolidus conditions is a common constituent of the disrupted mafic assemblages, and also occurs as clasts in submarine talus sequences (Cortesogno and others, 1975; Gianelli and Principi, 1979). Such talus sequences contain a wide range of mafic and ultramafic detritus and are covered by chert and pelagic mudstone and limestone. Where ultramafic rocks were the principal exposures in the ancient seascape, ophicalcite formed the main superjacent clastic assemblage. Serpentinite protrusions, in some cases, may have reached sea level, as suggested by pedogenic-type features and shallow water structures in some ophicalcite bodies (Folk and McBride, 1976, 1978). Such paleo-elevated sequences are also transgressed by pelagic deposits. Fracture zone ultramafic rocks uplifted above sea level are recognized at St. Pauls Rocks (Melson and others, 1972), and ridge areas underlain by ultramafic rocks of the Romanche fracture zone are known to have complex Neogene uplift and subsidence patterns (Bonatti and others, 1977). Thus, the interpretations presented by Folk and McBride (1976, 1978) for the northern Apennine ophicarbonates are not without possible modern analogues. Analogies between ophiolitic melange of the Apennines and modern oceanic fracture zones have been made by Abbate and others (1980) and Gianelli and Principi (1979). They stress the importance of transgressed pelagic sequences separating initial ophiolite disruption and olistostromal activity from further tectonic disruption and olistostromal reworking that occurred during nappe emplacement of the fracture zone assemblages.

Rocks of the Coastal Complex of western Newfoundland have also been interpreted as an on-land fracture zone assemblage (Karson and Dewey, 1978). Here, complex structural sequences of amphibolitic gabbro, mafic gneiss, deformed dike rock, serpentinized peridotite, and serpentinite-matrix melange are cut by non-deformed mafic dikes. Such cross-cutting relations are thought to represent migration of an active transform assemblage past an adjacent spreading-ridge segment (Karson and Dewey, 1978). Mafic breccias, ophicalcites, and off-axis alkalic volcanics represent deposits that accumulated along the ancient fracture trace. The Coastal Complex is locally part of the same thrust sheet as the Bay of Islands ophiolite (Karson and Dewey, 1978), but in places it is also apparently overthrust by the ophiolite sheet (Williams, 1973, 1975). Karson and Dewey (1978) suggest that obduction of the Coastal Complex-Bay of Islands sequence initiated along the ancient fracture zone trace.

Considering the distribution of fracture zones on the modern ocean floors, it is surprising that more fracture zone assemblages have not been recognized in on-land ophiolite belts. The Gibbs Island ultramafic complex in the South Shetland Islands of the Antarctic Cordillera is thought to be a fracture zone protrusion (De Witt and others, 1977). Sarwar and De Jong (this volume) interpret the Kanar melange, southern Pakistan, as having formed along a major transform juncture which evolved into a zone of convergence and ophiolite obduction. Moores and Vine (1971) identified a fossil transform fault in the Troodos ophiolite of Cyprus. This fault is quite small compared to most fracture zones and probably represents a second-order ephemeral transform. Such structures are common along modern spreading ridges and are considered to represent a structural response to minor variations in the pole of relative motion (Vogt and others, 1971; Johnson and Vogt, 1973; Vogt and Avery, 1974). In a general survey of the literature, I have found little more than the above accounts of on-land transform or fracture zone assemblages. This is a considerable problem if uniformitarian principles are to be applied in the comparison of ophiolites and modern oceanic crust.

The view expressed here is that the fracture zone assemblages are common in orogenic zones, but they occur primarily in the form of transported and remobilized ophiolitic melange and related ultramafic belts. Such belts are often interpreted to be a direct result of ocean basin closure and the suturing of collided crustal blocks. This interpretation is easily accepted, inasmuch as collisional sutures are, in part, defined by such belts. However, in Gansser's (1974) treatment of Tethyan belts, the relations between melange mixing and collisional suturing were not discussed, and the structural relations presented did not reflect mixing of ophiolitic crust during tectonic transport related to suturing. The transgressed melanges stand in direct contrast to such a mechanism and suggest that such ophiolitic melanges are accreted into collisional sutures in a pre-mixed state.

OPHIOLITIC MELANGE AND PLATE CONVERGENCE

Discussion on the relations between ophiolitic melange and plate convergence will focus first on the structural chronology that begins in the ocean floor along zones of primary structural weakness. Figure 2a shows the major structural elements of intersections between fracture zones and spreading ridges. Tectonic disruption of newly created crust begins in the intersection environment. Here, complex cross-fault patterns develop in response to transform motions and dip-slip displacements of both the rift axis and transform axial deep (Choukroune and others, 1978). Slumped mafic rubble, fault breccia, and pelagic ooze begin to accumulate in the axial deep immediately following crustal generation. The plutonic assemblages at the intersection may lack well-defined cumulate sequences and sheeted dikes, and are commonly deformed under protoclastic and hot subsolidus conditions. As the newly created crust migrates along the transform segment of the fracture, it undergoes progressive deformation by strike-slip transport and dip-slip adjustments resulting from the opposing plates being in different stages of their thermal decay as they slip past one another. As permeability and tectonic shuffling of the crust increase, the probability of ocean water interaction with upper mantle materials also increases. The onset of serpentinite protrusive activity may begin at a fairly early stage, particularly if the pole of relative motion varies with rift increments across the fracture (Van Andel and others, 1969; Thompson and Melson, 1972). With further migration of the cooling transform assemblage into the proximity of the adjacent ridge segment, substantial horizontal temperature gradients develop, and serpentinization fault processes may commence as envisaged by Francis (1981). At this stage, the most vigorous protrusive activity may begin. Thus, by the time major invasion of crustal materials by serpentinite begins, the mafic crust has already gone through a substantial history of faulting and tectonic brecciation in the transform environment. Such a deformation chronology is ideal for the development of serpentinite-matrix melange. Furthermore, the axial deep region has continued to fill with material slumped off the fracture walls, and thus, chaotic piles of surficial debris and pelagic ooze will be intergradational with tectonically mixed material. As the fracture zone complex migrates past the adjacent ridge segment, transform motions cease. The nontransform history of the fracture will also be characterized by prolonged dip-slip faulting related to thermal equilibration of the different ages of crust juxtaposed by the fracture (DeLong and others, 1977). Transgressive sequences of pelagic and/or flyschoidal sediments will cover the fracture as a function of ocean floor residence time and sediment source patterns.

Fracture zones constitute significant primary structural weaknesses in oceanic crust. Spreading-ridge segments also constitute zones of such weakness due to structural discontinuity and extremely high thermal gradients. Ridge-parallel serpentinization faults, when developed, may constitute second-order zones of primary structural weakness. The profound crust and upper mantle discontinuities represented by fracture zones and spreading ridges make these primary junctures ideal loci for the initiation of plate convergence in the event of a plate motion change.

Speculations on the behavior of primary ocean floor structures during a change to plate convergence within an actively spreading ocean basin are shown in Figure 2b. Obductive-type imbrication is shown initiating along transform and spreading ridge segments. The important point here is the nature of the initial structural sequence. Imbrication along ridge segments will result in dynamothermal aureoles along the thrust soles, as a result of primary heat within the overthrust sheet (Armstrong and Dick, 1974). Along fracture zone segments, pre-deformed sheets of ophiolitic melange and serpentinized peridotite will be imbricated with ophiolite sequences that may or may not carry a basal thermal aureole. With further tectonic transport, the initial imbricate sequence will be progressively dismembered and perhaps in part reworked as olistostromes. Such a structural chronology will leave high temperature metamorphic blocks and slices of ophiolitic melange scattered along the base of the ophiolitic nappe. The thrusting of ophiolitic nappes across fracture zone complexes or spreading ridges sets up an ideal situation for further transport by gravity sliding. This is suggested by the pronounced bathymetric variations across both spreading and transform plate junctures. Such gravity transport could conceivably lead to nappe emplacement across the edge of a deeply submerged rifted crustal block. This type of a gravity transport mechanism is especially viable along a boundary transform immediately adjacent to a rifted block, or in the case of a small ocean basin, where general proximity is maintained between initial rift edges and spreading ridge environments.

Ophiolitic nappes of the Oman belt lie structurally above metamorphic tectonic slices and local ophiolitic melange units, which together may represent disrupted vestiges of initial ocean floor imbricate sequences. Most notable are amphibolite slices derived from basaltic and pelagic sedimentary protoliths that were metamorphosed under high temperature-high strain conditions almost instantaneously after the ophiolite formed (Boudier and Coleman, 1981; Ghent and Stout, 1981; Lanphere, 1981; Tilton and others, 1981). Ophiolite obduction evidently initiated along the spreading axis, and then progressed by gravitational sliding across chaotic pelagic deposits and onto the deeply submerged edge of northeastern Arabia (Coleman, 1981). Some segments of the Oman belt are underlain by serpentinite and volcanic-matrix ophiolitic melange units that lack features indicative of deep-seated tectonic mixing (Coleman, 1981). These melange units lie in a structural position that is analogous with the amphibolite slices and may represent vestiges of the initial imbricate sequence, where obduction initiated along small fracture zone segments. The Oman ophiolite belt at present does not

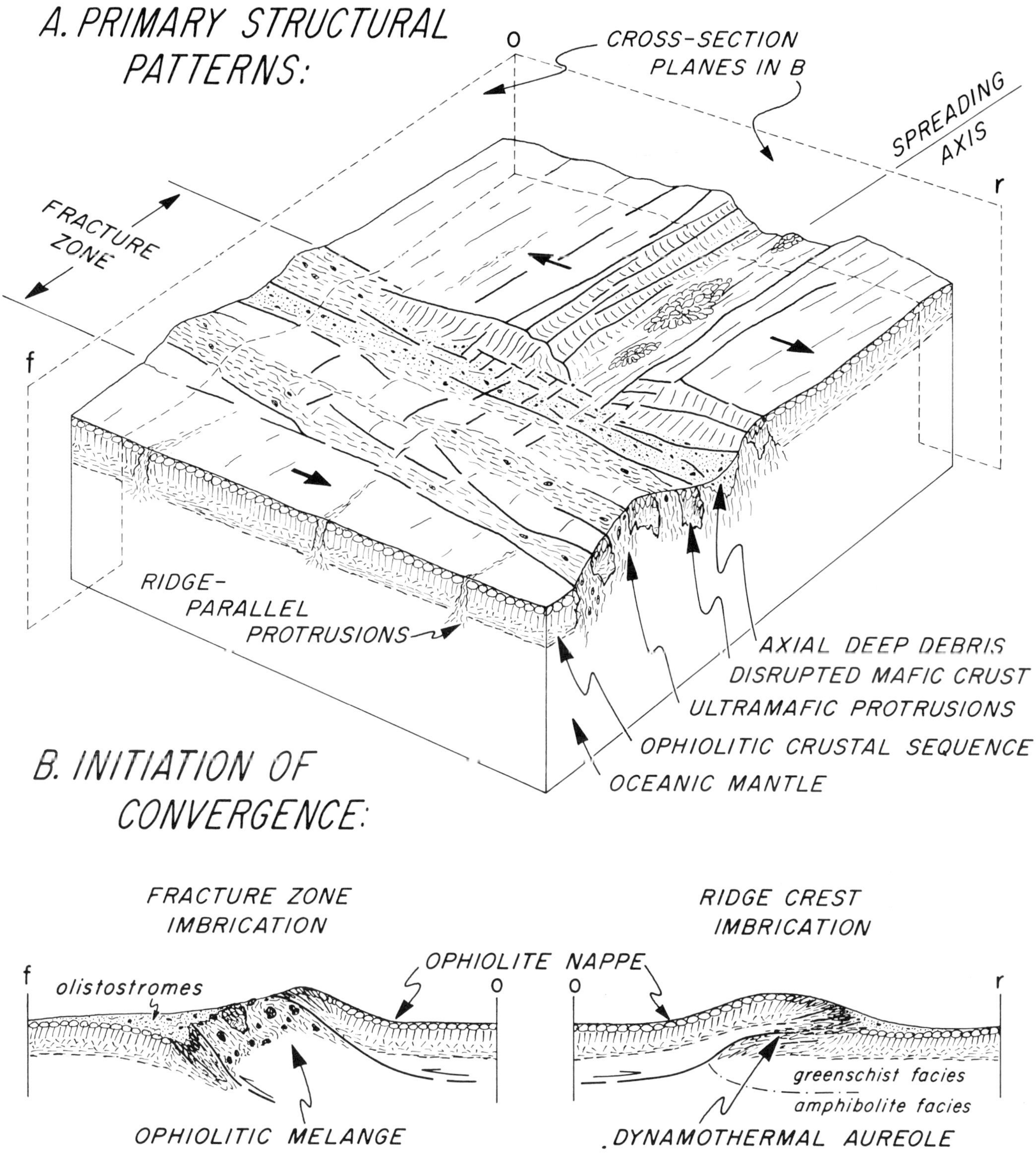

Figure 2. Schematic block diagram showing: (a) primary structural features that develop at and near fracture zone-spreading ridge intersections (after Bonatti and Honnorez, 1976; Fox and others, 1976; Choukroune and others, 1978); and (b) speculative ancillary sections of probable primary structural imbrication relations along fracture zone and spreading plate junctures upon initiation of convergence along primary junctures. Diagrammatic greenschist-amphibolite isograd depicts primary thermal structure near imbricated spreading ridge.

coincide with a collisional suture. Closure of the Gulf of Oman in the future may result in the accretion of the high-standing Oman nappes within an Arabian-Eurasia suture belt. However, it is clear that initial imbrication was an ensimatic process and that the present structural position of the nappes resting on Arabian crust does not mark a collisional suture.

Initial imbrication of oceanic crust is considered to be primarily a result of plate motion changes. Such imbrication along fracture zone and spreading ridge segments was predicted by the theoretical arguments of Menard and Atwater (1968) and Dewey (1975). Sufficient plate motion changes may occur during the closure of an ocean basin, with ensuing collisional suturing, or they may occur quite independently of such collisions. The critical point here is that once the initial imbricate sequence has formed, it will be susceptible to tectonic accretion, when and if the ocean basin closes or if it is by chance swept into a subduction zone. Such oceanic structural sequences and perhaps even large unmodified fracture zone massifs are likely to behave like oceanic plateaus with preferential accretion in zones of convergence (Ben-Avraham and others, 1981).

The utilization of primary structural weaknesses within oceanic crust for the accommodation of plate motion changes is shown for the Gorda plate in Figure 3. Here the Mendocino triple junction joins the Pacific, North American, and Gorda plates by a transform-transform-subduction triple junction (Silver, 1971). However, departure from ideal rigid plate tectonics is evident by discordant trends between the Mendocino and Blanco fracture zones, internal deformation of the Gorda plate, and limited convergence across the Mendocino fracture zone. Much of the area of interest is heavily sedimented, and, thus, extensive basement sampling by dredging is not feasible. However, the Mendocino fracture zone is still expressed as a major morphotectonic feature with a ridge and trough bathymetric profile. First motion studies suggest right-lateral motion along the fracture with a small southward underthrusting component (Bolt and others, 1968; Seeber and others, 1970). Reflection profiling across the fracture shows deformation and erosion of Pleistocene strata (Silver, 1971), which in conjunction with the dredging of rounded basaltic cobbles from the fracture ridge, suggests an emergent stage in the not too distant past (Krause and others, 1964). A system of ridge parallel faults is actively deforming the Gorda plate and its overlying sediments (Silver, 1971). These faults are shown in Figure 3 along with the trends of the Raff and Mason (1961) magnetic anomaly pattern. Analysis of marine geological data and first motion studies (McEvilly, 1968; Tobin and Sykes, 1968) led Silver (1971) to conclude that this fault system has a predominate left-slip movement pattern. The significant question that arises concerns the nature of the primary weakness that this fault system is following. Perhaps ridge-parallel serpentinization faults are controlling the manner in which the Gorda plate is deforming internally.

Oceanic crust is imbricated along the Macquarie Ridge and Island complex (Figure 4) as a result of a minor component of convergence across fracture zone trends. This system of ridges

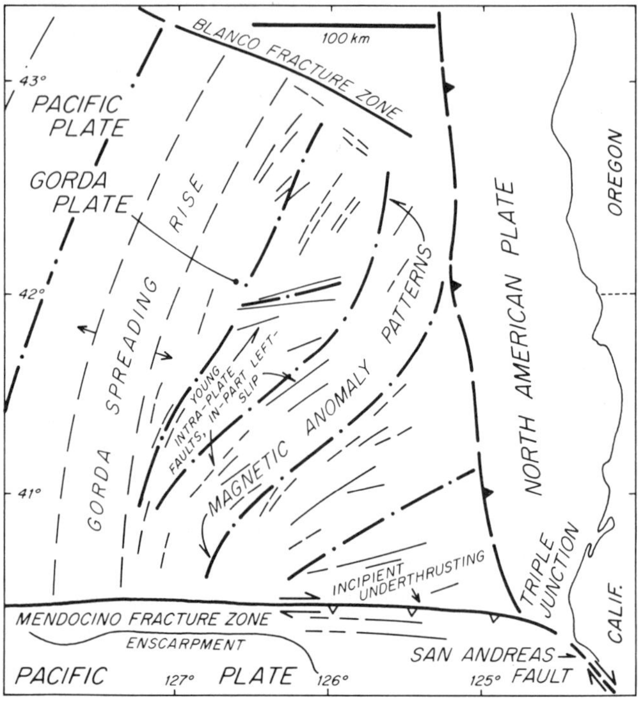

Figure 3. Map showing tectonic setting of the Mendocino triple junction and incipient internal deformation features of the Gorda plate (after Silver, 1971; Drummond and others, 1981).

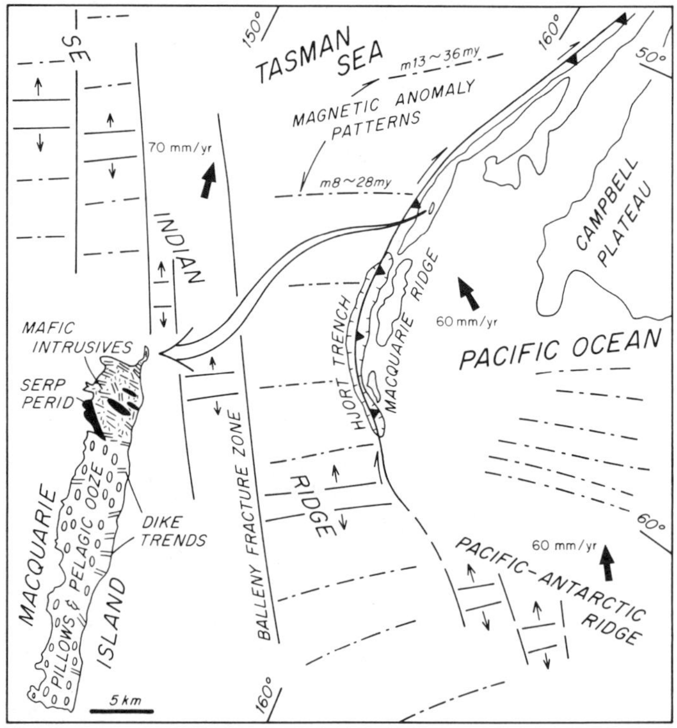

Figure 4. Map showing tectonic setting of Macquarie Ridge and Island complex emphasizing convergence component across fracture zone, and ancillary geologic map of Macquarie Island ophiolitic uplift (after Varne and Rubenach, 1972; Hayes and Talwani, 1972; Doutch and others, 1981; Ruff and others, 1982; and written commun., 1981). Heavy arrows = absolute motion.

and troughs is the morphological expression of the Pacific-Indian plate juncture in the extreme southwest Pacific. The pole of rotation between these two plates has changed progressively over about the last 40 m.y., resulting in an evolution from dextral-slip with minor extension along the Macquarie Ridge juncture to present-day dextral-slip with convergence (Walcott, 1978; Minster and Jordan, 1978). First motion studies on large and intermediate earthquakes along the ridge show dextral-slip along steeply-dipping faults and thrust movement on shallow east-dipping faults (Ruff and others, 1982; written commun., 1981) consistent with the theoretical pole of rotation. Also shown on Figure 4 are the trends of the magnetic anomaly patterns on the ocean floor adjacent to the ridge. The anomalies are trending at a high angle to the ridge, which further suggests that uplift is occurring along fracture zone trends. The ancillary map shows the generalized geology of Macquarie Island, which is the upper layers of an ophiolite being thrusted directly out of the ocean floor (Varne and others, 1969; Varne and Rubenach, 1972). Note the trend of the dike swarms and tabular bodies of serpentinized peridotite that cross the island and grossly parallel the magnetic anomaly trends on the adjacent ocean floor. Varne and others (1969) suggest that the structural patterns exposed on Macquarie Island are manifest directly on the ocean floor as the magnetic anomaly pattern. It has also been suggested that the serpentinized peridotite bodies of Macquarie Island are vestiges of ridge-parallel protrusions (Bonatti, 1976).

The Macquarie Ridge and Island complex exposes oceanic crust imbricated along a fracture zone as a result of a moderate change in the pole of rotation between the adjacent plates. One might suspect that lower plate rocks along this imbrication zone consist, at least in part, of fracture zone assemblages, quite likely ophiolitic melange. In the event that the Tasman Sea or the ocean basin between Antarctica and Australia was to close in the geologic future, the Macquarie Ridge complex would probably be accreted along a suture belt between the Campbell Plateau-New Zealand block and the Australian or Antarctic blocks. The suture belt ophiolitic assemblages in such a hypothetical case would clearly have undergone an imbrication event long before the collisional suturing event.

The manner and extent to which ocean floor is disrupted into ophiolitic melange within sediment-poor subduction zones must also be considered. For example, the Mariana trench is for the most part starved of terrigenous sediment (Karig and Sharman, 1975), and dredging along its inner wall has revealed disrupted ophiolitic sequences, ultramafic massifs, and abundant serpentinite (Yagi, 1960; Evans and Hawkins, 1979; Hawkins and others, 1979; Bloomer, 1981). These data point to the likely existence of ophiolitic melange within the inner-trench wall environment. However, if the developmental history of the Mariana arc-trench system is considered, the origin of such melange units appears ambiguous.

The major tectonomorphic features of the Mariana system are shown in Figure 5a (after Karig, 1971a, 1971b, 1972; Watts and others, 1977; Nishwaki and others, 1981). These features consist chiefly of parallel arcuate ridges and basins which lie inboard of the Mariana trench. Active elements of the system include the subducting trench, the Mariana Ridge volcanic arc, and the Mariana Trough, which is an inter-arc basin producing new oceanic crust. Inactive elements of the system include the West Mariana Ridge, which is a remnant arc that ceased activity upon growth of the Mariana Trough, and the Parece Vela Basin and Kyushu-Palau Ridge, which represent an inter-arc basin-remnant arc pair produced by rifting and ocean floor spreading in Early Miocene time. The West Philippine basin is thought to be a trapped fragment of the Pacific plate. Reconstructions of the Pacific plate's developmental history by Hilde and others (1976) suggest that during Late Cretaceous and Early Tertiary time the west Pacific basin was characterized by east-west trending spreading ridges with large north-south fracture zones (Figure 5b). A major change in Pacific plate motion in mid-Tertiary time resulted in a strong westward component (Clague and Jarrard, 1973) that led to convergence with Asia across fracture zone trends. Initiation of the Mariana trench along a large fracture zone led to the entrapment of a large fragment of the Pacific plate between the proto-Mariana trench and the eastern Asia margin (Figure 5c). The trapped oceanic crust contains the remnants of its Pacific plate spreading history in its generally east-west magnetic anomaly patterns and in a fossil spreading ridge that ceased activity at about the time of the major change in plate motions. The Hilde and others (1976) model suggests that fracture zone crust may constitute part of the Mariana arc basement and inner-trench wall. The view taken here is that the proto-Mariana trench may have resembled the imbricate sequence of Figure 2b, where plate convergence is depicted as initiating along a fracture zone. Such a view poses basic ambiguities in the interpretation of disrupted ophiolitic rocks within the Mariana inner-trench wall.

Franciscan melanges of the California Coast Ranges are the most closely studied rocks considered to have been mixed within a subduction setting. However, in a survey of the literature and from my own field observations, it appears that Franciscan serpentinite melange units rarely, if at all, contain block assemblages that reflect the mixing of complete ophiolite sequences. In contrast, many of the serpentinite melange units contain blocks with a wide range of protoliths and metamorphic overprint histories. Such compositional variations are suggestive of protrusive serpentinite masses that rose through an actively growing accretionary prism. Slabs of pillow basalt ± chert may represent the upper intervals of ophiolite sections that were sliced off during subduction, but nowhere can they be seen intermixed with blocks of dike complex, high-level gabbro, cumulates, and ultramafic tectonites.

Remnants of Jurassic ophiolite sit in an upper thrust plate position along the eastern edge of much of the Franciscan Complex (Bailey and others, 1970; Hopson and others, 1981). However, there are few signs of the upper plate ophiolite having been sliced off and mixed into ophiolitic melange during the Cretaceous construction of the Franciscan Complex. Serpentinite-matrix ophiolitic melange occurs as a major part of the northern segment of the Jurassic ophiolite terrane (Hopson and others, 1981); how-

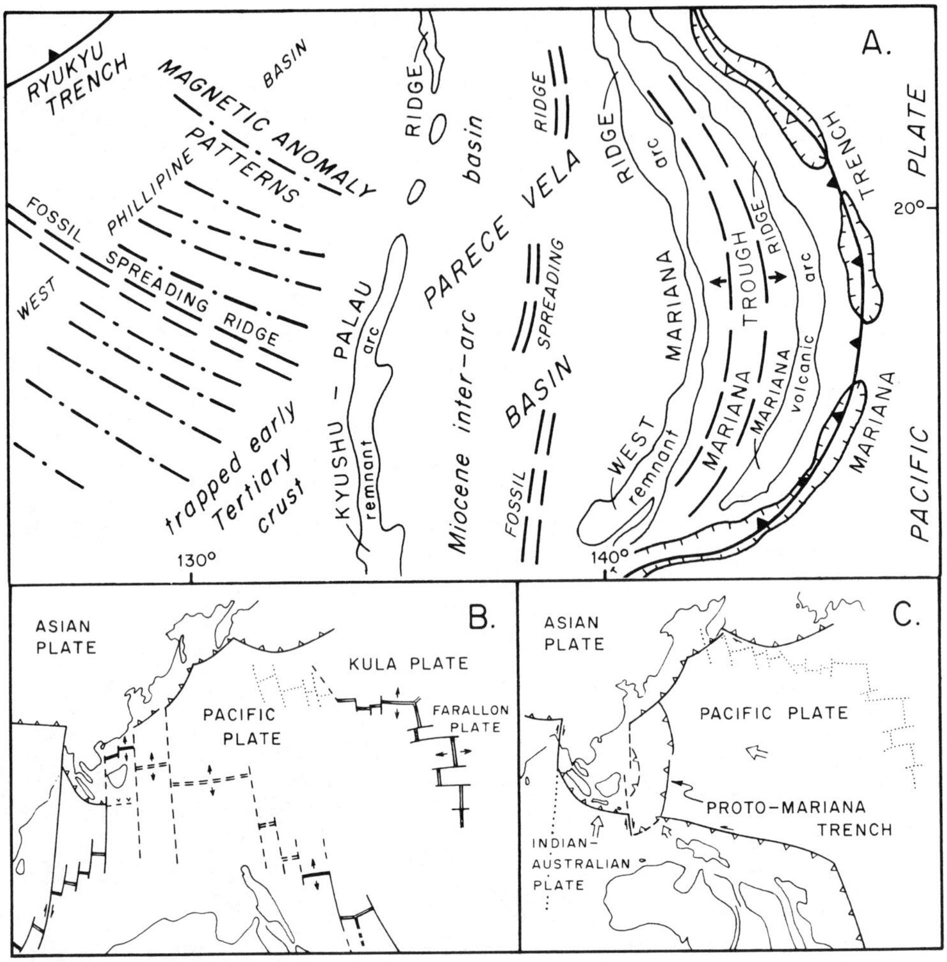

Figure 5. Map showing major tectonomorphic features of Mariana arc-trench system and the West Philippine Basin (after Karig, 1971a, 1971b, 1972; Watts and others, 1977; Nishiwaki and others, 1981). B and C show Hilde and others (1977) model for inception of the Mariana system by a major mid-Tertiary change in Pacific plate motion.

ever, intermixing with Franciscan rocks is localized along post-ophiolitic melange high-angle faults. Semi-intact ophiolite slabs associated with the melange contain upper stratigraphic sequences of locally derived volcanic and plutonic submarine talus breccias interbedded with and above pillow basalt and chert. Thus, tectonic disruption commenced on the ocean floor during and immediately following ophiolite genesis. The disrupted ophiolite was subsequently transgressed by distal flysch of the Great Valley sequence prior to juxtaposition with the Franciscan Complex (Maxwell, 1974; Hopson and others, 1981). Such relations suggest a fracture zone history. Furthermore, the trend of the disrupted ophiolite belt coincides with the regional trend of the Franciscan Complex and its boundary with the Great Valley sequence, considered to be a reflection of the Franciscan subduction zone trend (Bailey and others, 1970; Dickinson, 1971). Perhaps Franciscan subduction initiated along fracture zones within Jurassic oceanic crust now represented by the Coast Range ophiolite fragments.

Substantial volumes of basement-derived serpentinite protrusions and olistostromes within the lower Great Valley sequence indicate that serpentinite from the northern Coast Range ophiolitic melange continued its mobility when in the continental margin regime (Carlson, 1981; Phipps, this volume). Such a history is similar to the older pre-Nevadan development of the Kaweah melange. Carlson (1981) and Phipps (this volume) also suggest that lower Great Valley serpentinitic olistostromes, carrying metamorphic and ophiolitic blocks, moved westward into the Franciscan trench. This is consistent with statements made above on the character of Franciscan serpentinite melange units.

ADDED COMPLEXITIES OF BOUNDARY TRANSFORMS AND POLYGENETIC IGNEOUS SUITES

The views presented above consider many (perhaps most) ophiolitic melanges as distinct petrotectonic assemblages with heritages stemming from the primary development of oceanic crust. The tectonic evolution of such assemblages may be considerably more complex than expressed above. For example, in an intra-oceanic arc-trench system, part of the initial imbricate sequence will form the primitive arc basement, and is likely to end up as a basement element in remnant arc terranes created by inter-arc rifting and ocean floor spreading. Such is probably the case with basement elements of the Kyushu-Palau, West Mariana, and Mariana Ridges, and the respective forearc region (Figure 5a). In such a system the disrupted ophiolitic basement assemblages may ultimately be juxtaposed with or cross-cut by younger inter-arc basin and intra-arc ophiolitic assemblages. Such polygenetic suites of ophiolitic rocks are widespread in the western Sierra Nevada and Klamath Mountains of California (Snoke, 1977; Saleeby, 1982; Saleeby and others, 1982) and pose considerable, but subtle, problems in their petrotectonic analysis.

Ophiolitic assemblages formed along boundary transforms are likely to have additional structural complexities, as well as possible polygenetic developmental histories. Boundary transforms extend into transform edges and marginal offsets in rifted crustal blocks bounding the respective spreading ocean basin (LePichon and Hayes, 1971). Such boundary structures define the main strike-slip domains of the rift system that initially opened the basin. A spectacular example is the Romanche fracture zone of the equatorial Atlantic, which partly coincides with major marginal offsets in the western African and eastern South American coast lines. The critical points concerning boundary transforms are as follows: 1) They are likely to be more structurally complex than ridge-ridge transforms. They must first accommodate the complexities of initial rifting, which will be affected by pre-existing structures in the rifted blocks. Then, once the active transform is bounded only by oceanic crust, the pole of relative motion may shift to a preferred ocean basin spreading geometry, resulting in additional deformation along the transform. 2) Boundary transforms provide a pre-weakened zone between oceanic and rifted continental or transitional crust, resulting in a preferred convergence initiation site. Ocean floor obduction may be favored in the case of significant subsidence of the rifted crustal block. Boundary transforms within marginal or inter-arc basins may border remnant arc or active arc crustal blocks instead of, or in addition to, continental blocks.

One of the main difficulties in working with inter-arc boundary structures lies in the recognition of distinctly different dynamothermal metamorphic events, in addition to the problem of resolving the different igneous ages of ophiolitic protolith within the polygenetic suite. Dynamothermal aureoles in such environments may form not only during the imbrication of young warm crust, but may also form along boundary structures, where hot mantle and mafic magma are intruded and deformed against older ophiolitic assemblages. Examples of such dynamothermal aureoles are given for polygenetic assemblages of the Kings River and American River areas of the western Sierra Nevada (Saleeby, 1982). Here, melange and metamorphic tectonites were thermally overprinted and redeformed as coherent units during the emplacement of substantially younger juvenile ophiolitic materials. Protoclastic and hot-subsolidus deformation features of the younger ophiolitic assemblages bear structural geometries that are seemingly continuous with structures developed during the dynamothermal overprint of the bounding ophiolitic melange.

CONCLUSION

There can be little question that ocean floor serpentinites are common and that they coincide with zones of primary weakness in oceanic crust. Nor can there be any question that serpentinites are readily available for mobilization or remobilization by gravitational or tectonic forces under a wide range of pressure and temperature conditions, and in a wide range of tectonic environments. The basic point of this paper is that the primary ocean floor structures along which abundant serpentinite forms and the physical properties of serpentinite profoundly affect the manner in which oceanic crust behaves during orogenic deformations.

Minor to moderate changes in the pole of rotation between adjacent plates should be considered as part of the broad spectrum of orogenic processes. Such changes spawn significant deformations along primary ocean floor plate junctures and may lead to the growth of an oceanic orogenic welt by ocean floor imbrication. Where imbrication initiates along spreading ridge segments, dynamothermal aureoles develop along the soles of ophiolitic nappes. Fracture zone initiation results in the imbrication of ophiolite sheets and ophiolitic melange. Such oceanic welts, along with little-modified high-standing fracture zone massifs, should be considered "fair game" for *en masse* accretion in zones of convergence and ocean basin closure, just as are oceanic plateaus, island arcs, and remnant arcs. Many of the world's suture belt ophiolite nappes, ultramafic massifs, and ophiolitic melanges may represent previously stacked oceanic welts that underwent later tectonic accretion into their respective suture belts during ocean basin closure.

It is not clear that intact ocean floor sequences are routinely disrupted into ophiolitic melange in subduction zones. Such intact sequences are likely to be subducted. Such is not the case for fracture zone assemblages, which are likely to be sliced and/or protruded off of descending oceanic crust and incorporated into accretionary prism assemblages. Inasmuch as ophiolitic melange is an important constituent of fracture zone crust, the interpretation of ophiolitic melange in uplifted accretionary prisms is ambiguous.

The physical properties of serpentinite favor its long-term involvement in orogenic processes and a tendency toward continued migration to high structural levels. The fact that such behavior begins at a large scale on the ocean floor suggests such a heritage for many orogenic serpentinites. In conclusion, geologists working in melange zones and orogenic terranes are rightfully intent on relating melange-mixing and tectonic accretionary processes to modern plate tectonic phenomena. However, until the full significance of ocean floor serpentinites and the structural complexities of primary oceanic plate junctures are realized, our concepts of melange mixing and tectonic accretion will be severely limited.

ACKNOWLEDGMENTS

Field and geochronological investigations in ophiolitic melange and tectonite terranes of California, southern Oregon, and Baja California Sur have been supported by Geological Society of America Penrose funds; the Alfred P. Sloan Foundation; N.S.F. grants EAR-7708691, EAR-7925998, and EAR-8206382; and the U.S. Geological Survey. Critical reviews of an earlier draft of this paper by R. G. Coleman and L. B. Platt are gratefully acknowledged.

REFERENCES CITED

Abbate, E., Bortolotti, V., and Principi, G., 1980, Apennine ophiolite: A peculiar oceanic crust: Centro Studi Geologia Apennino in rapporto alle Geosenclenali Mediterranee, C.N.R., Firenze, Publication 90, 38 p.

Amrstrong, R. L., and Dick, H.J.B., 1974, A model for the development of thin overthrust sheets of crystalline rock: Geology, v. 2, p. 35–40.

Aumento, F., and Loubat, H., 1971, The Mid-Atlantic Ridge near 45°N. XVI. serpentized ultramafic intrusions: Canadian Journal of Earth Science, v. 8, p. 631–663.

Bailey, E. H., Blake, N. C., and Jones, D. L., 1970, On-land Mesozoic oceanic crust in California Coast Ranges: U.S. Geological Survey Professional Paper 700-C, p. C70–C81.

Ben-Avraham, Z., Nur, A., Jones, D. L., and Cox, A., 1981, Continental accretion: From oceanic plateaus to allochthonous terranes: Science, v. 213, p. 47–54.

Bloomer, S., 1981, Marianas forearc ophiolite—structure and petrology: EOS (Transactions of the Geophysical Union), v. 62, p. 1086.

Bolt, B. A., Lominitz, C., and McEvilly, T. V., 1968, Seismological evidence on the tectonics of central and northern California and the Mendocino Escarpment: Seismological Society of America Bulletin, v. 58, p. 1725–1767.

Bonatti, E., 1976, Serpentinite protrusions in the oceanic crust: Earth and Planetary Science Letters, v. 32, p. 107–113.

——1978, Vertical tectonism in oceanic fracture zones: Earth and Planetary Science Letters, v. 37, p. 369–379.

——Emilioni, C., Ferrara, G., Honnorez, J., and Rydell, H., 1974, Ultramafic-carbonate breccias from the equatorial Mid-Atlantic Ridge: Marine Geology, v. 16, p. 83–102.

——and Honnorez, J., 1976, Sections of the Earth's crust in the equatorial Atlantic: Journal of Geophysical Research, v. 81, p. 4104–4116.

——,——, and Ferrara, G., 1971, Peridotite-gabbro-basalt complex from the equatorial Mid-Atlantic ridge: Philosophical Transactions of the Royal Society of London, Ser. A, v. 268, p. 385–402.

——, —— and Gartner, Jr., S., 1973, Sedimentary serpentinites from the Mid-Atlantic Ridge: Journal of Sedimentary Petrology, v. 43, p. 728–735.

——Sarnthein, M., Boersma, A., Gorini, M., and Honnorez, J., 1977, Neogene crustal emersion and subsidence at the Romanche fracture zone, equatorial Atlantic: Earth and Planetary Science Letters, v. 35, p. 369–383.

Boudier, F., and Coleman, R. G., 1981, Cross section through the peridotite in the Samail ophiolite, southeastern Oman Mountains: Journal of Geophysical Research, v. 86, p. 2573–2592.

Cann, J. R., 1968, Geological processes at mid-ocean ridge crests: Geophysical Journal of the Royal Astronomical Society, v. 15, p. 331–341.

——1974, A model for oceanic crustal structure development: Geophysical Journal of the Royal Astronomical Society, v. 39, p. 169–187.

Carlson, C., 1981, Sedimentary serpentinite of the Wilbur Springs area—a possible Early Cretaceous link between the Franciscan Complex and the Great Valley sequence (M.S. thesis): Stanford, California, Stanford University, 105 p.

Casey, T.A.L., and Dickinson, W. R., 1976, Sedimentary serpentinite of the Miocene Big Blue Formation near Cantua Creek, California, *in* Fritsche, A. E., ed., Neogene Symposium volume: Pacific Section Society of Economic Mineralogists and Paleontologists, Los Angeles, California, p. 65–74.

Choukroune, P., Francheteau, J., Le Pichon, X., 1978, In situ observations along Transform Fault A in the FAMOUS area, Mid-Atlantic Ridge: Geological Society of America Bulletin, v. 89, p. 1013–1029.

Christensen, N. I., 1972, The abundance of serpentinites in the oceanic crust: Journal of Geology, v. 80, p. 709–719.

Church, W. R., 1972, Ophiolite: Its definition, origin as oceanic crust, and mode of emplacement in orogenic belts, with special reference to the Appalachians, *in* Irving, E., ed., The ancient oceanic lithosphere: Canada Department of Energy, Mines and Resources Earth Physics British Publication, v. 42, p. 71–85.

Clague, D. A., and Jarrard, R. I., 1973, Tertiary Pacific plate motion deduced from Hawaiian-Emperor chain: Geological Society of America Bulletin, v. 84, p. 1135–1154.

Coleman, R. G., 1971a, Petrologic and geophysical nature of serpentinites: Geological Society of America Bulletin, v. 82, p. 897–918.

——1971b, Plate tectonic emplacement of upper mantle peridoites along continental edges: Journal of Geophysical Research, v. 76, p. 1212–1222.

——1977, Ophiolites, ancient oceanic lithosphere: New York, Springer-Verlag, 240 p.

——1981, Tectonic setting for ophiolite obduction in Oman: Journal of Geophysical Research, v. 86, p. 2497–2508.

Cortesogno, L., Gianelli, G., and Piccardo, G. B., 1975, Preorogenic metamorphic and tectonic evolution of the ophiolite mafic rocks (Northern Apennine Tuscany): Bollettin della Societe Geologica Italiana, v. 94, p. 291–327.

Cowan, D. S., and Mansfield, C. F., 1970, Serpentinite flows on Joaquin Ridge, southern Coast Ranges, California: Geological Society of America Bulletin, v. 81, p. 2615–2628.

Davies, H. L., 1971, Peridotite-gabbro-basalt complex in eastern Papua: An overthrust plate of oceanic mantle and crust: Australian Bureau of Mineral Resources, Geology and Geophysics Bulletin; v. 128, p. 1–48.

DeLong, S. E., Dewey, J. F., and Fox, P. J., 1977, Displacement history of oceanic fracture zones: Geology, v. 5, p. 199–202.

Dewey, J. F., 1975, Finite plate motion: Some implications for the evolution of rock masses at plate margins: American Journal of Science, 275-A, p. 260–284.

——and Bird, J. M., 1971, Origin and emplacement of the ophiolite suite: Appalachian ophiolites in Newfoundland: Journal of Geophysical Research, v. 76, p. 3179–3266.

DeWit, M. J., Dutch, S., Kligfield, R., Allen, R., and Stern, C., 1977, Deformation, serpentinization, and emplacement of a dunite complex, Gibbs Island, South Shetland Islands: Possible fracture zone tectonics: Journal of Geology, v. 85, p. 745–762.

Dickinson, W. R., 1966, Table Mountain serpentinite extrusion in California Coast Ranges: Geological Society of America Bulletin, v. 77, p. 451–472.

——1971, Clastic sedimentary sequences deposited in shelf, slope and trough settings between magmatic arcs and associated trenches: Pacific Geology, v. 3, p. 15–30.

Doutch, H. F., and others, 1981, Plate tectonic map of the Circum-Pacific region, southwest quadrant: American Association of Petroleum Geologists, Tulsa, Oklahoma.

Drummond, K. J., and others, 1981, Plate tectonic map of the Circum-Pacific region, northeast quadrant: American Association of Petroleum Geologists, Tulsa, Oklahoma.

Eittreim, S., and Ewing, J., 1975, Vema fracture zone transform fault: Geology, v. 3, p. 555–558.

Engel, C. G., and Fisher, R. L., 1975, Granitic to ultramafic rock complexes of the Indian Ocean ridge system, western Indian Ocean: Geological Society of America Bulletin, v. 86, p. 1553–1578.

Evans, C., and Hawkins, J., 1979, Mariana arc-trench system: Petrology of "seamounts" on the trench-slope break: EOS (Transactions of the American Geophysical Union), v. 46, p. 968.

Fisher, R. L., and Engel, C. G., 1969, Ultramafic and basaltic rocks dredged from the nearshore flank of the Tonga Trench: Geological Society of America Bulletin, v. 80, p. 1373–1378.

Folk, R. L., and McBride, E. F., 1976, Possible pedogenic origin of Ligurian ophicalcite: A Mesozoic calichified serpentinite: Geology, v. 4, p. 327–332.

——and McBride, E. F., 1978, Radiolarites and their relation to subjacent "oceanic crust" in Liguria, Italy: Journal of Sedimentary Petrology, v. 48, p. 1069–1102.

Fox, P. J., Pittman, W. C., III, and Shepard, F., 1969, Crustal plates in the central Atlantic: Evidence for at least two poles of rotation: Science, v. 165, p. 487–489.

——Schreiber, E., Rowlett, H., and McCamy, K., 1976, The geology of the oceanographer fracture zone: A model for fracture zones: Journal of Geophysical Research, v. 811, p. 4117–4128.

Francis, T.J.G., 1981, Serpentinization faults and their role in the tectonics of slow spreading ridges: Journal of Geophysical Research, v. 86, p. 11,616–11,622.

Gansser, A., 1974, The ophiolite melange, a world wide problem on Tethyan examples: Ecologae Geologicae Helvetiae, v. 67, p. 479–507.

Gass, I. B., and Smewing, J. D., 1973, Intrusion, extrusion, and metamorphism at constructive margins: Evidence from the Troodos massif, Cyprus: Nature, v. 242, p. 26–29.

Ghent, E. D., and Stout, M. Z., 1981, Metamorphism at the base of the Samail ophiolite, Southeastern Oman Mountains: Journal of Geophysical Research, v. 86, p. 2557–2571.

Gianelli, G., and Principi, G., 1979, Northern Apenpine ophiolite: An ancient transcurrent fault zone: Bollettin della Societe Geologica Italiana, v. 96, p. 53–58.

Gregory, R. T., and Taylor, H. P., 1981, An oxygen isotope profile in a section of Cretaceous oceanic crust, Samail ophiolite, Oman: Evidence for $\delta^{18}O$ buffering of the oceans by deep (>5 km) seawater-hydrothermal circulation at mid-ocean ridges: Journal of Geophysical Research, v. 86, p. 2737–2755.

Hawkins, Jr., J. W., Bloomer, S., Evans, C., and Melchoir, J., 1979, Mariana arc-trench system: Petrology of the inner trench wall: EOS (Transactions of the American Geophysical Union), 46, p. 968.

Hayes, D. E., and Talwani, M., 1972, Geophysical investigation of the Macquarie Ridge complex: Science, v. 166, p. 211–234.

Herron, T. J., Ludwig, W. J., Stoffa, P. L., Kan, T. K., and Buhl, P., 1978, Structure of the east Pacific rise crest from multichannel seismic reflection data: Journal of Geophysical Research, v. 83, p. 798–804.

Hess, H. H., 1962, History of ocean basins, *in* Engel, A.E.J., James, H. L., and Leonard, B. K., eds., Petrologic Studies: A Volume in Honor of A. F. Buddington: Geological Society of America, p. 599–620.

Hilde, T.W.C., Uyeda, S., and Kroenke, L., 1977, Evolution of the western Pacific and its margin: Tectonophysics, v. 38, p. 145–165.

Hopson, C. A., Mattinson, J. M., and Pessagno, Jr., E. A., 1980, Coast Range ophiolite, western California, *in* Ernst, W. G., ed., The Geotectonic Development of California, Rubey Volume 1: Englewood Cliffs, N.J., Prentice-Hall, p. 419–510.

Johnson, A. M., 1970, Physical processes in geology: San Francisco, California, Freeman, Cooper and Co., 577 p.

Johnson, G. L., and Vogt, P. R., 1973, Mid-Atlantic ridge from 47° to 51° north: Geological Society of America Bulletin, v. 84, p. 3443–3462.

Karig, D. E., 1971a, Structural history of the Mariana Island arc system: Geological Society of America Bulletin, v. 82, p. 323–344.

——1971b, Origin and development of marginal basins in the western Pacific: Journal of Geophysical Research, v. 76, p. 2542–2561.

——and Sharman, G. F., III, 1975, Subduction and accretion in trenches: Geological Society of America Bulletin, v. 86, p. 377–389.

Karson, J., and Dewey, J. F., 1978, Coastal complex western Newfoundland: An early Ordovician oceanic fracture zone: Geological Society of America Bulletin, v. 89, p. 1037–1049.

Krause, D. C., Menard, H. W., and Smith, S. M., 1964, Topography and lithology of the Mendocino Ridge: Journal of Marine Research, v. 22, p. 236–249.

Lanphere, M. A., 1981, K-Ar ages of metamorphic rocks at the base of the Samail ophiolite, Oman: Journal of Geophysical Research, v. 86, p. 2777–2782.

LePichon, X., and Hayes, D. E., 1971, Marginal offsets, fracture zones, and the early opening of the South Atlantic: Journal of Geophysical Research, v. 76, p. 6283–6293.

Lockwood, J. P., 1972, Possible mechanisms for the emplacement of alpine-type serpentinite: Geological Society of America Memoir 132, p. 273–287.

Macdonald, K. C., Kastens, K., Spiess, F. N., and Miller, S. P., 1979, Deep tow studies of the Tamayo transform fault; Marine Geophysical Research, v. 4, p. 37–70.

Maxwell, J. C., 1974, Anatomy of an orogen: Geological Society of America Bulletin, v. 85, p. 1195–1204.

McEvilly, T. V., 1968, Sea floor mechanics north of Cape Mendocino, California: Nature, v. 220, p. 901–903.

Melson, W. G., Hart, S. R., and Thompson, G., 1972, St. Pauls Rocks, equatorial Atlantic: Petrogenesis, radiometric ages, and implications on sea-floor spreading: Geological Society of America Memoir 132, p. 241–272.

——and Thompson, G., 1971, Petrology of a transform fault zone and adjacent ridge segments: Philosophical Transactions of the Royal Society of London, Series A, v. 263, p. 423–442.

Menard, H. W., and Atwater, T., 1968, Changes in the direction of sea floor spreading: Nature, v. 219, p. 463–467.

—— and ——1969, Origin of fracture zone topography: Nature, v. 222, p. 1037–1040.

Minster, J. B., and Jordan, T. H., 1978, Present-day plate motions: Journal of Geophysical Research, v. 83, p. 5331–5354.

Moores, E. M., 1970, Ultramafics and orogeny, with models of the U.S. Cordillera and the Tethys: Nature, v. 228, p. 837–842.

——and Vine, F. J., 1971, The Troodos massif, Cyprus and other ophiolites as oceanic crust: Evaluations and implications: Philosophical Transactions of the Royal Society of London, Series A, v. 268, p. 443–466.

Nishiwaki, C., 1981, Plate tectonic map of the Circum-Pacific region, northwest quadrant: American Association of Petroleum Geologists, Tulsa, Oklahoma.

Orcutt, J. A., Kennett, B.L.N., and Dorman, L. M., 1976, Structure of the East Pacific Rise from an ocean bottom seismometer survey: Geophysical Journal of the Royal Astronomical Society, v. 45, p. 305–320.

Oxburgh, E. R., and Turcotte, D. L., 1968, Mid-ocean ridges and geotherm distribution during mantle convection: Journal of Geophysical Research, v. 73, p. 2643–2661.

Phillips, J. D., and Fleming, H. S., 1978, Multi-beam sonar study of the Mid-Atlantic Ridge rift valley, 36°-37°N, Geological Society of America MC-19.

Phipps, S. P., 1984, Ophiolitic olistostromes in the basal Great Valley sequence, Napa County, northern California Coast Ranges, *in* Raymond, L. A., ed., Melanges: Their nature, origin, and significance: Geological Society of America Special Paper 198 (this volume).

Raff, A. D., and Mason, R. G., 1961, Magnetic survey off the west coast of North America, 40°N latitude to 52°N latitude: Geological Society of America Bulletin, v. 72, p. 1267–1270.

Raleigh, C. B., and Paterson, M. S., 1965, Experimental deformations of serpentinite and its tectonic implications: Journal of Geophysical Research, v. 70, p. 3965–3985.

Ruff, L., Given, J., and Sanders, C., 1982, The tectonics of the Macquarie Ridge, New Zealand: New Evidence of strike-slip motion from the earthquake of May 25, 1981, Mw = 7.7: EOS (Transactions of the American Geophysical

Union), v. 63, p. 384.
Saleeby, J. B., 1978, Kings River Ophiolite, Southwest Sierra Nevada Foothills, California: Geological Society of America Bulletin, v. 89, p. 617–636.
——1979, Kaweah Serpentinite Melange, Southwest Sierra Nevada Foothills, California: Geological Society of America Bulletin, v. 90, p. 29–46.
——1982, Polygenetic ophiolite belt of the California Sierra Nevada, geochronological and tectonostratigraphic development: Journal of Geophysical Research, v. 87, p. 1803–1824.
——Harper, G. D., Snoke, A. W., and Sharp, W. D., 1982, Time relations and structural-stratigraphic patterns in ophiolite accretion, west-central Klamath Mountains, California: Journal of Geophysical Research, v. 87, p. 3831–3848.
Sarwar, G., and De Jong, K. A., 1983, Composition and origin of the Kanar melange, southern Pakistan, *in* Raymond, L. A., ed., Melanges: Their nature, origin, and significance: Geological Society of America Special Paper 198 (this volume).
Schweickert, R. A., and Cowan, D. S., 1975, Early Mesozoic tectonic evolution of the western Sierra Nevada, California: Geological Society of America Bulletin, v. 86, p. 1329–1336.
Sclater, J. G., Dick, H., Norton, I. O., Woodroffe, D., 1978, Tectonic structure and petrology of the Antarctic plate boundary near the Bouvet triple junction: Earth and Planetary Science Letters, v. 37, p. 393–400.
Seeber, L., Barazangi, M., and Nowroozi, A. A., 1970, Microearthquake seismicity and tectonics of coastal northern California: Seismological Society of America Bulletin, v. 60, p. 1669–1699.
Shand, S. J., 1949, Rocks of the Mid-Atlantic Ridge: Journal of Geology, v. 57, p. 89–92.
Silver, E. A., 1971, Tectonics of the Mendocino Triple Junction: Geological Society of America Bulletin, v. 82, p. 2965–2978.
Snoke, A. W., 1977, A thrust plate of ophiolitic rocks in the Preston Peak area, Klamath Mountains, California: Geological Society of America Bulletin, v. 88, p. 1641–1659.
Thompson, G., and Melson, W. G., 1972, The petrology of the oceanic crust across fracture zones in the Atlantic Ocean: Evidence of a new kind of sea-floor spreading: Journal of Geology, v. 80, p. 526–538.
Tilton, G. R., Hopson, C. A., and Wright, J. E., 1981, Uranium - Lead isotopic ages of the Samail ophiolite, Oman, with applications to Tethyan ocean ridge tectonics: Journal of Geophysical Research, v. 86, p. 2763–2775.
Tobin, D. G., and Sykes, L. R., 1968, Seismicity and tectonics of the northeast Pacific Ocean: Journal of Geophysical Research, v. 73, p. 3821–3845.
van Andel, T. H., Phillips, J. D., and von Herzen, R. P., 1969, Rifting origin for the Vema fracture zone in the north Atlantic: Earth and Planetary Science Letters, v. 5, p. 296–300.
——von Herzen, R. P., and Phillips, J. P., 1971, The Vema fracture zone and the tectonics of transverse shear zones in oceanic crustal plates: Marine Geophysical Researches, v. 1, p. 251–283.
Varne, R., Gee, D. R., and Quilty, P.G.J., 1969, Macquarie Island and the cause of oceanic linear magnetic anomalies: Science, v. 166, p. 230–233.
——and Rubenach, M. J., 1972, Geology of Macquarie Island and its relationship to oceanic crust, *in* Hayes, D. E., Antarctic Oceanology II, The Australian-New Zealand Section: American Geophysical Union, p. 251–266.
Vogt, P. R., and Avery, O. E., 1974, Detailed magnetic surveys in the northeast Atlantic and Labrador Sea: Journal of Geophysical Research, v. 79, p. 363–388.
——Johnson, G. L., Holcombe, T. L., Gilg, J. G., and Avery, O. E., 1971, Episodes of sea-floor spreading recorded by the North Atlantic basement: Tectonophysics, v. 12, p. 211–234.
Walcott, R. I., 1978, Present tectonics and late Cenozoic evolution of New Zealand: Geophysical Journal of the Royal Astronomical Society, v. 52, p. 137–164.
Watts, A. B., Weissel, J. K., and Larson, R. L., 1977, Sea-floor spreading in marginal basins of the western Pacific: Tectonophysics, v. 37, p. 167–181.
Williams, H., 1973, Bay of Islands map-area, Newfoundland: Canada Geological Survey Paper 72-34, 7 p.
——1975, Structural succession, nomenclature, and interpretation of transported rocks in western Newfoundland: Canadian Journal of Earth Science, v. 9, p. 1216–1229.
Yagi, K., 1960, A dolorite block dredged from the bottom of the Vityoz Deep, Mariana trench: Japan Academy Proceedings, v. 36, p. 213–216.

Manuscript Accepted by the Society July 12, 1984
California Institute of Technology Contribution No. 3891

Printed in U.S.A.

Geological Society of America
Special Paper 198
1984

Epilogue
Melanges: Future studies

Loren A. Raymond
Department of Geology
Appalachian State University
Boone, North Carolina 28608

The general character of melanges is well known as a result of hundreds of studies of melange terranes throughout the world. In this volume, new descriptions of melanges from California to Pakistan have been added to the literature. However, while many studies reveal general similarities and differences between melanges, fewer are unveiling clues essential to uniquely characterizing melanges formed in different ways within various paleotectonic environments. In addition, editing this volume revealed anew that the same data examined by different workers lead to conflicting interpretations.

Albeit complete agreement on a melange definition will probably never be reached, some uniformity in application of the terminology will facilitate future studies. Evidently some views are widely held. Notwithstanding the observations of a few workers who suggest that locally within some melanges there are zones that have no matrix (e.g., J. Desmons, personal communication, 1978; Yilmaz and Maxwell, 1984), nearly all workers agree that *melanges consist of blocks in a matrix.* Bodies of fragmented and mixed rock that lack a matrix perhaps should be assigned another name. Likewise, other kinds of units similar to melanges should be assigned specific appellations. The view that *fragmentation and mixing are the essential processes in melange formation* is also widely accepted. These and other views should be formally adopted by a "working group" on melanges, a step that would clarify future communications.

Future studies should focus on the formative processes for each type of melange and for similar appearing or related rock bodies. Important questions include, but are not limited to: how do fragmentation and mixing occur, do different processes result in similar internal fabrics, how can one distinguish fabrics produced in different ways, what are the relative roles of soft-sediment and post-lithification deformation in melange formation, what are the effects of deformation on unlithified or poorly lithified sediments, how do the different materials typical of melanges respond to progressive deformation, what roles do flow and folding play in melange formation, what are the relative roles of brittle and ductile deformation in melange genesis, and how are exotic blocks introduced into melanges? One or more of these problems have been addressed recently by several workers including Moore and his coworkers (Moore and Geigle, 1974; Moore and Wheeler, 1978; Moore, 1978), Cowan and his coworkers (Cowan and Page, 1975; Cowan and Silling, 1978; Cowan, 1978, 1982), Karig (1980), Bosworth and Vollmer (1981; Vollmer and Bosworth, 1984-this volume), Naylor (1981, 1982), Nelson (1982), Byrne (1982, 1984-this volume), and Cloos (1982, 1984-this volume). Yet much remains to be learned.

The types of studies that lead to an understanding of melange-forming processes also help to clarify definitive criteria for distinguishing melanges formed in different environments. Clarification is especially important in studies of polygenetic melanges. Polygenetic melanges like those discussed in this volume by Jacobi (1984), Saleeby (1984), Sarwar and DeJong (1984), Yilmaz and Maxwell (1984), and Vollmer and Bosworth (1984) are widely distributed in the geologic record. Did they have an early sedimentary or tectonic history? As an aid in resolving such questions, several workers have noted criteria for distinguishing olistostromes and Naylor (1982) has recently provided a list. However, where soft sediments are deformed in tectonically active environments, distinctions between sedimentary and tectonic processes become blurred. Consequently, paleogeographic analysis retains a central role in evaluations of paleoenvironments of melange petrogenesis.

At present, a detailed analysis is required for recognition of the significance of melanges in the geologic record. The days of assuming that melanges represent subduction zones are past. The works of Page and Suppe (1981), Phipps (1984), Saleeby (1984), Saleeby and others (1978), and Naylor (1982), for example, suggest that even the association of melange and ophiolite does not significantly narrow the choice of environment. Associated melange and ophiolite have been reported from fore-arc and back-arc basins, transform fault zones, subduction zones, and base of slope sites on continental margins. As is the case with granites, there are melanges and melanges. Future studies should clarify details of the nature of melanges, specify the processes and history of fragmentation and mixing, and, consequently, result in a clearer understanding of the significance of each melange type in the geologic record.

ACKNOWLEDGMENTS

I wish to thank F. K. McKinney and an anonymous reviewer for reviews of earlier versions of this manuscript, Susan Wilson for typing the manuscript, and Appalachian State University for support.

REFERENCES CITED

Bosworth, W., and Vollmer, F. W., 1981, Structures of the medial Ordovician flysh of eastern New York: Deformation of synorogenic deposits in an overthrust environment: Journal of Geology, v. 89, p. 551–568.

Byrne, T., 1982, Structural evolution of coherent terranes in the Ghost Rock Formation, Kodiak Island, Alaska, *in* Leggett, J., ed., Trench-Forearc geology: Sedimentation and tectonics on modern and ancient active plate margins: Geological Society of London Special Publication 10, p. 229–242.

—— 1984, Early deformation in melange terranes of the Ghost Rocks Formation, Kodiak Islands, Aslaska, *in* Raymond, L. A., ed., Melanges: Their nature, origin, and significance: Geological Society of America Special Paper 198 (this volume).

Cloos, M., 1982, Flow melanges: Numerical modeling and geologic constraints on their origin in the Franciscan subduction complex, California: Geological Society of America Bulletin, v. 93, p. 330–345.

—— 1984, Flow melanges and the structural evolution of accretionary wedges, *in* Raymond, L. A., ed., Melanges: Their nature, origin, and significance: Geological Society of America Special Paper 198 (this volume).

Cowan, D. S., 1978, Origin of blueschist-bearing chaotic rocks in the Franciscan Complex, San Simeon, California: Geological Society of America Bulletin, v. 89, p. 1415–1423.

—— 1982, Deformation of partly dewatered and consolidated Franciscan sediments near Piedras Blancas Point, California, *in* Leggett, J., ed., Trench-Forearc geology: Sedimentation and tectonics on modern and ancient active plate margins: Geological Society of London Special Publication 10, p. 429–457.

—— and Page, B. M., 1975, Recycled Franciscan material in Franciscan melange west of Paso Robles, California: Geological Society of America Bulletin, v. 86, p. 1089–1095.

—— and Silling, R. M., 1978, A dynamic, scaled model of accretion at trenches and its implications for the tectonic evolution of subduction complexes: Journal of Geophysical Research, v. 83, p. 5389–5396.

Jacobi, R. D., 1984, Modern submarine sediment slides and their implications for melange and the Dunnage Formation in north-central Newfoundland, *in* Raymond, L. A., ed., Melanges; Their nature, origin, and significance: Geological Society of America Special Paper 198 (this volume).

Karig, D. E., 1980, Material transport within accretionary prisms and the "knocker" problem: Journal of Geology, v. 88, p. 27–39.

Moore, J. C., 1978, Orientation of underthrusting during latest Cretaceous and earliest Tertiary time, Kodiak Islands, Alaska: Geology, v. 6, p. 209–213.

—— and Geigle, J. E., 1974, Slaty cleavage: incipient occurrences in the deep sea: Science, v. 183, p. 509–510.

—— and Wheeler, R. L., 1978, Structural fabric of a melange Kodiak Islands, Alaska: American Journal of Science, v. 278, p. 739–765.

Naylor, M. A., 1981, Debris flow (olistostromes) and slumping on a distal passive continental margin: The Palombini limestone-shale sequence of the northern Apennines: Sedimentology, v. 28, p. 837–852.

—— 1982, The Casanova Complex of the northern Appennines: A melange formed on a distal passive continental margin: Journal of Structural Geology, v. 4, no. 1, p. 1–18.

Nelson, K. D., 1982, A suggestion for the origin of mesoscopic fabric in accretionary melange, based on features observed in the Chrystalls Beach Complex, South Island, New Zealand: Geological Society of America Bulletin, v. 93, p. 625–634.

Page, B. M., and Suppe, J., 1981, The Pliocene Lichi melange of Taiwan: Its plate-tectonic and olistostromal origin: American Journal of Science, v. 281, p. 193–227.

Phipps, S. P., 1984, Ophiolitic olistostromes in the basal Great Valley sequence, Napa County, northern California Coast Ranges, *in* Raymond, L. A., ed., Melanges: Their nature, origin and significance: Geological Society of America Special Paper 198 (this volume).

Saleeby, J., 1984, Tectonic significance of serpentinite mobility and ophiolitic melange, *in* Raymond, L. A., ed., Melanges: Their nature, origin, and significance: Geological Society of America Special Paper 198 (this volume).

—— Goodin, S. E., Sharp, W. D., and Busby, C. J., 1978, Early Mesozoic paleotectonic-paleogeographic reconstruction of the southern Sierra Nevada region, *in* Howell, D. G., and McDougall, K. A., eds., Mesozoic paleogeography of the western United States: Los Angeles, Pacific Section Society of Economic Paleontologists and Mineralogists, p. 311–336.

Sarwar, G., and DeJong, K. A., 1984, Composition and origin of the Kanar melange, southern Pakistan, *in* Raymond, L. A., ed., Melanges: Their nature, origin, and significance: Geological Society of America Special Paper 198 (this volume).

Vollmer, F. W., and Bosworth, W., 1984, Formation of melanges in a foreland basin overthrust setting: Example from the Taconic orogen, *in* Raymond, L. A., ed., Melanges: Their nature, origin, and significance: Geological Society of America Special Paper 198 (this volume).

Yilmaz, P. O., and Maxwell, J. C., 1984, An example of an obduction melange: The Alakir Cay unit, Antalya Complex, SW Turkey, *in* Raymond, L. A., ed., Melanges: Their nature, origin, and significance: Geological Society of America Special Paper 198 (this volume).

MANUSCRIPT ACCEPTED BY THE SOCIETY JULY 12, 1984

Typeset by WESType Publishing Services, Inc., Boulder, Colorado
Printed in U.S.A. by Malloy Lithographing, Inc., Ann Arbor, Michigan